Notations of the Principal Crystalline Structures

Strukturbericht	Schoenflies	Type	System
A1	O_h^5	Cu	fcc
A4	O_h^7	C (diam)	fcc
A7	D_{3d}^5	As	rhombohedral
A8	D_3^4	Se	trigonal
B1	O_h^5	NaCl	fcc
B2	O_h^1	CsCl	cubic
B3	T_d^2	ZnS (bl)	fcc
B4	6_{6v}^4	ZnS (w)	hexagonal
B8$_1$	D_{6h}^4	NiAs	hexagonal
B8$_2$	D_{6h}^4	Ni$_2$In	hexagonal
B13	C_{3v}^5	NiS (mill)	rhombohedral
B17	D_{4h}^9	PtS (coop)	tetragonal
B18	D_{6h}^4	CuS (cov)	hexagonal
B29	D_{2h}^{16}	SnS	orthorhombic
B31	D_{2h}^{16}	MnP	orthorhombic
B34	C_{4h}^2	PdS	tetragonal
C1	O_h^5	CaF$_2$	fcc
C2	T_h^6	FeS$_2$(pyr)	cubic
C4	D_{4h}^{14}	TiO$_2$	rutile
C6	D_{3d}^3	CdI$_2$	trigonal
C7	D_{6h}^4	MoS$_2$	hexagonal
C18	D_{2h}^{12}	FeS$_2$ (marc)	orthorhombic
C19	D_{3d}^5	CdCl$_2$	rhombohedral
C22	D_3^2	Fe$_2$P	trigonal
C40	D_6^4	CrSi$_2$	hexagonal
DO$_2$	T_h^5	CoAs$_3$	bcc
DO$_3$	O_h^5	BiF$_3$	fcc
DO$_9$	O_h^1	ReO$_3$	cubic
DO$_{11}$	D_{2h}^{16}	Fe$_3$C (cem)	orthorhombic
DO$_{19}$	D_{6h}^4	Ni$_3$Sn	hexagonal
D5$_1$	D_{3d}^6	α-Al$_2$O$_3$ (cor)	rhombohedral
D5$_2$	D_{3d}^3	La$_2$O$_3$	trigonal
D5$_3$	T_h^7	Mn$_2$O$_3$	bcc
D7$_2$	O_h^7	Co$_3$S$_4$	fcc
D7$_3$	T_d^6	Th$_3$P$_4$	bcc

CRYSTAL CHEMISTRY AND SEMICONDUCTION

in Transition Metal Binary Compounds

CRYSTAL CHEMISTRY AND SEMICONDUCTION

in Transition Metal Binary Compounds

J. P. SUCHET

Centre National de la Recherche Scientifique, Paris

 1971

ACADEMIC PRESS New York and London

Copyright © 1971, by Academic Press, Inc.
ALL RIGHTS RESERVED
NO PART OF THIS BOOK MAY BE REPRODUCED IN ANY FORM,
BY PHOTOSTAT, MICROFILM, RETRIEVAL SYSTEM, OR ANY
OTHER MEANS, WITHOUT WRITTEN PERMISSION FROM
THE PUBLISHERS.

ACADEMIC PRESS, INC.
111 Fifth Avenue, New York, New York 10003

United Kingdom Edition published by
ACADEMIC PRESS, INC. (LONDON) LTD.
Berkeley Square House, London W1X 6BA

Library of Congress Catalog Card Number: 78-137636

PRINTED IN THE UNITED STATES OF AMERICA

Contents

PREFACE ix
ACKNOWLEDGMENTS xiii
SYMBOLS, ABBREVIATIONS, AND PHYSICAL CONSTANTS xv

PART ONE
REVIEW AND DISCUSSION OF SOME USEFUL THEORETICAL BASES

Chapter 1 From the Atom to the Molecule

1.1.	Atomic Orbital Functions	3
1.2.	H_2 and H_2^+ Molecules	9
1.3.	Heteronuclear Molecules	13
1.4.	Extension of the VB Method	18
	References	23

Chapter 2 From the Molecule to the Crystal

2.1.	Covalent Aspect of Bonds	24
2.2.	VB Approach (Crystallochemical Model)	29
2.3.	MO Approach (Band Model)	33
2.4.	Electrical Conductibility	39
	References	47

Chapter 3 **The Magnetic Crystal**

3.1.	Crystal Field Theory	49
3.2.	Magnetic Interactions	53
3.3.	Metal-to-Metal Transfers	61
3.4.	Magneton–Electron Interactions	68
	References	74

PART TWO

BIBLIOGRAPHICAL DIGEST (1947–1967)

Chapter 4 **IIIB, IVB, and VB Metalloid Compounds**

4.1.	Main Structures	79
4.2.	Interstitial Compounds	88
4.3.	Silicides, Germanides, Stannides	95
4.4.	Phosphides, Arsenides, Antimonides	103
	References	115

Chapter 5 **Oxides of the Metals Ti, V, Cr, Mn, and Homologues**

5.1.	TO, T_3O_4, and T_2O_3 Oxides	119
5.2.	Rutile TiO_2	130
5.3.	Other Dioxides	139
5.4.	T_2O_5 and TO_3 Oxides	148
	References	155

Chapter 6 **Oxides of the Metals Fe, Co, Ni, Cu, and Homologues**

6.1.	Mössbauer Effect	161
6.2.	TO Oxides	167
6.3.	T_3O_4 Oxides	177
6.4.	T_2O_3 Oxides	184
	References	191

Chapter 7 **Transition Metal Chalcogenides**

7.1.	Main Structures	196
7.2.	Sulfides	204
7.3.	Selenides	213
7.4.	Tellurides	219
	References	229

Chapter 8 Compounds of Rare Earths and Similar Elements

8.1.	Main Structures	235
8.2.	IIIB, IVB, and VB Compounds	239
8.3.	Oxides	246
8.4.	Chalcogenides	257
	References	268

PART THREE
SOME PRESENT PROBLEMS AND POSSIBLE APPLICATIONS

Chapter 9 Various Magnetoelectric Effects

9.1.	Magnetic Scattering	277
9.2.	Magnetoresistance	281
9.3.	Astrov Effect	288
9.4.	Possible Applications	292
	References	296

Chapter 10 Hall Magnetoelectric Effects

10.1.	Ordinary Hall Effect	299
10.2.	Extraordinary Hall Effect	303
10.3.	Experimental Research	308
10.4.	Possible Applications	313
	References	320

Chapter 11 Electro- and Magnetooptical Effects

11.1.	Forced Birefringencies	322
11.2.	Faraday Effect	325
11.3.	Experimental Research	329
11.4.	Possible Applications	335
	References	341

Appendix 343

INDEX OF FORMULAS 345
INDEX OF STRUCTURES 354
AUTHOR INDEX 355
SUBJECT INDEX 374

Preface

Several generations of semiconductor materials can be distinguished:

(1) The elements Ge and Si, which brought about a scientific and industrial revolution in the fifties.

(2) The "classic" binary, ternary, etc. compounds. Some of these are already being used industrially (InSb, GaAs), and many are fairly familiar (III V, II VI). Research undoubtedly will continue for many years because of their very large number. Although original devices have been constructed from these materials, their impact has not been revolutionary.

(3) The "special" compounds: transition metal or rare earth compounds, organic semiconductors, vitreous or liquid substances, etc. Here we are faced with different semiconduction mechanisms which are still obscure, and combinations of possibly unexpected properties.

The presently emerging generation of semiconductor crystals containing magnetic atoms can be traced to research on oxides of mixed valency and the development of magnetic ferrites in the immediate post-war years. Since these crystals combine high permeability with transparency to electromagnetic waves, they have recently been used in hyperfrequency devices. Such uses, in fact, constitute the first practical application of magnetic semiconductors.

This book covers more generally the substances whose transport properties are not typically metallic, and in which atomic magnetic moments exist. It will be shown, in fact, that it is impossible to draw such neat boundaries between semiconductor compounds and metallic alloys of transition elements as for compounds of alkaline or alkaline-earth elements. In addition, there is scarcely any difference between the effect of a ferromagnetic order and that of an antiferromagnetic order on the transport properties. What is more, this order always ends at a certain temperature, and it is very difficult to ignore phenomena occurring at higher temperature (i.e., in the paramagnetic region).

Oxide materials with a high energy gap have been studied fairly systematically, although recent discoveries, such as that of the ferroelectric material $Ni_3B_7O_{13}I$, may be opening up further fields for research. In contrast, little is known about antimonides, selenides, and tellurides. Their chemical properties and crystallographic structure are usually obvious, and in many cases, their magnetism is defined. However, there is still lively controversy about their transport properties, and the very definition of the semiconductor seems to be far less precise. The wide range of phase homogeneity and the effect of the stoichiometric ratio in all these substances give them a versatility not inherent in conventional materials. However, scientists are handicapped in dealing with these substances because of the absence of any book combining, in a comprehensible form, the essential chemical and physical information and a detailed analysis of existing experimental results. One of my aims has been to fill this gap.

In Part Two all the experimental work on the electrical conductibility of compounds of transition metals, rare earths, or actinides published since the war is analyzed and summarized. I have tried to provide those embarking on scientific research with a guide that may make it easier for them to approach a body of writing that can sometimes be heavy going.

Part Three deals briefly with some areas in which applications may be expected. It would be premature to devote any greater space to this. The theoretical concepts needed for the construction of approximate models to estimate the properties of new compounds are given in a condensed and fairly original form in Part One.

To make clear the spirit in which this book has been written, I shall briefly review the three separate activities that ensure the advancement of science: *understanding* (for the research scientist), *application* (for the engineer), and *teaching* (for the teacher). The research scientist gathers experimental data, correlates them, deduces partial laws, and tries to construct a working model. These procedures involve a certain amount of trial-and-error work, and frequent backward steps. It is, therefore, pioneering work. Tough industrial competition forces the engineer to apply the results of research immediately, without awaiting their verification. The teacher, in more basic courses, cannot run the risk of presenting concepts that may be subject to radical modification. The scientific instruction that he provides for most of his pupils is therefore based on research conducted 5 to 10 years earlier. In this very rough scheme, this book is concerned with in the first stage—*understanding*. It is intended primarily for young research workers wishing to enter into this field, and secondarily for research engineers, who are investigating the construction of new devices. Finally, this book will be of use to teachers, who will themselves assess what they can include in their courses without too much risk.

Solid state chemistry is a new branch of science, which came into existence a few years ago in the United States as a result of the need to solve the numerous practical problems connected with the synthesis and crystallization of pure substances. Its rapid expansion has been due to close contact between pure and applied research. It also involves a certain reaction against the esoteric tendencies that often appear in solid state physics and nuclear physics. One of the most enthralling aspects is the hope that one day it may be possible to carry out, to order, the synthesis of compounds with the properties required for a given application. However, this goal implies significant advances in some fields of inorganic chemistry, and wider training of theoreticians.

Acknowledgments

I should like to express my gratitude to my colleague, Francis Bailly, Chargé de Recherche at the C.N.R.S., for his active collaboration in the preparation of Part One of this book.

I should like also to thank Dr. J. B. Goodenough, Group Leader in the Lincoln Laboratory of M.I.T., Cambridge, Massachusetts, and Professor A. Wold of Brown University, Providence, Rhode Island, for the comments they were kind enough to make about the manuscript.

I must also thank Denis Mahaffey for his help in translating the original French manuscript.

Symbols, Abbreviations, and Physical Constants

AB   antibonding level
AO   atomic orbital
A   actinide element
B   bonding level
c   speed of light, 2.998×10^{10} cm sec^{-1}; number of Lewis pairs formed by one atom; height of the elementary cell (NiAs structure)
C_1, C_2, C_3, C_4 normation coefficients
d   electron with azimuthal quantum number 2
D   electronic density; Madelung constant
e   electron charge, 1.602×10^{-20} cgs or 10^{-19} C
e_d, e_a height of forbidden band gap (extrinsic excitation mechanism)
e_g fused $d_{x^2-y^2}$ and d_{z^2} MOs
E_A activation energy (transfer mechanism)
E_F Fermi level energy
E_G height of forbidden band gap (intrinsic excitation mechanism)
EPR   electronic paramagnetic resonance
f   electron with azimuthal quantum number 3
G   Avogadro's number, 6.02×10^{23}
h   Planck's constant, 4.14×10^{-15} eV sec $= 6.62 \times 10^{-27}$ erg sec or 10^{-34} J sec
H_h hyperfine field (Mössbauer effect)
I   magnetization intensity
J_e exchange integral of two electrons
k   Boltzmann's constant, 8.62×10^{-5} eV °K^{-1} $= 1.38 \times 10^{-16}$ erg °K^{-1} or 10^{-23} J °K^{-1}
k   wave vector with module $2\pi/\lambda$

K electron with principal quantum number 1
L electron with principal quantum number 2
L rare earth element (Sc, Y, or Ln)
LCOA linear combination of atomic orbitals
Ln lanthanide element (rare earth)
m mass of the electron "at rest," 9.109×10^{-28} gm; covalent "charge" of the atom
m^*/m effective mass of a carrier
M symbol for a metal
M electron with principal quantum number 3
MO molecular orbital method
n principal quantum number; ionic charge
N number of electrons per cubic centimeter
N electron with principal quantum number 4
NMR nuclear magnetic resonance
p electron with azimuthal quantum number 1
p any integer
P number of positive holes per cubic centimeter; oxygen pressure; electric polarization
q effective charge of the atom "at rest"
r polar coordinate (radial); ionic radius
R_H Hall coefficient
R_0 ordinary Hall coefficient
R_1 extraordinary Hall coefficient
s electron with azimuthal quantum number 0
S atomic component of spin quantum numbers
t_{2g} fused d_{xy}, d_{yz}, and d_{zx} MOs
T symbol for a transition metal
T temperature
v velocity of a particle
VB valence bond method
x trirectangular coordinate; fraction of crystallographic sites
X symbol for a metalloid
y trirectangular coordinate
z trirectangular coordinate
ϕ polar coordinate (geographical longitude)
α Seebeck coefficient
α, β spin functions
$\alpha, \beta, \gamma, \delta, \varepsilon$ phases of a diagram
δ asphericity of the electronic distribution; isomeric shift (Mössbauer effect)
Δ energy difference between the d sublevels
ε dielectric constant; quadrupolar interaction (Mössbauer effect)
φ amplitude of an AO wave function
λ wavelength; ionicity parameter
λ_0 equilibrium ionicity of a bond
μ Hall mobility of carriers
μ_B Bohr magneton

Symbols, Abbreviations, and Physical Constants

μ_D drift mobility of carriers
ν frequency of an electromagnetic wave or associated wave; hopping frequency (transfer mechanism)
π_g, π_u types of bond with a symmetry plane (MO approach)
ρ electrical resistivity
σ electrical conductivity, $1/\rho$
σ_g, σ_u types of bond with a symmetry axis (MO approach)
θ polar coordinate (geographical latitude); Curie–Weiss parameter
ψ amplitude of an MO wave function
ψ_+, ψ_- bonding and antibonding functions
χ magnetic susceptibility
I, II, III, etc.   elements in the corresponding columns of the periodic table

---

$$1 \text{ eV} = 8068 \text{ cm}^{-1} = 23.063 \text{ kcal} = 1.602 \times 10^{-19} \text{ J or } 10^{-12} \text{ erg}$$

CRYSTAL CHEMISTRY AND SEMICONDUCTION

in Transition Metal Binary Compounds

Part One | REVIEW AND DISCUSSION OF SOME USEFUL THEORETICAL BASES

Chapter 1 | **From the Atom to the Molecule**

1.1. ATOMIC ORBITAL FUNCTIONS

To avoid overburdening this book unnecessarily, it is assumed that the reader is familiar with the basic principles of chemistry, i.e., the structure of the atom and the periodic table. Those who need some reminder of these points should refer to the classic works on the subject, such as the one by Moore [1.1]. Beginners might use the extremely simple and straightforward little book written by Seel [1.2].

It is known that a particle (photon) has to be associated with an electromagnetic wave. The momentum (mv) attributed to a photon of energy E is, according to Maxwell, $(mv) = E/c = hv/c$ (where c is the speed of light, h Planck's constant, and v the frequency). This gives us an expression for the wavelength

$$\lambda = h/(mv) \tag{1.1}$$

In 1924, de Broglie proposed a generalization of this equation for electromagnetism, postulating that it also defined a wave of a new type associated with any material particle of momentum mv (m mass and v velocity). Three years later, experiments in electronic diffraction confirmed this bold concept.

The stationary state of a system vibrating in one dimension, such as a vibrating cord, is described by a wave with an amplitude $\varphi(x)$, a wavelength λ, and is a solution to the differential equation

$$(d^2\varphi/dx^2) + (4\pi^2/\lambda^2) = 0 \tag{1.2}$$

If one accepts that this differential equation, well known in conventional mechanics, applies to the associated wave imagined by de Broglie (Schrödinger's postulate), it becomes a simple matter to deal with the movement of an electron. Let us assume, for instance, that the electron moves along the x-axis in a zero field. All its energy is kinetic, $E = \frac{1}{2}mv^2$, giving $\lambda = h/mv = h(2mE)^{-1/2}$. The differential equation is written as

$$(d^2 \varphi/dx^2) + (8\pi^2 m/h^2) E\varphi = 0 \tag{1.3}$$

In the simple case of a single dimension and purely kinetic energy, it is the basic equation obtained by Schrödinger in 1926. Let us assume that the electron rebounds elastically ($\varphi = 0$) for $x = 0$ and l. Its energy is $E = p^2 h^2/8ml^2$. It has not just any value, but is equal to one of the elements of a discrete series arising from the integer p. In other words, the energy of the electron is *quantized*. This results from the existence of conditions at the limits.

The simplest atom is the hydrogen atom, in which one electron rotates round one proton. This involves a three-dimensional problem, and the solution of Schrödinger's equation requires its separation into three equations, each of which depends only on one of the three variables, for instance the polar coordinates r (proton–electron distance), θ and ϕ (the electron's geographical latitude and longitude in relation to the proton). In addition, the total energy E includes a term $V = -e^2/r$, due to the proton's coulombic potential ($e =$ elementary charge), so that the difference $(E-V)$ appears in place of E in (1.3). Finally, one finds the following solutions (principal, azimuthal, and magnetic quantum numbers in parentheses on the left) [1.3]

(1, 0, 0)	K, s	$\varphi = (\pi a^3)^{-1/2} \exp(-r/a)$	(1.4)
(2, 0, 0)	L, s	$\varphi = \frac{1}{4}(2\pi a^3)^{-1/2}(2-r/a)\exp(-r/2a)$	(1.5)
(2, 1, 0)	L, p_z	$\varphi = \frac{1}{4}(2\pi a^3)^{-1/2}(r/a)\exp(-r/2a)\cos\theta$	(1.6)
(2, 1, +1)	L, p_x	$\varphi = \frac{1}{4}(\pi a^3)^{-1/2}(r/a)\exp(-r/2a)\sin\theta\cos\phi$	(1.7)
(2, 1, −1)	L, p_y	$\varphi = \frac{1}{4}(\pi a^3)^{-1/2}(r/a)\exp(-r/2a)\sin\theta\sin\phi$	(1.8)

with

$$a = (h^2/4\pi^2)me^2 = 0.53 \text{ Å} \tag{1.9}$$

which indicates the radius of the first orbit in Bohr's model. These are the *orbital functions* of the hydrogen atom.

It can be seen straightaway that the first two functions have a spherical symmetry. This is the case for all s functions and for them alone (azimuthal quantum number is zero). Figure 1.1 shows, in terms of r, φ (1.4), $|\varphi|^2$, and $4\pi r^2 |\varphi|^2$. The second curve shows that the density or probability of presence in the center (in other words, as near the proton as the repelling forces will allow) is not nil. This is another characteristic property of s functions. The third curve shows that the probability is maximum in the shell, $r = 0.53$ Å, in accordance with Bohr's model. But a systematic uncertainty remains concerning simultaneous awareness of the position of the electron and its momentum. This unusual fact should not cause any surprise. One realizes that phenomena taking place on a very different scale from our own may obey other laws. The electron is moving at an average velocity of 2000 km sec^{-1}, and, at this speed, it traverses the hydrogen atom in 10^{-17} sec!

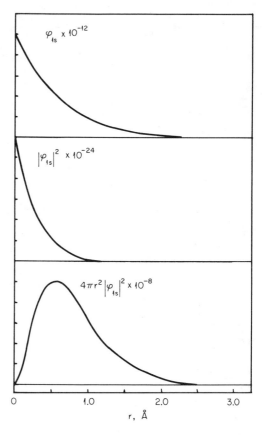

Fig. 1.1. In terms of the proton–electron distance r: Top, φ function corresponding to the fundamental state (1.4). Middle, square of modulus $|\varphi|^2$ representing the electron density at a point located at a distance r from the proton. Bottom, radial function $4\pi r^2 |\varphi|^2$ representing the probability of presence of the electron in a spherical shell of radius r and negligible thickness dr.

In the functions (1.6)–(1.8), it is clear that the electronic density can no longer be distributed spherically because of the intervention of the angles θ and ϕ. The graphic representation of $\cos \theta$ in polar coordinates is a combination of two circles. The angular factor of p_z is thus represented by two spheres centered on the z-axis, the upper part representing the positive values of the cosine and the lower part the negative values. In contrast to the s functions, the function p_z is antisymmetrical in relation to the horizontal plane;

in other words, its two parts are obtained by symmetry in relation to this plane but with a change in the sign. The electronic distribution, represented by $|\varphi|^2$, assumes a closely related form and is naturally symmetrical. The functions p_x and p_y have the same form and characteristics but their directions lie along the x- and y-axis, respectively. Figure 1.2 shows the usual diagrammatic representation of these functions.

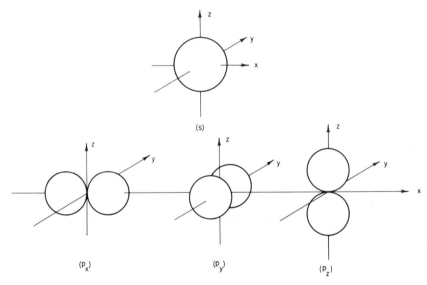

Fig. 1.2. Diagrammatic representation of the angular distribution of electrons in the s and p orbital functions.

The energy of the electron on the orbits represented by the functions mentioned above is given, as in Bohr's model, by the expression

$$E = \tfrac{1}{4}e^2/n^2 a^2 \tag{1.10}$$

$n = 1$ corresponds to the fundamental state and the following values to excited states. It should be noted that when $n = 2$, the energy is the same for the s and p functions. This result is valid only for the hydrogen atom.

For atoms other than hydrogen, which involves several electrons, strict calculation of orbital functions is generally much more complicated, and use is made of their property of being able to be put in the form of a product of a radial term by an angular term. This angular term is fortunately independent of the individual atom, depending only on the azimuthal quantum number. There is only one angular distribution for all the s electrons, three for all the p electrons, five for all the d electrons, seven for all the f electrons, and so on. Mention has already been made, in connection with the hydrogen atom, of the

1.1. Atomic Orbital Functions

spatial distributions of the charge in the s and p orbitals. Something should now be said of the charge in the d orbitals, which have a particularly important role in this book.

Whereas the three p functions clearly evoked the three directions of a trirectangular trihedron, it is much more difficult to give a simple representation of the d functions, even by associating them with one another in the form of linear combinations. The best solution is to retain the independence of the different functions, by introducing an orbital d_{z^2} which has a different spatial distribution from that of the orbitals d_{xy}, d_{yz}, d_{zx}, and $d_{x^2-y^2}$ (Fig. 1.3). The difference is only apparent, however. The orbital d_{z^2} can in fact be written as a linear combination of the functions $(1/\sqrt{3})\, d_{z^2-x^2}$ and $(1/\sqrt{3})\, d_{z^2-y^2}$ with the same spatial distribution as the other d orbitals, but which are no longer independent of it. Its exact expression would be $d_{(3z^2-r^2)/\sqrt{3}}$.

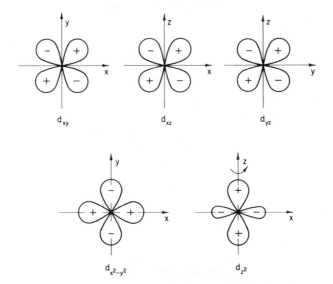

Fig. 1.3. Diagrammatic representation of the angular distribution of electrons in the d orbitals. The orbital, referred to by the symbol d_{z^2}, presents symmetry of rotation around the z-axis.

As far as the radial term is concerned, strict calculation of it is usually impossible, and two main approximations are used to evaluate it. The first, and older, consists of using the so-called "hydrogenoid" orbitals, with the same form as the hydrogen atom orbitals given in (1.4) to (1.8), but in which the atomic number Z and the parameter $\rho = Zr/a$ intervene. Here are the results for the first two layers, for example (quantum numbers in parentheses on the left) [1.4]

(1, 0, 0)	K, s	$\varphi = Z^{3/2}(\pi a^3)^{-1/2}\exp(-\rho)$ (1.11)
(2, 0, 0)	L, s	$\varphi = \tfrac{1}{4}Z^{3/2}(2\pi a^3)^{-1/2}(2-\rho)\exp(-\rho/2)$ (1.12)
(2, 1, 0)	L, p_z	$\varphi = \tfrac{1}{4}Z^{3/2}(2\pi a^3)^{-1/2}\rho\exp(-\rho/2)\cos\theta$ (1.13)
(2, 1, +1)	L, p_x	$\varphi = \tfrac{1}{4}Z^{3/2}(2\pi a^3)^{-1/2}\rho\exp(-\rho/2)\sin\theta\cos\phi$ (1.14)
(2, 1, −1)	L, p_y	$\varphi = \tfrac{1}{4}Z^{3/2}(2\pi a^3)^{-1/2}\rho\exp(-\rho/2)\sin\theta\sin\phi$ (1.15)

The other possible approximation is Hartree–Fock's, which allows the radial term to be estimated numerically.

In any case, the simplicity of the calculations mentioned for the hydrogen atom quickly disappears when the atomic number of the element rises. It would be an exaggeration to see a simple application of mathematics in chemistry, as Dirac predicted in 1929, in the enthusiasm of the first years of quantum mechanics. However, this method of presenting the problem seems essential for its understanding, and it has the additional advantage of providing a few points of reference for readers intending to go on later to more theoretical works.

Finally, it might be mentioned that certain linear combinations of the orbital functions of an atom are particularly important. These are the *hybrid* orbital functions, the energy of which is lower than those of their component orbital functions. The main ones are those combining the atomic orbital functions (AO) s, p_x, p_y, and p_z of the highest level in the elements C, Si, Ge, and Sn. They can be written, after a change of axes [1.5] as

$$\begin{aligned}\varphi_1 &= \tfrac{1}{2}(s + p_x + p_y + p_z)\\ \varphi_2 &= \tfrac{1}{2}(s + p_x - p_y - p_z)\\ \varphi_3 &= \tfrac{1}{2}(s - p_x - p_y + p_z)\\ \varphi_4 &= \tfrac{1}{2}(s - p_x + p_y - p_z)\end{aligned} \quad (1.16)$$

It will be seen in Section 1.3 that this hybridization is the only way of explaining the structure of many molecules, for example CH_4. The electrons involved in such orbital functions are no longer either s or p electrons. Seel [1.2] has suggested the general name of "q electrons." However, the custom has remained of referring to the orbital on the basis of its composition: here sp^3. The angular distribution of the charge shows four maximum points of electronic density in four directions, forming an angle of 109° 28′ between them.

The main hybrid orbitals are s, p_x, p_y (sp^2, 120° plane); $d_{x^2-y^2}$, d_{z^2}, s (d^{2}s, 120° plane); d_{z^2}, p_x, p_y (dp^2, 120° plane); $d_{x^2-y^2}$, d_{z^2}, p_z (d^{2}p, 90°); s, p_x, p_y, p_z (sp^3, 109°28′); d_{xy}, d_{yz}, d_{zx}, s (d^{3}s, 109°28′); $d_{x^2-y^2}$, s, p_x, p_y, (dsp^2, 90° plane); $d_{x^2-y^2}$, d_{z^2}, s, p_x, p_y, p_z (d^{2}sp^3, 90°); and s, p_x, p_y, p_z, d_{xy}, d_{yz}, d_{zx}, f_{xyz} (sp^{3}d^{3}f, 70°29′).

1.2. H_2 AND H_2^+ MOLECULES

The way in which two atoms are associated in a molecule was explained satisfactorily for the first time by G. N. Lewis in 1916 (see the facsimile reprint [1.6]). He showed that the stability of the electronic octet, already known for ions, could also be attained by the pooling of a *pair* of electrons. The simplest example of such a pair is provided by the hydrogen molecule H_2. It is the covalent chemical bond. By means of the mathematical language of quantum mechanics, its stability can be interpreted and its energy calculated.

It can be shown easily from Schrödinger's equation, by separating the variables, that the wave function describing the whole of two systems without interaction, separately described by $\varphi_1(x_1, y_1, ...)$ and $\varphi_2(x_2, y_2, ...)$, is simply the product $\varphi_1 \varphi_2$, and that the total energy is the sum of the energies of each of them. It is also known that, if φ_I and φ_{II} are two possible solutions for the same system, the linear combination $a\varphi_I + b\varphi_{II}$ is also a solution. Finally, of the various solutions giving various energies for the ground state of a system, the best is always the one giving the lowest energy. These general principles provide guidelines in the search for a molecular wave function.

Let us take two identical hydrogen atoms, in which, for the convenience of argument, we shall refer to the protons as A and B, and the electrons as 1 and 2. The wave functions describing the electrons of each system separately are $\varphi_A(1)$ and $\varphi_B(2)$. In accordance with the first principle mentioned above, the total wave function is $\varphi_A(1)\varphi_B(2)$. Let us assume that this function remains a valid approximation when the atoms approach each other and form a molecule. The electrons can no longer be distinguished, and the same molecule would have been obtained from two atoms in which the electrons would have been described by $\varphi_A(2)$ and $\varphi_B(1)$, in other words a total wave function $\varphi_A(2)\varphi_B(1)$. These two products are thus both possible solutions. In accordance with the second principle mentioned above, their linear combination $a\varphi_A(1)\varphi_B(2) + b\varphi_A(2)\varphi_B(1)$ is also a solution and, for reasons of symmetry due to indistinguishability, $a = \pm b$. If one replaces each of the atomic orbital functions (AO) φ by its value in the fundamental state (1.4), one finds that the energy of the total wave function varies depending on the distance AB, as indicated in Fig. 1.4. The linear combination is thus the best solution.

The wave function common to the two electrons in the molecule

$$\psi = (1/\sqrt{2})[\varphi_A(1)\varphi_B(2) \pm \varphi_A(2)\varphi_B(1)] \qquad (1.17)$$

was found by Heitler and London [1.7] in 1927. The factor $1/\sqrt{2}$ is introduced by the fact that the probability of finding an electron (1 or 2) in the whole space is necessarily one (condition of normalization leaving out overlap integrals). Numerical calculation gives the minimum of the curve ψ_+ an energy of

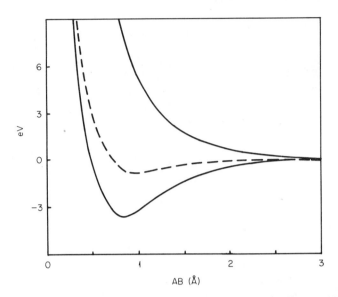

Fig. 1.4. Variation in the energy, in electron volts, in relation to the distance AB between protons. The solid line is the combination $\varphi_A(1)\varphi_B(2) \pm \varphi_A(2)\varphi_B(1)$ with the signs $+$ (ψ_+, above) and $-$ (ψ_-, below). The dotted line is the simple product $\varphi_A \varphi_B$.

72.5 kcal mole^{-1} (3.2 eV) and a distance between the protons of 0.80Å. There is by no means a perfect correspondence, since the experimental values are 109.5 kcal mole^{-1} (4.72 eV) and 0.74 Å, but this calculation was none the less a revelation at the time since it provided a resounding confirmation of G. N. Lewis' ideas. In addition, an important density of charge $(\varphi_A + \varphi_B)^2$ appears halfway between A and B in the function ψ_+ and explains the bonding of the two protons by the attraction of a common negative charge. Such behavior is bound up with the existence of de Broglie's associated wave. The discrepancy in relation to experimental results can be eliminated by various improvements, such as the intervention of an effective nuclear charge, ionic terms $H_A^+ H_B^-$ and $H_A^- H_B^+$ and the interaction between the two electrons. But here we are interested only in understanding the essential phenomena and will leave these refinements aside.

The method used by Heitler and London to form the wave function of the two electrons in the hydrogen pair can be extended to other molecules. It is known as the *valence bond* method (VB) or sometimes the mesomerism method, in allusion to the linear combination of the two solutions $\varphi_A(1)\varphi_B(2)$ and $\varphi_A(2)\varphi_B(1)$ which, in less simple cases than that of the hydrogen molecule, correspond to separate mesomer formulas. This method provides a natural expression of a chemist's usual concepts according to which molecules consist of atoms linked by the chemical bond but still retaining their individuality

14   *From the Atom to the Molecule*

another, one faces the question of finding out how bonding can occur between AOs of different types. It is shown that the zones of orbital functions which overlap must have the same analytical symmetry around the axis joining the two nuclei [1.8], and the greater the overlap, the more stable the bond will be [1.5]. Figure 1.6 illustrates cases where bonding is possible (a and b) or impossible (c and d). This shows the *directional* character of the covalent bond, and the close link existing between the AO involved on the one hand, and the angle of the bonds established by one atom with its neighbors, on the other. Consideration of a few simple molecules, resulting from the combination of atoms of hydrogen with those of elements in columns IV, V, and VI of the periodic table, will make these points clearer.

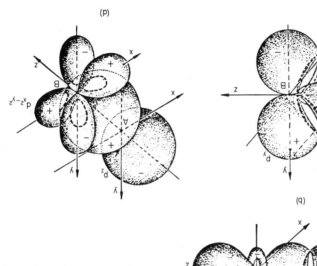

Fig. 1.6. Overlapping of some AO functions: s–p_z (a) and p_z–d_{z^2} (b) result in the establishment of a bond along the z-axis, while s–p_y (c) and p_z–$d_{x^2-y^2}$ (d) cannot allow along it since the overlap zones are equal and have opposite signs on each side of the z-axis. [Based on G. Pannetier, "Chimie générale, atomistique, liaisons chimiques." Masson, Paris, 1966.]

1.3. Heteronuclear Molecules

The principal MOs are, in increasing order of energy: $1s\sigma_g$, $1s\sigma_u$, $2s\sigma_g$, $2s\sigma_u$, $2p_x\sigma_g$, $2p_y\pi_u$, or $2p_z\pi_u$, $2p_y\pi_g$ or $2p_z\pi_g$, $2p_x\sigma_u$. [1.1]. Calculation of the energy corresponding to the expression (1.20) gives two different values. The MO containing the $+$ signs has the lower energy and is called the bonding orbital, while the one with the $-$ signs has the higher energy and is called the antibonding orbital. In the case of diatomic molecules of elements, they have symmetries corresponding to the indices g and u, respectively.

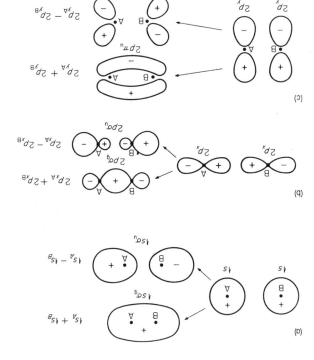

Fig. 1.5. Main notations for the MOs obtained by LCOA. (a) $s\sigma$, (b) $p\sigma$, and (c) $p\pi$. [Based on Walter J. Moore, "Physical Chemistry," 3rd ed., Fig. 13.7, p. 531. © 1962. By permission of Prentice-Hall, Inc.]

1.3. Heteronuclear Molecules

Whenever a bond is formed between two different atoms, only external electrons (valence electrons) can take part in the Lewis pair. It can happen that their AO functions are of the same type (both s or p_x, for example), but this is not usually the case. In view of the spatial distribution of the electron density in the different types of AO and the arrangement of the atoms in relation to one

12 *From the Atom to the Molecule*

is a linear combination of atomic orbitals (LCAO). As above, the factor $1/\sqrt{2}$ results from the normalization condition. This way of obtaining a mono-electron molecular orbital is very common, and will be discussed again later.

Let us consider the electron system described in (1.19), which we shall call 1, as a first system to which we shall add another system, made up of an electron 2, assumed to be independent of 1. The whole function will consist of the product of the functions of both systems (where the $+$ and $-$ signs are not independent):

$$\psi = (1/\sqrt{2})[\phi_A(1) \pm \phi_B(1)][\phi_A(2) \pm \phi_B(2)] \quad (1.20)$$

This wave function provides another representation of the H$_2$ neutral molecule, different from (1.17).

The result of the numerical calculations based on this function is even less satisfactory than for Heitler and London's, giving 61.8 kcal mole^{-1} (2.64 eV) and 0.85 Å [1.1]. This can easily be understood by writing (1.20) in the form

$$\psi_+ = (1/\sqrt{2})[\phi_A(1)\phi_B(2) + \phi_A(2)\phi_B(1) + \phi_A(1)\phi_A(2) + \phi_B(1)\phi_B(2)] \quad (1.21)$$

The first two terms in the bracket are identical with function (1.16), and the following two represent the ionic configurations H$_A^-$H$_B^+$ and H$_A^+$H$_B^-$, which therefore are as important as the covalent bond. The initial mistake is obvious. This method is of little use for the H$_2$ molecule, but, on the other hand, it has been extended successfully to other molecules under the name of the *molecular orbitals* method (MO). Its mathematical developments are often simpler than those for the VB method, and its drawback for hydrogen (undue weight of the ionic terms) turns out to be an advantage for certain heteronuclear molecules. However, it must be pointed out that functions similar to (1.19) do not include, in more complex molecules, any localization of the electron between two given atoms. The Lewis pair is completely lost sight of, and the electrons are assumed to move round each nucleus without distinction.

This method involves certain particular notations: MOs rotating symmetrically in relation to the axis of the nuclei are referred to by the symbol σ while most of the others, symmetrical in relation to the zero electron density plane, are referred to by the symbol π. These are accompanied by the subscripts g (from the German gerade, even) or u (ungerade, uneven), depending on whether the function retains or changes its sign after inversion in relation to the middle point on the line of the nuclei (Fig. 1.5). All the electrons in the molecule are divided up into two per MO, as would be done for the AOs of an atom, for example,

$$H_2^+ = (1s\sigma_g)^1, \quad H_2 = (1s\sigma_g)^2, \quad He_2^+ = (1s\sigma_g)^2(1s\sigma_u)^1$$

$$Li_2 = (1s\sigma_g)^2(1s\sigma_u)^2 \quad \text{or} \quad KK(2s\sigma_g)^2$$

1.2. H_2 and H_2^+ Molecules

despite the few valency electrons transferred or pooled. It shows that there are always two possible wave functions for the Lewis pair, referred to as *bonding* (ψ_+) and *antibonding* (ψ_-) functions, because of their high or zero density of charge halfway between the protons. Similarly, their energies are divided into two distinct levels, bonding (B) and antibonding (AB). At the equilibrium distance, 0.74 Å in the hydrogen molecule, the energy separating them is around 200 kcal mole^{-1} (9 eV).

So far we have not considered the electron's own magnetic moment since the magnetic interaction energy is insignificant compared with that of the coulombic interaction. Two functions, α and β, corresponding to the spin quantum numbers $+\frac{1}{2}$ and $-\frac{1}{2}$, are usually introduced. For a two-electron system, there are four possible functions, three symmetrical (sign retained by the exchange of electrons 1 and 2) and one antisymmetrical (sign reversed). The total wave function of a two-electron system is the product of the space function, just described above, and the spin function. The Pauli exclusion principle, according to which no two electrons in the same atom can have the same quantum numbers, limits the number of possible combinations of space and spin functions to four:

$$[\phi_A(1)\phi_B(2) + \phi_A(2)\phi_B(1)][\alpha(1)\beta(2) - \alpha(2)\beta(1)]$$
symmetrical   antisymmetrical

singlet state (bonding), anti-parallel spins   (1.18)

$$[\phi_A(1)\phi_B(2) - \phi_A(2)\phi_B(1)]\begin{cases}\alpha(1)\alpha(2)\\ \beta(1)\beta(2)\\ [\alpha(1)\beta(2) + \alpha(2)\beta(1)]\end{cases}$$
antisymmetrical   symmetrical

triplet state (antibonding), parallel spins

When electrical discharges occur in a hydrogen atmosphere, the ionized molecule H_2^+ is revealed by the characteristic lines of its emission spectrum. The heat of dissociation and the distance of the protons are thus known. If one looks for the wave function of the single electron in this molecule, one realizes immediately that each of the AO functions ϕ_A and ϕ_B is a possible solution since the system can always be regarded as the juxtaposition of a hydrogen atom and a proton. The linear combination $a\phi_A + b\phi_B$ is thus another solution, and, for reasons of symmetry due to indistinguishability, $a = \pm b$. For the functions containing the signs $+$ and $-$, one finds energy curves in terms of the distance between protons of the same type as those in Fig. 1.4. For the former, the calculation gives 40.7 kcal mole^{-1} (1.75 eV) and 1.21 Å. Experimental results are 63.8 kcal mole^{-1} (2.75 eV) and 1.06 Å [1.3].

The wave function of the single electron in the molecule

$$\psi = (1/\sqrt{2})(\phi_A \pm \phi_B) \tag{1.19}$$

1.3. Heteronuclear Molecules

An atom of carbon, the electronic formula for which is $2s^2p^2$, seems to be able to form only two bonds, with a nonbonding doublet $2s^2$ and two bachelor electrons, $2p_x$ and $2p_y$, for example, each of which can participate in a Lewis pair. The molecule of methane CH_4 and much of organic chemistry prove that this is by no means the case. The four atoms of hydrogen are strictly equivalent, and are situated at the corners of a tetrahedron with the carbon at its center. The question is sufficiently familiar not to need further discussion here. In Section 1.1, the existence of sp^3 hybrid AOs on the carbon atom was mentioned. Each of the four bonds will therefore be written as sp^3/s (Fig. 1.7). The nitrogen atom, with the formula $2s^2p^3$, can form three bonds, with one nonbonding doublet $2s^2$ and three bachelor electrons $2p_x$, $2p_y$, and $2p_z$, ready to enter into a Lewis pair. The oxygen atom, with the formula $2s^2p^4$, can form two bonds, with two nonbonding doublets $2s^2$ and $2p_x^2$ (for instance) and two bachelor electrons $2p_y$ and $2p_z$. In these last two cases, each of the bonds will be written as p/s, although the bond angles involved are more than 90° [1.3]:

CH_4	109° 28′	NH_3	106° 45′	H_2O	104° 27′
		PH_3	93° 50′	H_2S	92° 30′
		AsH_3	91° 35′	H_2Se	91° 00′
		SbH_3	91° 30′	H_2Te	89° 30′

We shall return later to the interpretation of these differences.

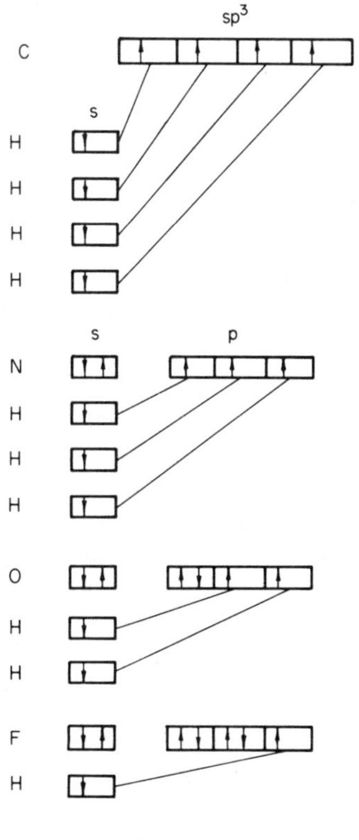

Fig. 1.7. Diagrammatic representation of the Lewis pairs in CH_4, NH_3, H_2O, and FH molecules.

The attempt to find a wave function by the VB method for the Lewis pairs in these molecules leads to function (1.17) where φ_B, for example, will represent a φ hybrid AO of element IV or a 2p AO of element V or VI. In the case of nitrogen and oxygen, some authors have suggested forming partially hybridized orbitals by the linear combination of the 2s and 2p AOs of the metalloid, and substituting them in φ_B for the 2p orbital. This mathematical artifice is, in fact, taken from the MO method. For instance, the modulus of the MO of NH_3 is kept in a rotation of $2\pi/3$ round the bisectrix of the trihedron formed by the three H atoms round the N corner, and one begins by replacing the AOs of the three hydrogen atoms h_0, h_1, and h_2 by

$$\varphi_0 = h_0 + h_1 + h_2$$
$$\varphi_\pm = h_0 + h_1 \exp(\pm 2\pi i/3) + h_2 \exp(\pm 4\pi i/3) \tag{1.22}$$

The LCOA of each of these functions, with the AOs of nitrogen (2p more or less hybridized with 2s) finally gives the MOs.

The attempt to find a wave function by the MO method for the electrons in these molecules does not always lead to function (1.20). It is usually done by replacing the AO of the metal (or metalloid) with LCOAs which respect the symmetry of the molecule and in which one tries to lower the energy of the fundamental state of the MO in order to increase the overlap and thus improve the function. This method, which is more adaptable to mathematical calculations, theoretically allows the exact bonding angles to be taken into account, by means of the artifice of hybridizations among AOs. Figure 1.8 shows the correspondence between angles and s, p, and d percentages. For instance, calculation gives angles of 91°30′ (instead of 91°) for H_2Se or 92°6′ (instead of 92°12′) for H_2S, but the angles of 93°30′ and 92°42′ obtained for H_2O and NH_3 are much too small [1.5]. It should be remembered that this method ignores the concept of Lewis pairs and that diagrams like those in Fig. 1.7 are ruled out (and even impossible). Complete delocalization of the MOs is usually a handicap, and the so-called "localized MO" method is often substituted, with appropriate LCOAs localizing the electron densities of the MOs on the atoms of a given element in the molecule. Their form in this case differs from (1.20), since $a \neq b$ in the expression $a\varphi_A + b\varphi_B$. But the concept of energy attached to the orbital disappears here, and this is a serious defect because, as will be seen in Chapter 3, one has to be able to classify the energy levels of d electrons (given by the crystal field theory) and of bonding and antibonding electrons in relation to one another.

Enough has been said to show that neither of the two methods, VB or MO, is perfect. This has long been known. As Ketelaar [1.3] wrote in 1958, "The present position is as follows: for trustworthy quantitative calculations, one would always use the MO method; in interpreting the results, however, one will base one's reasoning on concepts that in fact correspond to the ideas of the

1.3. Heteronuclear Molecules

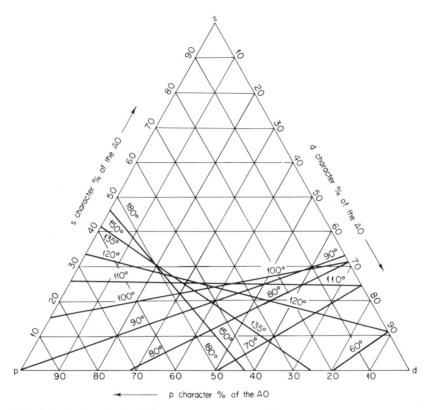

Fig. 1.8. Angle calculated between two maximum electron density directions in the spatial distribution of the charge of s–p–d hybrid AOs (based on Kasha [1.9]).

VB method. For qualitative interpretations, the VB method is still very widely used. We calculate by the MO method, and talk the VB language." Moore [1.1] put it more precisely in 1963, "For most molecules, the simple MO method considerably overestimates the ionic terms, whereas the simple VB method considerably underestimates them. The true structure is usually some compromise between these two extremes, but the mathematical treatment of such a compromise is more difficult. It is necessary to add further terms to the expressions for the wave functions, for instance, ionic terms to the VB function." Here the tendency will be to use the VB approach, partly because our aim is to understand the phenomena and not calculate them, and partly because a remedy is provided for this main defect in the next section.

A number of authors [1.10, 1.11] have considered separating the purely covalent and purely ionic aspects of the bond into two separate wave functions:

$$\psi = a\psi_{\text{cov}} + a'\psi_{\text{ion}} \tag{1.23}$$

The ratio a'/a provides an evaluation of the ionic character of the bond. One can then adopt a function of the Heitler and London type for ψ_{cov} and confine oneself to adopting the most likely ionic configuration for ψ_{ion}. This is exactly the improvement suggested by Moore. It is worth recalling one of the first attempts made in this connection by Karo and Olson [1.12] on an LiH molecule. The 2s and 2p AO functions of lithium produce fairly low overlap charges. They therefore tried to substitute an LCOA of these two functions varying the respective weights of the 2s and 2p orbitals so as always to obtain the minimum energy for the function at the equilibrium distance of the two nuclei in the molecule. The best result was obtained for $a = 0.88$ and $a' = 0.47$, giving an idea of the ionic character of the molecule. The complexity of such an attempt, even on a very simple molecule, shows the need for a more general approach, however.

In 1960, the author [1.13, 1.14] suggested that an ionicity parameter could be worked out on the basis of the weights of the covalent and ionic functions, representing the Lewis pair and the nonbonding doublet of an anion respectively. The value of such a parameter lay in the certainty of its involvement in the lattice-related properties of crystals. The recently published work by Suchet and Bailly [1.15] provides a solid basis for these ideas and backs up the semi-empirical estimate previously given by a closer calculation, using cohesion energies.

1.4. Extension of the VB Method

According to most authors [1.1, 1.3, 1.16], the most general wave function that can be given for the two electrons in a Lewis bonding pair between two nuclei i and j is

$$\psi = C_1 \psi_1 + C_2 \psi_2 + C_3 \psi_3 + C_4 \psi_4$$
$$\psi_1 = \varphi_i(1) \varphi_j(2), \quad \psi_3 = \varphi_i(1) \varphi_i(2) \quad (1.24)$$
$$\psi_2 = \varphi_i(2) \varphi_j(1), \quad \psi_4 = \varphi_j(1) \varphi_j(2)$$

In the pure covalent bond, represented by Heitler and London's function, $C_1 = C_2 = C_0$ and $C_3 = C_4 = 0$. In the pure ionic bond, $C_1 = C_2 = 0$ and C_3 or C_4 is zero, depending on whether the strongly electronegative nucleus (in other words with a strong power of attraction on the electrons) is j or i. In the simple MO method, $C_1 = C_2 = (C_3 C_4)^{1/2}$. As already pointed out, most bonds have a partial ionic character, so that the terms with indices 1 and 2 are insufficient to describe them. The MO method gives the terms with indices 3 and 4 a weight that is not necessarily correct. The following essential remark should be made: one of the terms, C_3 or C_4, is always negligible at a first approximation, either because the ionic terms are both very slight in a covalent bond, or because the probability of a metal forming an anion is infinitesimal

1.4. Extension of the VB Method

in an ionic bond. We shall therefore assume that it is j which tends to produce an anion, and adopt the simplified function

$$\psi = C_0(\psi_1 + \psi_2) + C_4\psi_4 \qquad (1.25)$$

Let us continue to ignore the overlap integrals $\int \varphi_i \varphi_j^* \, d\tau$ and $\int \varphi_j \varphi_i^* \, d\tau$ where φ^* is the conjugated complex wave function of φ, and $d\tau$ is the volume change. Incidentally, this approximation is made by all authors, and rejection of it would make any simple reasoning impossible. The condition of normalization is written [1.15] as

$$2C_0^2 + C_4^2 = 1 \qquad (1.26)$$

C_4^2 represents the possibility of finding the two bonding electrons in a j doublet, in other words the probability of a pure ionic configuration. We shall take this quantity as a parameter of the ionicity of the bond, and refer to it by the symbol λ. We therefore have $C_4 = \lambda^{1/2}$, $C_0 = (1-\lambda)^{1/2}/(2)^{1/2}$ and

$$\psi = (1-\lambda)^{1/2}[\varphi_i(1)\varphi_j(2) + \varphi_i(2)\varphi_j(1)]/(2)^{1/2} + \lambda^{1/2}\varphi_j(1)\varphi_j(2) \qquad (1.27)$$

or

$$\psi = [(1-\lambda)^{1/2}\varphi_i(1)/(2)^{1/2} + \lambda^{1/2}\varphi_j(1)/2]\varphi_j(2)$$
$$+ [(1-\lambda)^{1/2}\varphi_i(2)/(2)^{1/2} + \lambda^{1/2}\varphi_j(2)/2]\varphi_j(1) \qquad (1.28)$$

One can verify that for $\lambda = 1$ (ionic configuration), $\psi = C_4\psi_4$ and that for $\lambda = 0$ (covalent configuration), one again meets with $C_0(\psi_1 + \psi_2)$.

Using expression (1.28), it is easy to calculate a distribution of the bonding electrons between the two atoms. For example, let us consider atom i. In the first term, one can see that a fraction $(1-\lambda)/2$ of electron 1 can be considered as being attached to atom i. In the second term, the same applies to a fraction $(1-\lambda)/2$ of electron 2. A total fraction of $(1-\lambda)$ can therefore be attached to atom i. Since we are considering a set of two bonding electrons, there remains a fraction $(1+\lambda)$ attached to atom j. The same result would have been reached by using the even simpler function which leaves out the permutation of 1 and 2 [1.13, 1.14]:

$$\psi = C_1\psi_1 + C_4\psi_4 \qquad (1.29)$$

or

$$\psi = (1-\lambda)^{1/2}\psi_{\text{cov}} + \lambda^{1/2}\psi_{\text{ion}} \qquad (1.30)$$

The definition of the number of electrons attached to each atom involves plotting somewhere between the nuclei i and j a boundary surface dividing in two the electron cloud corresponding to the Lewis pair. This has two major consequences. The first is that such a definition is arbitrary in character involving a choice that can be justified only after the event in the light of the practical usefulness of the image. The second is that the two halves of the electron cloud each contain a certain charge, which may be assumed to be a point charge, fixed on the nuclei i and j. One thus finds effective charges, which naturally have the same arbitrary character as the number of electrons attached to the atoms. We shall return later to these charges.

Schrödinger's equation can be written in the form

$$H\psi = E\psi \tag{1.31}$$

where H is in quantum mechanics an operator, indicating the performance of a complex operation on the symbol following it. Let us multiply by the conjugated function and integrate in the whole space.

$$\int \psi^* H\psi \, d\tau = \int \psi^* E\psi \, d\tau \tag{1.32}$$

E, being a constant, can be removed from the integral [1.1]:

$$E = \left[\int \psi^* H\psi \, d\tau\right] \Big/ \int \psi^* \psi \, d\tau \tag{1.33}$$

If we continue to ignore the overlap integrals, the denominator is equal to 1. Let us replace ψ by the expression (1.30) and ψ^* by the conjugated quantity

$$E = (1-\lambda) \int \psi^*_{\text{cov}} H\psi_{\text{cov}} \, d\tau + \lambda \int \psi_{\text{ion}} H\psi_{\text{ion}} \, d\tau$$
$$+ 2\lambda^{1/2}(1-\lambda)^{1/2} \int \psi_{\text{ion}} H\psi_{\text{cov}} \, d\tau \tag{1.34}$$

We now reach a stage at which, having ignored the overlap integrals, it is harder to keep our reasoning as rigorous. It is worth working out semi-phenomenological expressions which have to be tested by experience. Let us use the terms E_{cov} and E_{ion} to refer to the bonding energies corresponding to the functions ψ_{cov} (bonding) and ψ_{ion}. One can write

$$E = (1-\lambda) E_{\text{cov}} + \lambda E_{\text{ion}} + 2\lambda^{1/2}(1-\lambda)^{1/2} E_{\text{res}} \tag{1.35}$$

taking the last integral as a resonance energy. It was seen in Section 1.2 that the best of the various solutions to Schrödinger's equation was the one corresponding to the lowest energy. Let us give various values to λ and take the one corresponding to the minimum energy. The equation $dE/d\lambda = 0$ allows one to calculate E_{res} in terms of E, E_{cov}, and E_{ion}, giving [1.17]

$$E = [(1-\lambda) E_{\text{cov}} - \lambda E_{\text{ion}}]/(1-2\lambda) \tag{1.36}$$

and

$$\lambda = (E_{\text{cov}} - E)/(E_{\text{cov}} + E_{\text{ion}} - 2E) \tag{1.37}$$

We therefore have a way of calculating the parameter λ in terms of the quantities E_{cov}, E_{ion}, and E, the latter corresponding to the actual bonding energy as it can be obtained from thermochemical results. We shall do this calculation for hydracid molecules. The average of the simple bonding energies of hydrogen and halogen is equal to E_{cov} [1.5]. $E_{\text{ion}} = I + A - C$ ($I =$ the ion-

1.4. Extension of the VB Method

ization energy of hydrogen, A = electronic affinity of halogen, C = coulombic interaction). Let us take

$$C = (e^2/d)(1 - 1/k) \tag{1.38}$$

where e is the elementary charge, d the interatomic distance, and $(1-1/k)$ allows for the repulsive forces with the usual value $k=9$; E is the formation energy of the gas molecule [1.15]. The results of the calculation correspond very satisfactorily with the percentage ratio of the dipolar moment measured and the moment that would be produced by two point charges carried by the nuclei to their equilibrium distance (values used in defining the "ionic character" (i.c.) by Pauling [1.5]).

HF	$\lambda = 0.57$	i.c. = 0 60
HCl	0.18	0.19
HBr	0.10	0.11
HI	0.02	0.04

As already mentioned, the two electrons in the Lewis pair are distributed $(1-\lambda)$ on i and $(1+\lambda)$ on j. Let us assume that the molecule is strongly "dilated" so that the atoms are in their neutral state. Let λ_0 be the particular value of the parameter in this situation, and c and c' the number of bond pairs in which each atom participates. In a purely ionic bond, the nucleus i is completely denuded and transfers all the $c(1-\lambda_0)$ valency electrons it contained to its j neighbors. Let $+n$ be the ionic charge which it then takes on, equal to the valency of the i element. We have

$$c(1-\lambda_0) = n \tag{1.39}$$

Similarly, in a purely covalent bond, the nucleus j retains only c' bond electrons (one per pair). Of the $c'(1+\lambda_0)$ which it contains, it therefore has to transfer $c'\lambda_0$ to its i neighbors. Let $+m$ be the charge corresponding to this transfer (covalent charge). We have

$$c'\lambda_0 = m \tag{1.40}$$

From Eqs. (1.39) and (1.40), we obtain the value of the parameter λ_0 in the particular situation of a dilated molecule made up of neutral atoms

$$\lambda_0 = m/c' = 1 - n/c \tag{1.41}$$

From this we deduce the relation between covalent and ionic charges when $c = c'$,

$$n + m = c \tag{1.42}$$

Let us now return to the real molecule with its atoms at their equilibrium distance. Let us assume that while atoms i and j are approaching one another, the former have transferred q electrons per j nucleus to the latter (or the latter

q' electrons per i nucleus to the former). It is obvious that the extreme charges n and m set the limits for such transfers,

$$q \leqslant n \quad \text{and} \quad q' \leqslant m \tag{1.43}$$

We shall call q and q' the *effective charges* supported by the atoms i and j, in contrast to the charges n and m, which involve ideal cases seldom met with in practice. The number of electrons retained by the isolated, neutral atom j was found to be equal to $c'(1+\lambda_0)$. When we moved from the dilated molecule to the real molecule, it was increased by the q electrons transferred to it by its i neighbors. The valency electrons contained by it in the real molecule are therefore

$$c'(1+\lambda) = c'(1+\lambda_0) + q \tag{1.44}$$

which gives [1.13]

$$\lambda = \lambda_0 + q/c' \tag{1.45}$$

We shall call λ_0 the *equilibrium ionicity* since it corresponds to a balance between the two opposing electron transfers bound up with the formation of pure covalent and ionic configurations and preserving for the atoms the number of electrons they had in their neutral, isolated state. The ionicity of the bond is therefore the result of the addition to the equilibrium ionicity of a polarity correction: the number of electrons transferred per bond. By definition, λ, λ_0 and q are probabilities of presence or electron densities extended to a certain volume, in other words, numbers without dimensions. However, q can represent an electrostatic charge expressed in electronic units.

In the case of hydracids, as in that of many simple molecules, $c = c' = 1$, $n = 1$, and $m = 0$. The equilibrium ionicity λ_0 is therefore zero, and $\lambda = q$. The effective charges borne by the halogens are thus equal to the values given above for λ. It can be seen that the upper signs in H^+X^- are wrong, even for HF, which is the most ionic molecule. The values of the effective charges carried by hydrogen atoms in a few simple molecules are given below

NH_3	+0.09		H_2O	+0.21
PH_3	+0.01		H_2S	+0.04
AsH_3	0		H_2Se	0
SbH_3	0		H_2Te	0

If one takes the bond H—Se ($\lambda = 0$) as purely covalent, it is possible to calculate the angle H—O—H, corresponding to the equilibrium between the forces of molecular rigidity and coulombic repulsion. Bailly [1.18] has done an approximate calculation, giving a result of around 105°, corresponding satisfactorily with experimental results. It is worth mentioning that the tetrahedral arrangement of the hydrogen atoms in the CH_4 molecule is the one ensuring maximum distance H—H for a given bond length C—H.

REFERENCES

[1.1] W. J. Moore, "Physical Chemistry." Longmans, Green, New York and London, 1963.
[1.2] F. Seel, "Atombau und chemische Bindung." F. Enke, Stuttgart, 1956.
[1.3] J. A. A. Ketelaar, "Chemical Constitution." Elsevier, Amsterdam, 1958.
[1.4] H. Eyring, J. Walter, and G. E. Kimball, "Quantum Chemistry." Wiley, New York, 1944.
[1.5] L. Pauling, "The Nature of the Chemical Bond." Cornell Univ. Press, Ithaca, New York, 1950.
[1.6] G. N. Lewis, "Valence and the Structure of Atoms and Molecules." Dover, New York, 1966.
[1.7] W. Heitler and F. London, *Z. Phys.* **44**, 455 (1927).
[1.8] G. Pannetier, "Chimie générale, atomistique, liaisons chimiques." Masson, Paris, 1966.
[1.9] M. Kasha, quoted by Moore [1.1].
[1.10] Y. K. Syrkin and M. E. Dyatkina, "Structure of Molecules." Butterworth, London and Washington, D.C., 1950.
[1.11] C. A. Coulson, "Valence." Oxford Univ. Press (Clarendon), London and New York, 1952.
[1.12] A. M. Karo and A. R. Olson, *J. Chem. Phys.* **30**, 1232 (1959).
[1.13] J. P. Suchet, *Proc. Int. Conf. Semicond. Phys.* (*Prague 1960*), *Czech. J. Phys.* 904 (1961).
[1.14] J. P. Suchet, *J. Phys. Chem. Solids* **21**, 156 (1961).
[1.15] J. P. Suchet and F. Bailly, *Ann. Chim.* (*Paris*) **10**, 517 (1965).
[1.16] A. Julg, "Chimie théorique." Dunod, Paris, 1964.
[1.17] F. Bailly, *J. Phys.* (Paris) **27**, 335 (1966).
[1.18] F. Bailly, Personal communication, 1967.

Chapter 2 | **From the Molecule to the Crystal**

2.1. COVALENT ASPECT OF BONDS

In this chapter, knowledge of the elements of crystallography and the structure of the main mineral compounds is assumed. What interests us in this structure, however, is the arrangement of the different atoms, their interatomic distances and bond angles. The symmetry classes so dear to the mineralogists are only of secondary interest here. Readers wishing to go more thoroughly into certain points can refer to the classic digests, such as Wells' [2.1, 2.2].

The importance of the covalent aspects of bonds in mineral crystals was brought into light in the fifties, mainly through the systematic work of Krebs [2.3–2.5]. Here is a brief analysis. Let MX be the crystal in an equiatomic binary compound. One can distinguish two main categories of crystal arrangements. The first arises from the A4 structure of the diamond, in which one carbon atom in every two is replaced by a metal atom M and the other by a metalloid atom X (Fig. 2.1). The second arises from the A7 structure of arsenic, in which the same alternate substitution has been done (Fig. 2.2). Let us consider them in turn, attaching the main derivative structures to each, particularly that of rock salt for the second.

In the first category of arrangement, the ideal example of which is the B3 structure of zinc blende ZnS, each atom is surrounded by four neighbors, arranged in the form of a tetrahedron, and the bond results from the formation of pairs of electrons each occupying the quantum state of an sp^3 hybrid (Fig. 2.3a). These AOs lie almost exclusively in the direction of the partner. Since, in addition, any two bonds MX and MX' always form the same angle $109°28'$, the part of the orbital opposite the partner has no possible overlap

2.1. Covalent Aspect of Bonds

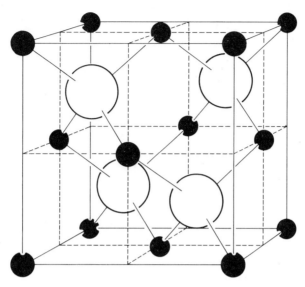

Fig. 2.1. Structure of a binary compound derived from the diamond. The black circles represent the metal atoms. Each atom has four neighbors.

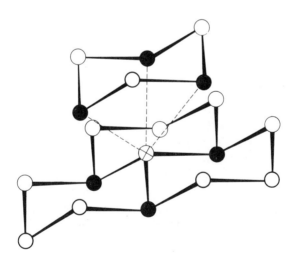

Fig. 2.2. Structure of a binary compound derived from arsenic. The dotted lines show the possibility of acquiring six neighbors instead of three, for the atom marked with a cross. (Based on Krebs [2.3] *Physica* **20**, 1125 (1954), Fig. 4, p. 1128. By permission of North-Holland Publishing Company.)

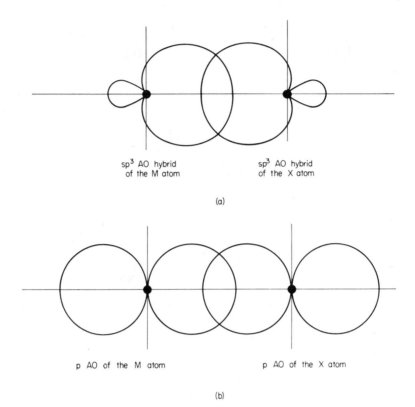

Fig. 2.3. Wave functions of valency electrons in bond pairs: (a) structure derived from A4 and (b) structure derived from A7. (Based on Krebs [2.3] *Physica* **20**, 1125 (1954), Fig. 2, p. 1127. By permission of North-Holland Publishing Company.)

with any one of the other three orbitals in the same atom. A pair is therefore more or less localized between two atoms M and X. The partially covalent aspect of the B3 structure has never been brought into question, and measurements of electron density, for ZnS crystals for instance, show clearly that it remains high halfway between the atoms. What is more, this structure, like the similar B4 structure of wurtzite, involves a low-density packing in which the stability can be due only to strongly directed bonds. All the compounds crystallizing there have much lower interatomic distances than the sum of the ionic radii: Pauling [2.6] has produced a set of tetrahedral "radii" which makes verification easy. The contraction is 9% for ZnS.

Among structures derived from zinc blende, let us look briefly at the C_1 structure of fluorite CaF_2, in which all the tetrahedral sites between the metal atoms are occupied (instead of only half in the B3 structure). This structure may include vacancies in the case of a trivalent metal (Y_2O_3 and mixed

2.1. Covalent Aspect of Bonds

crystals CeO_2–Y_2O_3). It can also accept mixed crystals (CaF_2–YF_3) in which the excess fluorine atoms occupy interstitial sites, contradicting the rule of electrostatic valencies. In the lattice of Bi_2OF_4, the fluorine even occupies octahedral sites. Finally, fluorite presents cleavages parallel to the sites of an octahedron, which can be explained only by the existence of a partially covalent aspect [2.7]. If one considers a sublattice similar to zinc blende, with tetrahedral bonds formed from hybrid AOs, one can see that the formula CaF will have one valency electron too many and that the remaining fluorine, in the anion state, would be excluded. But since all the fluorine sites are identical, it would be just as logical to adopt another formula in which their roles are permutated

$$[CaF_I]^+ \; F_{II}^- \longleftrightarrow [CaF_{II}]^+ \; F_I^- \tag{2.1}$$

Only this resonance between two mesomer states can explain the remarks above. In particular, in the case of Bi_2OF_4, in which the zinc blende sublattice would have two electrons too many, the presence of an extra fluorine anion in an interstitial site ensures the equilibrium of the electron formula [2.5].

In the second category of arrangement, the ideal structure would consist of lattices of superimposed hexagons, spatially distorted in such a way that the angles are 90°. One obtains a series of double atomic layers, without bonds. Figure 2.2 thus corresponds approximately to the B29 structure of SnS. The real bond angles are usually higher than 90°, but approach it as the atomic number increases. Simultaneously, the distance between double layers falls, so that each atom, in addition to its three neighbors at a distance r_1 (pyramidal trigonal arrangement), acquires three others on the next layer at a distance r_2 (octahedral arrangement) (Fig. 2.2):

As	97°00′	$r_1 = 2.51$ Å	$r_2 = 3.15$ Å	$r_2/r_1 = 1.25$
Sb	95°35′	2.91	3.35	1.15
Bi	95°29′	3.07	3.53	1.15

There is thus a trend toward 90° and $r_1/r_2 = 1$, which would correspond to the B1 structure of rock salt. Only the three p AOs available on these atoms can explain these bonds (Fig. 2.3b), and their high value in two opposing directions on either side of the atom explains the resonance between two succeeding double layers (Fig. 2.4).

The octahedral arrangement of the B1 structure is thus defined, from the standpoint of its partial covalent aspect, as a derivative of the A7 structure. Let us compare the compounds SrS and PbS, for example. Strontium has an ionization energy of only 17 eV, compared with 23 eV for lead. It thus loses its valency electrons more easily. In addition, it has an electron formula of $5s^2$ compared with $6s^2p^2$ for lead. The formation of a pair of electrons each occupying a quantum state p thus first requires the exciting of an s electron in a

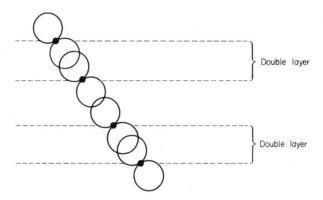

Fig. 2.4. Overlapping of the p electron AOs of bismuth in the direction of a symmetry axis of fourth order. (Based on Krebs [2.3] *Physica* **20**, 1125 (1954), Fig. 5, p. 1129. By permission of North-Holland Publishing Company.)

p state and will be more difficult. It may be expected that the covalent aspect of the structure, although present in all compounds, will be more important in the case of PbS. The interatomic distance, what is more, shows a contraction of 3% in relation to the sum of the ionic radii, and the energy of the lattice is 4% higher than the calculated value for an ionic crystal. The resonance system is thus particularly significant between the two mesomer states [2.4]

$$Pb_I-S_I \quad Pb_{II}-S_{II} \quad Pb_{III}- \longleftrightarrow -Pb_I \quad S_I-Pb_{II} \quad S_{II}-Pb_{III} \qquad (2.2)$$

This remark shows the essential difference between this and the tetrahedral type of arrangement, where the electron pairs were more strictly localized.

Following on from this work, the author [2.8] in 1959 recalled Bradley and Hume–Rothery's rule, according to which each atom of an element in column N of the periodic table possesses $8-N$ next neighbors, emphasizing that it merely expresses a tendency to complete the electron octet through the effect of *covalent* bonds. If $a = N$, the number of valency electrons in the atom, and $b = 8-N$, the number of covalent bonds formed with neighbors, we are tempted to write

$$a+b = 8 \qquad (2.3)$$

But this equation assumes that all the valency electrons are liable to participate in Lewis pairs (the case of tetrahedral bonds of the first type of arrangement with $N = $ IV). If only the p electrons intervene, $a = N-2$, and the second term is reduced to 6 (the case of pyramidal trigonal bonds of the second type of arrangement with $N = $ V). If, in addition, there is a nonbonding p pair, one drops to 4 (a chain of diagonal bonds with $N = $ VI), and with two nonbonding p pairs to 2 (diatomic molecules with $N = $ VII). The type pf arrangement depends basically on the number of valency electrons, and so it is preserved by isoelectronic substitutions:

$$\begin{array}{l} sp^3: \quad IV\ (A4) \longrightarrow III\ V \longrightarrow II\ VI \longrightarrow I\ VII\ (B3) \\ p^3: \quad\ \ V\ (A7) \longrightarrow IV\ VI\ or\ II\ VI \longrightarrow I\ VII\ (B1) \end{array} \quad (2.4)$$

Naturally, in the series above, the covalent aspect diminishes as the difference in electronegativity among the component elements increases. When, for instance, two Ge atoms are replaced by one In atom and one Sb atom, the bond per pair of electrons in the sp^3 state would remain strictly identical if one could write $In^{(sp^3)}Sb^{(sp^3)}$ or In^-Sb^+. In fact, the electrons are more strongly bound to Sb, and the center of gravity of the Lewis pair shifts towards the antimony atom, approaching the electron distribution of a neutral bond In^0Sb^0. For ZnS or CuBr, there is even a tendency towards the ionic bond $Zn^{2+}S^{2-}$ or Cu^+Br^-. The wave function has to allow for this fact by including an ionic term and a heteropolarity coefficient [2.4]. Here we find a problem similar to the one involving molecules in Section 1.4. We shall return to it in the next section.

Few compounds of transition metals or rare earths adopt the first type of arrangement. The main example is chalcopyrite $CuFeS_2$, the structure of which is derived from B3 by alternate substitution on the metal sites. In the second type, MnO, MnS, MnSe, FeO, CoO, NiO, EuO, EuS, EuSe, and EuTe are to be found in the B1 structure, but most of the equiatomic binaries crystallize in the $B8_1$ structure of nickel arsenide NiAs, which is derived from the preceding one by a slight distortion, and the B31 structure of MnP, which is more distorted. Elimination of half the metal sites produces the C19 structure of $FeCl_2$. One also finds many mixed types of arrangement, such as the C2 structure of pyrites FeS_2 and the distorted C18 structure of $FeAs_2$ (octahedral for Fe and tetrahedral for S or As), the C4 structure of rutile TiO_2 (octahedral for Ti, plane trigonal for O), the C6 structure of PdS_2 (octahedral for Pd, pyramidal trigonal for S), the B17 structure of PtS (plane tetragonal for Pt, tetrahedral for S) and the $D7_3$ structure of Th_3P_4 (cubic for Th, octahedral for P). The presence of a partially filled d level naturally complicates the bonds and the octahedral arrangement around the metal often results from d^2sp^3 hybrid orbitals (C2, C18, C4, C6, C19). These structures will be examined in greater detail in Part 2.

2.2. VB Approach (Crystallochemical Model)

The problems raised by the extension of the Heitler and London function to the crystal were defined clearly in 1953 in a paper read by Krebs and Schottky [2.9] to the German Physics Society. A simple general solution in the form of a product of functions for the different pairs in the cell can be found only if these pairs are independent, namely in the case of the tetrahedral arrangement. If all the atoms are identical (A4 structure), the fundamental state is written in the form of the product

$$\psi = \prod^N (\psi_\text{I}) \prod^N (\psi_\text{II}) \prod^N (\psi_\text{III}) \prod^N (\psi_\text{IV}) \tag{2.5}$$

where N is the number of elementary cells and the indices I to IV refer to the four pairs formed by each atom:

$$\psi_\text{I} = \varphi_i(\alpha)\varphi_j(\beta) + \varphi_i(\beta)\varphi_j(\alpha) \quad \text{etc.} \tag{2.6}$$

$\varphi_i(\alpha)$ and $\varphi_j(\beta)$ are the AOs for two successive atoms i and j contributing two electrons with opposite spins α and β to the bond pair. The total energy is the sum of the energies of all the bond pairs. One can also define an excited state with higher energy corresponding to the antibonding Heitler and London functions. Although it is possible to try to find a more accurate solution [2.10], this expression does not allow one to move to a type of representation in which the same function would represent each elementary cell ("homotypical" representation).

On the other hand, some aspects of the bond can be investigated more thoroughly. Let us return to the point of view expressed by Suchet and Bailly [2.11] as described in Section 1.4, and calculate the ionicity of a crystal from the bonding energies. Now E, E_ion, and E_cov will refer to the real energy, the energy associated with an ionic configuration, and the energy associated with a purely covalent configuration, respectively. As a reference state, we shall choose that of the crystal broken down into its gas elements in the atomic state. The real energy is then negative, and can be worked out from thermochemical data:

$$E = -[S_\text{M} + S_\text{X}] + \Delta H_\text{f} \tag{2.7}$$

S_M and S_X are the sublimation energies of the metal M and the metalloid X. ΔH_f is the energy required to form the crystal from elements in the solid state (standard formation heat). The energy E_ion is equal here to $I + A - Dn^2 C$, where I and A, supplied by tables or estimated by analogy, refer to the ionization of the M and X atoms in M^{n+} and X^{n-}. The coulombic energy (1.38) is supplied by bringing the ions together; D is the Madelung constant.

It is less simple to calculate E_cov, since this energy does not generally correspond to the average simple bonding energies of the elements. It was seen, at the end of the previous section, that formation of the covalent configuration of a crystal (such as $\text{In}^{(sp^3)}\text{Sb}^{(sp^3)}$) implies an electron transfer, a problem which did not arise for hydracid molecules. Let us take the case of tetrahedral arrangement. This transfer of m electrons involves an affinity A' and an ionization energy I' and gives $M^{(s^2p^2)}$ and $X^{(s^2p^2)}$. Their exciting in the states

2.2. VB Approach (Crystallochemical Model)

$M^{(sp^3)}$ and $X^{(sp^3)}$ requires the hybridization energies H_M and H_X. The excited atoms are then brought together, producing the restitution energy R caused by the coupling of the electron orbits and a coulombic energy C' caused by the existence of the "covalent" charges m. Finally, one obtains [2.11]

$$E_{cov} = A' + I' + H_M + H_X - R - C' \tag{2.8}$$

Figure 2.5 shows the quantity $-\Delta H_f$ in terms of the ionicity λ, for different series of compounds crystallizing in the B3 structure. It will be noted that this quantity reaches a minimum level around 0.25 for the III V compounds (InSb) and 0.50 for the II VI compounds (HgTe). It is generally accepted that in the expression (2.7), ΔH_f corresponds approximately to the polar interactions in the crystal while the square bracket corresponds to the juxtaposition of neutral atoms. The minimum observed therefore probably corresponds to the ionicity of a compound in which there is neither any transfer of electrons between the isolated atoms during the formation of the crystal nor, consequently, any polar interactions. This viewpoint is confirmed by various physical properties of InSb and HgTe. What is more, if one calculates the equilibrium ionicity λ_0 defined in Section 1.4, Eq. (1.41) gives precisely 0.25 for the III V, 0.50 for the II VI and 0.75 for the I VII compounds ($c = 4$). In the octahedral arrangement due to the p orbitals, it would give 0.33 for the IV VI or II VI and 0.67 for the I VII compounds ($c = 3$). Finally, it is worth noting that the quantity $c\lambda_0$ is equal to the difference in electronegativities given by Pauling [2.6] for hypothetical binary compounds combining elements from the first period of the table.

Effective charges q were defined in Section 1.4. They result from two antagonistic phenomena. On the one hand, the electronegativity difference between the M and X atoms tends to transform them into ions ($q = +n$). On the other hand, the polarization of each ion in its partner's field causes interpenetration of electron clouds. This constitutes a return to a covalent configuration and would tend to transform a doublet of the anion into a shared Lewis pair ($q = -m = n - c$). One can therefore write:

$$q = n - F \quad (0 < F < c) \tag{2.9}$$

Approximate reasoning shows that estimation is possible using the expression [2.11, 212]:

$$q = n[1 - 0.01185(Z/r' + Z'/r)] \tag{2.10}$$

If the ionic charge n' of the anion is not identical with n, one obtains for the X atom an effective charge $q' \neq -q$, which it is easy to calculate. These charges concern the atoms at rest whereas the effective charges defined earlier by other authors referred to vibrating atoms. They are not independent of one another [2.11].

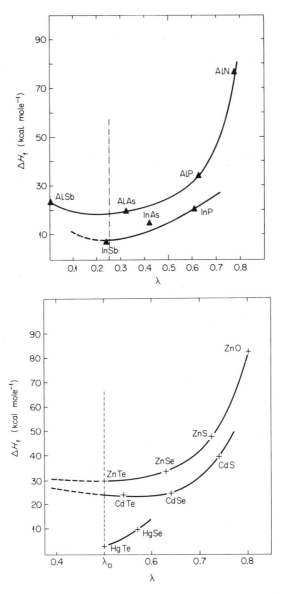

Fig. 2.5. Variation in the standard formation heat $-\Delta H_f$ in relation to the ionicity of the bonds: (a) III V compounds, (b) II VI compounds (based on Suchet et al. [2.11]).

Calculation of λ from the bonding energies requires experimental data which do not exist for most crystals. It is therefore very important to point out that calculation from the fundamental expression (1.45) and the Eq. (2.10) gives the same result to within 0.05 as the calculation based on the bonding energies for III V and II VI compounds in the B3 structure. We will use these expressions for the main compounds of transition metals T and rare earths L. Table 2.1 gives the effective charge q borne by the metal atom in the crystal and the ionicity λ of a bond pair T—X (in a structure with vacancies, this differs from the average ionicity). Negative values for λ are obviously impossible, by definition. It can be shown that they constitute a strong presumption in favor of metal properties [2.11]. The formulas in parentheses refer to non-stoichiometric phases. Taking distorted structure bonds, such as B31 or C19, as being the same as those for the ideal B1 structure, naturally provides a reliable approximation.

2.3. MO Approach (Band Model)

Krebs and Schottky [2.9] have also discussed the MO approach to the A4 structure, referring to the molecular wave functions with a single electron proposed by Hund [2.13]:

$$\psi = \prod^N \psi_{I\alpha+} \prod^N \psi_{I\beta+} \prod^N \psi_{II\alpha+} \cdots \prod^N \psi_{IV\beta+} \qquad (2.11)$$

N always refers to the number of elementary cells and $\psi_{\alpha+}$ and $\psi_{\beta+}$ to the LCOAs

$$\psi_{\alpha+} = \varphi_i(\alpha) + \varphi_j(\alpha), \qquad \psi_{\beta+} = \varphi_i(\beta) + \varphi_j(\beta) \qquad (2.12)$$

The approximation obtained is not so good as with the VB approach, but the use of equal coefficients for the Heitler–London term and the ionic term introduces a worthwhile simplification by producing a homotypical representation. Without going into mathematical explanations, let us merely mention that such a situation allows the electrons to be described by normal Bloch functions $\exp(i\mathbf{k}\mathbf{x}) u(\mathbf{x})$, where $\mathbf{k} = 2\pi/\lambda$ is the wave number of the wave associated to the electron. Occupation of all the elementary cells then shows all the values of $\mathbf{k}$, and leads to the representation of the crystal by completely filled energy "bands," corresponding to the insulating properties of the diamond. Empty energy bands, or bands with much higher energy, would have appeared if we had taken the antibonding molecular functions into account.

This representation is that of the band model, the detailed mathematical formulas of which meets an engineer's need to calculate all the physical effects in semiconductor devices using the germanium, silicon, etc. The numerical coefficients for a given body, however, can be found only after measuring a large number of parameters to establish the "band structure" of the crystal. Unfortunately, neither this nor the parameters on which it is based can be worked out from atomic and crystallographic data alone. In these circum-

TABLE 2.1

Rock salt B1 (resonating bonds)

L(III) V ($\lambda_0 = 0$)

LaN	$q = 1.57$	$\lambda = 0.52$	GdAs	$q = 0.77$	$\lambda = 0.26$
LaP	1.54	0.51	GdSb	0.23	0.08
LaAs	1.02	0.34	LuN	1.21	0.40
LaSb	0.55	0.18	LuP	1.17	0.39
GdN	1.38	0.46	LuAs	0.53	0.18
GdP	1.35	0.45	LuSb	−0.05	−0.02

T(II) VI or L(II) VI ($\lambda_0 = 0.33$)

MnO	$q = 1.31$	$\lambda = 0.77$	EuO	$q = 0.75$	$\lambda = 0.58$
MnS	1.17	0.72	EuS	0.84	0.61
MnSe	0.66	0.55	EuSe	0.51	0.50
FeO	1.28	0.76	EuTe	0.21	0.40
CoO	1.26	0.75	YbO	0.62	0.54
NiO	1.23	0.74	YbS	0.71	0.57
			YbSe	0.35	0.45
			YbTe	0.02	0.34

Nickel arsenide B8$_1$ and manganese phosphide B31 (resonating bonds)

T(III) V ($\lambda_0 = 0$)   T(II) VI ($\lambda_0 = 0.33$)

MnAs	$q = 0.70$	$\lambda = 0.23$	CrS	$q = 1.21$	$\lambda = 0.74$
FeAs	0.63	0.21	FeS	1.13	0.71
CoAs	0.58	0.19	CrSe	0.72	0.57
CrSb	−0.09	−0.03	FeSe	0.41	0.47
MnSb	−0.24	−0.08	(CrTe)	0.24	0.41
FeSb	−0.33	−0.11	MnTe	0.15	0.38
CoSb	−0.40	−0.13	(FeTe)	0.05	0.35
NiSb	−0.45	−0.15	(CoTe)	0.00	0.33
MnBi	−1.96	−0.65	(NiTe)	−0.02	0.31

Defect nickel arsenide (resonating bonds)

□ T$_2$(III) VI$_3$ ($\lambda_0 = 0$)

(□Cr$_2$S$_3$)	$q = 1.67$	$\lambda = 0.56$	□Cr$_2$Te$_3$	$q = -0.12$	$\lambda = -0.04$
(□Fe$_2$S$_3$)	1.56	0.52	□Fe$_2$Te$_3$	−0.36	−0.12
(□Cr$_2$Se$_3$)	0.77	0.26	□Co$_2$Te$_3$	−0.43	−0.14
(□Fe$_2$Se$_3$)	0.59	0.20	□Ni$_2$Te$_3$	−0.49	−0.16

TABLE 2.1 (continued)

CADMIUM CHLORIDE C19 (RESONATING BONDS)
$\square T(II) \ VII_2 \ (\lambda_0 = 0.33)$

$\square MnCl_2$	$q = 1.17$	$\lambda = 0.72$	$\square CoCl_2$	$q = 1.10$	$\lambda = 0.70$
$\square MnBr_2$	0.65	0.55	$\square CoBr_2$	0.54	0.51
$\square MnI_2$	0.15	0.38	$\square CoI_2$	0.00	0.33
$\square FeCl_2$	1.15	0.71	$\square NiCl_2$	1.07	0.69
$\square FeBr_2$	0.58	0.53	$\square NiBr_2$	0.50	0.50
$\square FeI_2$	0.05	0.35	$\square NiI_2$	−0.07	0.31

CASSITERITE C4 ($c = 6$, $c' = 3$)
$T(IV)O_2$ ($\lambda_0 = 0.33$)			$T(II)F_2$ ($\lambda_0 = 0.67$)		
TiO_2	$q = 2.70$	$\lambda = 0.78$	MnF_2	$q = 1.30$	$\lambda = 0.88$
VO_2	2.57	0.76	FeF_2	1.27	0.88
CrO_2	2.47	0.74	CoF_2	1.24	0.87
$\beta\text{-}MnO_2$	2.41	0.73	NiF_2	1.22	0.87

PYRITE C2 AND MARCASITE C18 ($c = 6$, $c' = 3+1$)
$T(IV)$ (V–V) ($\lambda_0 = 0.33$)			$T(II)$ (VI–VI) ($\lambda_0 = 0.67$)		
$CrSb_2$	$q = -0.93$	$\lambda = 0.18$	MnS_2	$q = 1.20$	$\lambda = 0.87$
FeP_2	1.91	0.68	$(MnSe_2)$	0.68	0.78
$FeAs_2$	0.34	0.39	$MnTe_2$	0.18	0.70
$FeSb_2$	−1.21	0.13	FeS_2	1.15	0.86
(CoP_2)	1.86	0.65	$FeSe_2$	0.61	0.77
$CoAs_2$	0.26	0.38	$FeTe_2$	0.08	0.68
$CoSb_2$	−1.32	0.11			
(NiP_2)	1.83	0.64			
$NiAs_2$	0.24	0.37			
$NiSb_2$	−1.33	0.11			

DEFECT FLUORITE C1 ($c = 4$, $c' = 2$, RESONATING BONDS)
$L_2(III) \ O_3 \square \ (\lambda_0 = 0.25)$

$Y_2O_3\square$	$q = 1.70$	$\lambda = 0.68$	$Gd_2O_3\square$	$q = 1.10$	$\lambda = 0.53$
...			...		
$La_2O_3\square$	1.32	0.58	$Er_2O_3\square$	0.94	0.49
...			...		
$Nd_2O_3\square$	1.22	0.55	$Lu_2O_3\square$	0.89	0.47
...					

TABLE 2.1 (continued)

Defect thorium phosphide $D7_3$ ($c = 4$, $c' = 3$, resonating bonds)
$\square_{0.25}L_2(III) VI_3$ or $\square_{0.25}A_2(III) VI_3$ ($\lambda = 0.25$)

	q	λ		q	λ
		$= 0.65$		$= 1.00$	$= 0.50$
$\square_{0.25}Y_2S_3$	1.62	0.65	$\square_{0.25}Lu_2S_3$	1.00	0.50
$\square_{0.25}Y_2Se_3$	0.98	0.49	$\square_{0.25}Lu_2Se_3$	0.40	0.35
$\square_{0.25}La_2S_3$	1.40	0.60	$\square_{0.25}Ac_2S_3$	0.76	0.44
$\square_{0.25}La_2Se_3$	0.92	0.48	$\square_{0.25}Ac_2Se_3$	0.31	0.33
$\square_{0.25}Ce_2S_3$	1.36	0.59	$\square_{0.25}Th_2S_3$	0.74	0.43
$\square_{0.25}Ce_2Se_3$	0.86	0.46	$\square_{0.25}Th_2Se_3$	0.28	0.32
...			...		
$\square_{0.25}Gd_2S_3$	1.20	0.55	$(\square_{0.25}U_2S_3)$	0.67	0.42
$\square_{0.25}Gd_2Se_3$	0.65	0.41	$(\square_{0.25}U_2Se_3)$	0.17	0.29
...			...		

stances, whatever its usefulness for electronics specialists, the model remains empirical, adapted individually to each compound. Research workers, trained as chemists, adopting as their main goal the forecasting of the approximate properties of new bodies before carrying out their synthesis, would find such a model useless. One has to speak out and say this. However, a summary of it will be given since it is essential to know the concepts and terminology involved, widely used to describe transport properties in semiconductor crystals. It also provides a point of view that is vital in investigating practical applications.

Among the numerous treatises devoted to it may be mentioned Smith's work of mathematical physics [2.14] and the briefer and more accessible one by Aigrain and Englert [2.15]. Readers not interested in the mathematical aspect and wishing simply to find out a few conventional facts about semiconductors may prefer the summary by Guillien [2.16]. The *valence band* refers to the fully occupied energy band mentioned above, corresponding to all the bonding MOs. The *conduction band* corresponds to the antibonding MOs.

The distribution of the electron in relation to the energy is extremely simple at absolute zero where the quantum states are occupied regularly one after the other beginning with the lowest energy states. The first empty level is called the *Fermi level* E_F. It is a characteristic property of the crystal concerned, and its position can easily be determined if one knows the number of available states in relation to the energy and the number of electrons. At temperatures above zero, thermal agitation disturbs this simple order, and the probability of finding an electron at the energy level is represented by the expression $1/[1 + \exp(E - E_F)/kT]$, which depends solely on the difference $E - E_F$. The probability reaches 50% at the Fermi level. In a pure semiconductor crystal

2.3. MO Approach (Band Model)

without defects E_F is located in the forbidden energy band, and the probabilities in the valence and conduction bands are 1 and 0, respectively, at absolute zero. They move gradually away from these values, on either side of the forbidden band, as the temperature rises. The electrons in the first band can become free and move into the second band if they acquire extra energy equal to the energy gap E_G separating them. This is the *intrinsic* conductibility. They leave behind them, in the valence band, a hole which behaves like a positive particle (Fig. 2.6). Their number is

$$N = N_0 \exp(-E_G/2kT) \qquad (2.13)$$

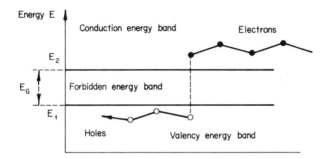

Fig. 2.6. Intrinsic conductibility in the band model.

The presence of E_F in a permitted energy band would mean that the band is partially filled even at absolute zero, in other words that a metal is involved.

When a crystal is not perfect, defects or impurities cause local disturbances in the Bloch waves and create discrete levels in the forbidden energy band, capable of giving the conduction band or accepting from the valence band a limited number of electrons. The extra energy needed e_d (or e_a) is generally lower here. Distribution of the electrons follows the same law as before, in

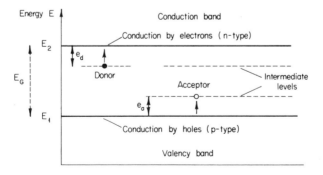

Fig. 2.7. Extrinsic conductibility in the band model.

relation to the temperature, but only one of the two bands is affected. This is *extrinsic* conductibility by N electrons (donor) or P holes (acceptor) (Fig. 2.7)

$$N = N_0 \exp(-e_d/kT) \tag{2.14}$$

This behavior is often provoked deliberately by doping, the addition of measured quantities of particular impurities to the pure crystal. At absolute zero, the Fermi level is between the donor level and the conduction band (or between the valence band and the acceptor level). As the temperature rises, however, it approaches the position it would have in intrinsic conductibility, with the influence of the levels of impurities diminishing. It can be seen that in the presence of a single intermediate level, assumed to be a donor, and if $e_d \ll E_G$, one can distinguish three successive regions in relation to T: (1) extrinsic conductibility, (2) region of exhaustion of the donor, (3) intrinsic conductibility (Fig. 2.8). The product of the densities of carriers per cubic centimeter is still, even in the extrinsic region,

$$NP = AT^3 \exp(-E_G/2kT) \tag{2.15}$$

where A is the characteristic constant of the crystal, which can be calculated if the band structure is known.

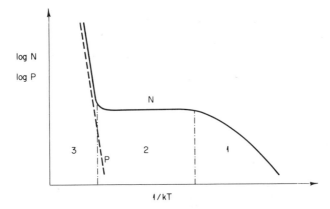

Fig. 2.8. Number of free carriers, electrons (N) or holes (P), in relation to the temperature in the case where $N > P$ (based on Aigrain *et al.* [2.15]).

The idea of forbidden band may be understood by assuming that the electrons of certain speeds can be reflected in the same way as X rays. Let us express their kinetic energy in terms of the associated wavelength (1.1):

$$E = mv^2/2 = h^2/2m\lambda^2 = (h^2/8\pi^2 m)|\mathbf{k}|^2 \tag{2.16}$$

A crystal arrangement with a single period a changes the parabola $E(\mathbf{k})$ into a series of distorted arcs in the region of the reflections (for $2a$ multiple of λ or $|\mathbf{k}|$ multiple of π/a). Let us take a straight line passing through the origin of the coordinates and give it all the possible orientations in turn. The values of $\mathbf{k}$, multiples of π/a, entered on this straight line, describe a series of concentric polyhedrons (Brillouin zones). Such a lattice, inverted in relation to the real lattice, is simply the crystallographer's reciprocal lattice (it is sometimes said that physicists, attached to the band model, reason in the space of the $\mathbf{k}$ numbers while chemists, attached to the VB approach, reason in the space of the x numbers). The E values corresponding to $\mathbf{k} = \pi/a$ at the limit of the first zone form the boundary of a forbidden band of a height $E_G = E_2 - E_1$. The upper limit E_1 of the valence band and the lower limit E_2 of the conduction band vary depending on the direction under consideration. If a minimum of E_2 exists in the same propagation direction as the maximum of E_1, the result is a minimum for E_G (direct transitions, the case of III V compounds). If the minimum for E_2 and the maximum for E_1 occur for two different directions, the situation is less straightforward (indirect transitions, the case of Si and Ge). Figure 2.9 illustrates these two types of transition.

The derivative of (2.16) can be written as

$$1/m^* = (4\pi^2/h^2)\delta^2 E/\delta k^2 \tag{2.17}$$

In the vacuum, one would have $m^* = m$, the rest mass of the electron. In a material medium, and from the curvature of the $E(\mathbf{k})$ arcs, this equation defines an *effective mass*, positive for electrons changing position at the bottom of the conduction band and negative for holes changing position at the top of the valence band. This is an artificial concept, according to which the electron is always subject to the imaginary forces that would result from the electrical field applied to the crystal and not the real forces resulting from the electrical polarization of the medium [2.16]. It is shown experimentally, in an extremely simple way, when $m^* \leqslant m$, by cyclotron resonance, in which an electron describes a spiral in a high-frequency alternating field. One finds, for instance, in InSb (direct transitions), an isotropic effective mass of $0.013m$. In Ge (indirect transitions), $m_x^* = m_y^* = 0.08\ m$ and $m_z^* = 1.3\ m$ for electrons and $m^* = -0.39m$ for holes (-0.044 for certain "light" holes). The existence of different types of carriers in a single crystal explains the complexity of the band structures in Fig. 2.9.

2.4. Electrical Conductibility

The MO method, in its initial form in which the electron is completely delocalized, thus leads to the essential concept of two energy bands, one filled and one empty, separated by an energy band whose occupation by a particle is theoretically impossible unless there is local disturbance. Electronic

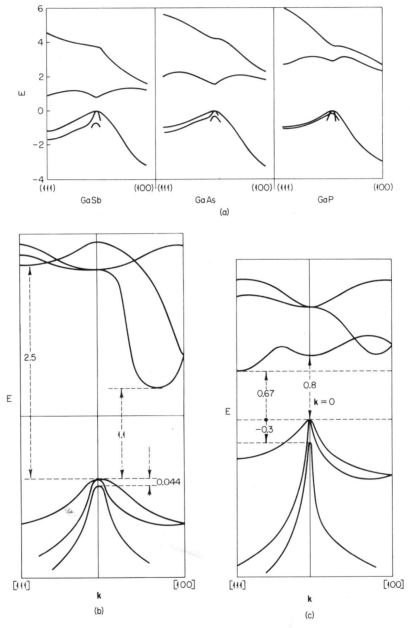

Fig. 2.9. $E(\mathbf{k})$ band structure between 0 and π/d in the [111] and [100] directions: (a) III V compounds of gallium (based on Hilsum [2.17]), (b) and (c) silicon and germanium (based on Aigrain et al. [2.18]).

2.4. Electrical Conductibility

formulation of the band model has much in common with the electronic theory of metals, and the behavior of these is linked with the existence of a partly filled band or, what comes to the same thing, the partial overlapping of the valence and conduction bands (Fig. 2.10). The VB method and its develop-

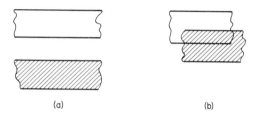

Fig. 2.10. Energy diagram of a crystal in the MO method (band model): (a) semiconductor and (b) metal.

ments, described in Sections 1.4 and 2.2, define two levels, bonding (B) and antibonding (AB) for each pair of electrons, with the passage to the metal state linked to the negative values of the ionicity parameter and the accompanying disappearance of these separate levels. Although, as has been seen, its extension to the crystal raises mathematical difficulties, one instinctively realizes that the juxtaposition of multiple electron pairs involves a widening of these levels for the crystal (Fig. 2.11).

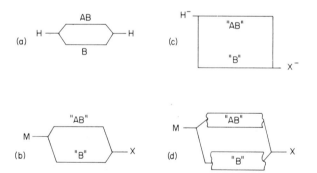

Fig. 2.11. Energy diagram of a crystal in the VB method (crystallochemical model): (a) covalent molecule ($\lambda = 0$), (b) possible state of an ionic molecule ($\lambda = 1$), (c) any molecule, and (d) crystal ($0 < \lambda < 1$).

The situation can also be described by explaining that there are three models for a crystal: (a) an ionic model, characteristic of an ideal state, (b) a multimolecular covalent model, obtained by the VB method, and (c) a crystal covalent model, obtained by the MO method (band model). Interpolation between (a) and (b) is possible (crystallochemical model described in Section

2.2), but not between (a) and (c) (Fig. 2.12). It can be seen in fact that there can be a common representation of these models, consisting of two systems of levels or two energy bands, the lower of which is completely occupied at absolute zero, and the upper of which is empty.

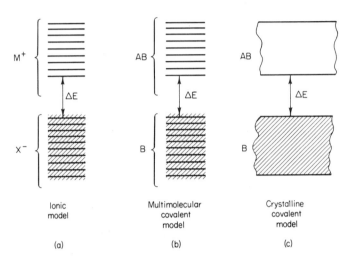

Fig. 2.12. Theoretical models for electrical conduction in crystal mineral compounds. (a) Ionic model, (b) multimolecular covalent model, and (c) crystalline covalent model.

Let us examine the experimental bases of these models. The existence of a forbidden band is deduced from optical, electrical, and electromagnetic phenomena. These all indicate the exciting of an electron from an energy E to an energy $E+E_G$. The optical transmission threshold links this excitation to the absorption by the electron of the energy quantum of an electromagnetic wave (photon). This can be effective only if $hv \geqslant E_G$ ($h =$ Planck's constant, $v =$ wave frequency), defining the threshold frequency. If $E_G \geqslant 3$ eV, the threshold is in the ultraviolet range and the crystal is transparent in the visible range. If $E_G \leqslant 1.5$ eV, the threshold is in the infrared range and the crystal is opaque in the visible range. In impure crystals where the intrinsic conductibility region is hard to reach, it is better to work out E_G from the optical spectrum than from electrical properties. In very impure crystals, a technique involving the dilution of powder in potassium bromide can give qualitative indications [2.19].* Suchet and Bailly [2.11] have estimated E_G in III V compounds by using the energy concepts described in Section 2.2. The results correspond sufficiently well with experimental values to justify identification of the widened levels (B) and (AB) with the valence and conduction bands of the

---

* See also Pearson [2.19a].

2.4. Electrical Conductibility

band model. In a series of compounds with the same structure, E_G varies simply as a function of the ionicity λ [2.20].

The electrical conductibility σ is obviously proportional to the total charge carried Ne (or Pe), where e is the elementary charge. The ratio σ/Ne shows the speed (centimeters per second) in an electric field of value one (V cm^{-1}). This is the *mobility* μ_n expressed in square centimeters per volt per second. Similarly, a hole mobility μ_p can be defined. The mobility in a series of compounds with the same structure varies simply as a function of the ionicity λ, and is at its maximum when the effective charge q is zero [2.20]. If, in exceptional circumstances, the carriers of the two types have comparable densities,

$$\sigma = Ne\mu_n + Pe\mu_p \tag{2.18}$$

A high mobility, an essential characteristic for many semiconductor applications, remains rare, however. Only three elements and a dozen or so compounds have a mobility of more than 1000 cm^2 V^{-1} sec^{-1} at room temperature. Measurement of the effective mass by cyclotron resonance is possible only in this small group, which does not include any transition element or rare earth compounds. In the presence of a magnetic induction B directed along $0x$ (the x-axis) and a current I along $0y$ (the y-axis), a voltage, known as the Hall voltage, appears along $0z$ (the z-axis) in a diamagnetic crystal:

$$V = RBI/a \tag{2.19}$$

where a is the thickness of the sample in the direction of the magnetic field and R is the Hall coefficient. When one type of carrier predominates, for instance electrons, it is shown that $R = 1/Ne$. The mobility can then easily be calculated:

$$\mu = \sigma R \tag{2.20}$$

Similar exponential laws, found experimentally for σ and R in the intrinsic regions

$$\sigma = \sigma_0 \exp(-E_G/2kT) \tag{2.21}$$

$$R = R_0 \exp(E_G/2kT) \tag{2.22}$$

or extrinsic

$$\sigma = \sigma_0 \exp(-e_d/kT) \tag{2.23}$$

$$R = R_0 \exp(e_d/kT) \tag{2.24}$$

show that the mobility of the carriers varies little in relation to temperature. It is not unaffected by it, however, and can follow several laws, depending on whether the movement of the charges is hindered mainly by the atom vibrations, the electrostatic charges they carry or the defects and impurities in the crystal. These laws contain T to a low power and are therefore negligible in relation to the exponential of the N or P density. They can however affect the curve in Fig. 2.8, in particular producing a minimum and maximum when the influence of the impurities is not predominant. The author [2.21] has calculated the position of the minimum, and shown that it varies rapidly with the density of impurities. This enables it to be distinguished from a minimum conductibility caused by another phenomenon, which will be mentioned in Section 3.4.

Something must be said of thermoelectric effects. A difference of potential $(\alpha_B - \alpha_A)(T_1 - T_2)$ appears at the terminals of the circuit when $T_1 > T_2$. This is the Seebeck effect, for which α_B and α_A V $°K^{-1}$ are the characteristic coefficients in relation to a reference substance (usually copper). Reciprocally, the passage of a current I in a weld AB introduces into the thermal exchanges between this weld and the environment the term $IT(\alpha_B - \alpha_A)$ with the same sign as I. This is the Peltier effect. It is shown that

$$\alpha = (E_F - E)/eT \qquad (2.25)$$

where E_F is the Fermi energy, E the average energy of the moving electrons, and e the elementary charge.

One conductibility mechanism, which will from now on be referred to as the *excitation mechanism*, is thus defined independently of the method of approach selected: breakage of a bond and exciting of an electron to a higher energy level across a forbidden band.

Hung and Gliessman [2.22], investigating the behavior of germanium at low temperatures, with higher densities of impurities than usual, have shown the existence of different mechanisms. At a few degrees Kelvin, σ and R are unaffected by the temperature and conductibility is metallic. At slightly higher temperatures, there is a region in which σ follows a law of type (2.23) while R remains fairly constant (Fig. 2.13). An explanation of this is that the carriers hop directly from one impurity atom to another, resulting in the expression *hopping* conductibility. If, as before, one forms the product σR, one finds that it increases exponentially as the temperature rises, with the intervention of a hopping activation energy E_A. It can, by extension, be called mobility, although its value is much lower and it applies to a different physical phenomenon from the one involved in the band model. Ioffe [2.23] has shown that this model cannot be applied to crystals in which μ is equal to or lower than around some cm^2 V^{-1} sec^{-1} and the effective mass is high. However, the expression "narrow band" is frequently, although incorrectly, used in such cases. The increase in σ in relation to frequency in alternating fields up to 9×10^9 Hz [2.24] has been attributed to the polarization caused by this phenomenon. Readers wishing to go more thoroughly into conductibility mechanisms in classic impure semiconductors may refer to Mott and Twose's digest [2.25]. They have been mentioned here only as a way of introducing the *transfer mechanism* observed in transition element compounds, which presents many similarities with them.

2.4. Electrical Conductibility

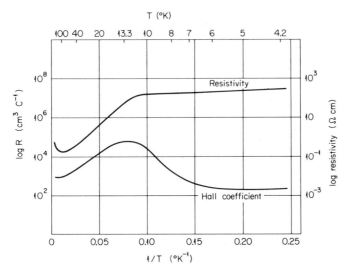

Fig. 2.13. Variation in resistivity and Hall coefficient in relation to temperature in a *p*-type germanium sample (based on Hung and Gliessman [2.22]).

The existence of multiple valencies in transition metal compounds is well known, and the discovery of a conductibility mechanism in which an electron spends an amount of time on a metal atom that is long in relation to the vibration period of the lattice resulted from the work of Wagner [2.26] and de Boer and Verwey [2.27]. This situation creates a potential barrier which the electron has to cross in order to move from one metal atom to the other, involving an activation energy E_A. Let us take the classic example of an oxide assumed to be strongly ionic. Any point defects in the crystal lattice are accompanied by an electrostatic disturbance, immediately corrected by the

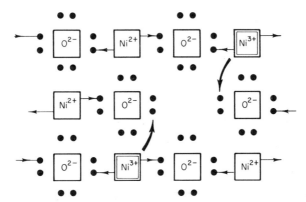

Fig. 2.14. Effect of a nickel vacancy in NiO.

exchange of a few electrons with the outside electronic layer of the neighboring ions. Less energy is needed to remove an electron from a divalent cation of a transition metal, making it pass to valency 3, than to extract it from the O^{2-} anion. If, for example, one has Ni^{2+} cations of formula $3d^8$ and O^{2-} anions of formula $2s^2p^6$, and if a nickel vacancy introduces a local deficit of two electrons, these will therefore be taken from two neighboring cations, which change to the state Ni^{3+} with formula $3d^7$ (Fig. 2.14).

If there are 12 neighboring cations (as in the B1 structure), they are all equally likely to supply an electron. A given configuration ($10Ni^{2+}$, $2Ni^{3+}$) therefore has no stability and can be replaced immediately by another, differing from the first one by the transfer of an electron between two neighboring cations, which immediately exchange their symbols Ni^{2+} and Ni^{3+} and their formulas $3d^8$ and $3d^7$. If nickel vacancies are distributed uniformly along the whole crystal, an electric field will cause the drift of a hole (Fig. 2.15a).

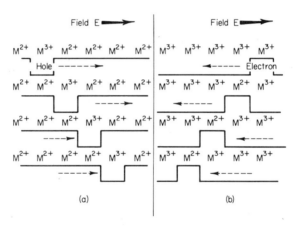

Fig. 2.15. Diagrammatic representation of the transfer mechanism: (a) induction of a higher valency and (b) induction of a lower valency. [Based on Suchet presented before the 8th session of the Société Française des Electriciens on 10 February 1955. *Bull. Soc. Elect.* (*Paris*) **5**, 274 (1955).]

This occurs whenever the induced valency is higher than the normal valency. In another case, such as an interstitial atom, where the passage is instead from M^{3+} to M^{2+}, the field would cause the drift of an electron (Fig. 2.15b). This occurs whenever the induced valency is lower than the normal valency [2.28]. These electron transfers obey two strict rules, defined by Verwey *et al.* [2.29]: (a) valency difference of one unit, (b) equivalent crystallographic sites; in addition, whenever the valency induction is obtained by adding a foreign body, (c) addition of a more stable valency cation differing by one unit, (d) solid solution between the addition and the original compound, and (e) preparation conditions favoring a change of valency.

The VB approach to the crystal is based on the Lewis electron pair. With this approach, the d electrons taking part in the transfer mechanism therefore occupy discrete levels similar to the impurity levels, even if conductibility is high (Fig. 2.16). The MO approach, on the other hand, describing each

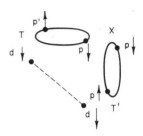

Fig. 2.16. Diagrammatic representation, in the VB method, of metal–metalloid and metal–metal interactions.

electron individually, applies to transfers. But since a charge can be carried only in an energy band, one is forced to speak of a partly filled d band, and consequently of metallic conductibility. In the particular case of insulating materials such as very pure MnO and NiO, this is a difficulty that was evaded by the concept of "narrow" band. In addition, as for any MO, one has to define bonding and antibonding states, involving an unfortunate confusion with the (B) and (AB) bands of metal–metalloid pairs in the VB method. We shall return to this.

If one wanted to represent the situation in an extremely schematic form, one could distinguish four cases:

(a) $N = N_0 \exp(-E_G/2kT)$ *excitation*, semiconductor.

(b) $N \sim C^{te}$, $E_G \neq 0$ *transfers*, $\mu = \mu_0 \exp(-E_A/kT)$, semiconductor.

(c) $N \sim C^{te}$, $E_G \neq 0$ *transfers*, $\mu \sim C^{te}$, metal.

(d) $N \sim C^{te}$, $E_G = 0$ metal.

Metals (c) and (d) do not behave identically. Some authors speak of first- and second-type metals or wide- and narrow-band conductors. We feel that only the (d) metals have been clearly defined up till now and have suggested using the term "pseudometal" in the case of (c). The experimental results are always complex, allowing a simple case to be identified only within a limited temperature interval. We shall return in greater detail to the transfer mechanism in Section 3.3, and can then define case (c) a little more closely.

REFERENCES

[2.1] A. F. Wells, *Solid State Phys.* **7**, 45 (1958).
[2.2] A. F. Wells, "Structural Inorganic Chemistry." Oxford Univ. Press (Clarendon), London and New York, 1962.
[2.3] H. Krebs, *Proc. Int. Conf. Semicond.* (*Amsterdam, 1954*) *Physica* **20**, 1125 (1954).
[2.4] H. Krebs, *Acta Crystallogr.* **9**, 95 (1956).

[2.5] H. Krebs, "Grundzüge der anorganischen Kristallchemie." Enke, Stuttgart, 1968; *Engl. transl.* McGraw-Hill, Maidenhead, 1968.
[2.6] L. Pauling, "The Nature of the Chemical Bond." Cornell Univ. Press, Ithaca, New York, 1950.
[2.7] A. Neuhaus, *Fortschr. Mineral.* **32**, 37 (1954).
[2.8] J. P. Suchet, *J. Phys. Chem. Solids* **12**, 74 (1959).
[2.9] H. Krebs and W. Schottky, *Halbleiterprobleme* **1**, 25 (1954).
[2.10] H. Hartmann, "Theorie der chemischen Bindung." Springer, Berlin, 1954.
[2.11] J. P. Suchet and F. Bailly, *Ann. Chim. (Paris)* **10**, 517 (1965).
[2.12] J. P. Suchet, *J. Phys. Chem. Solids* **21**, 156 (1961).
[2.13] F. Hund, *Z. Phys.* **51**, 759 (1928); **63**, 719 (1930).
[2.14] R. A. Smith, "Semiconductors." Cambridge Univ. Press, London and New York, 1959.
[2.15] P. Aigrain and F. Englert, "Les semiconducteurs." Dunod, Paris, 1958; *Engl. transl.* Methuen, London, 1965.
[2.16] R. Guillien, "Les semiconducteurs." Presses Univ., Paris 1963.
[2.17] C. Hilsum, *Proc. Int. Conf. Phys. Semicond. Paris, 1964*, p. 1127. Dunod, Paris, 1964.
[2.18] P. Aigrain and M. Balkanski, Table de constantes, "Semiconducteurs." Pergamon, Paris, 1961.
[2.19] P. Manca and G. Saut, *C. R. Acad. Sci.* **262**, 1621 (1966).
[2.19a] A. D. Pearson, *J. Phys. Chem. Solids* 316 (1958).
[2.20] J. P. Suchet, "Séminaires de chimie de l'état solide" (J. P. Suchet, ed.), vol. 1, p. 37. S.E.D.E.S., Paris, 1969.
[2.21] J. P. Suchet, *Ann. Phys. (Paris)* **8**, 285 (1963).
[2.22] C. S. Hung and J. R. Gliessman, *Phys. Rev.* **79**, 726 (1950).
[2.23] A. F. Ioffe, *J. Phys. Chem. Solids* **8**, 6 (1959).
[2.24] M. Pollak and T. H. Geballe, *Phys. Rev.* **122**, 1742 (1961).
[2.25] N. F. Mott and W. D. Twose, *Advan. Phys.* **10**, 107 (1961).
[2.26] C. Wagner, *Z. Phys. Chem. (B)* **32**, 439 (1936).
[2.27] J. H. de Boer and E. J. W. Verwey, *Proc. Phys. Soc. London* **49**, 59 (1937).
[2.28] J. P. Suchet, *Bull. Soc. Elect. (Paris)* **5**, 274 (1955).
[2.29] E. J. W. Verwey, P. W. Haayman, F. C. Romeyn and G. W. van Oosterhout, *Philips Res. Rep.* **5**, 173 (1950).

Chapter 3 | The Magnetic Crystal

3.1. Crystal Field Theory

Readers interested in a detailed analysis may refer to the numerous works on the subject, such as those by Ballhausen [3.1] and Cotton [3.2]. The summary given here is based on the briefer work by Orgel [3.3]. There are two ways of approaching this theory which is aimed at providing a relative classification of the energy levels of d electrons: by considering the spatial distribution of point charges, which may be assumed to occupy the different sites in an ionic crystal, or by considering the symmetry and compatibility of the AOs involved. The first method, which can be applied to partially covalent crystals as well, is an extension of the VB approach. The second completes the MO approach which would otherwise be inadequate in the case of transition metal compounds. These two viewpoints do not result in contradictory results so that one can subsequently argue on the basis of the energy level diagram obtained without worrying any more about the method employed to produce it.

The electrostatic theory developed by Bethe [3.4] and Van Vleck [3.5] has so far been applied only to ionic crystals. A transition metal atom therefore has to appear in it in the form of a cation carrying the $+n$ charge surrounded by anions each carrying the $-n$ charge. Let us consider the effect of the main possible arrangements on the angular distribution of the d electrons. The d AO functions and their distribution in space were described in Section 1.1. The main crystal arrangements were described in Section 2.1. Let us identify the x-, y-, and z-axes ($0xyz$) in Figs. 1.2 and 1.3 with the axes parallel to the edges of the cubic elementary cells of the B3 (tetrahedral arrangement) and B1 (octahedral arrangement) structures. In the first case, the electrons of the orbitals $d_{x^2-y^2}$ and d_{z^2}, which point in the direction of the axes, are subject to a smaller force of repulsion from the negatively charged anions than those of orbitals d_{xy}, d_{yz}, and d_{zx}, which point in the directions of their bissectrixes (Fig. 3.1a). The energy of the electrons will therefore increase less than that of the d_{xy}, d_{yz}, and d_{zx} electrons. In the second case (B1), the maximum repulsion is felt by the electrons in the orbitals $d_{x^2-y^2}$ and d_{z^2}, which will therefore have a higher energy level (Fig. 3.1b).

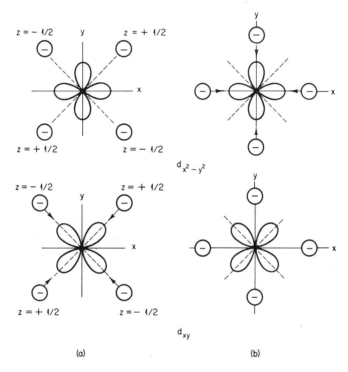

Fig. 3.1. Spatial distribution of the $d_{x^2-y^2}$ and d_{xy} orbitals and forces of repulsion exerted by the anions: (a) in a tetrahedral arrangement and (b) in an octahedral arrangement.

In this way one can classify the relative energies of these orbitals. The levels of the three orbitals d_{xy}, d_{yz}, and d_{zx}, on the one hand, and of the two orbitals $d_{x^2-y^2}$ and d_{z^2}, on the other, are shown artificially separated in Fig. 3.2, but what is in fact involved is triple and double orbitals, which are referred to either as $d\varepsilon$ and $d\gamma$, or, using a notation borrowed from the group theory, t_2 and e (according to this notation, the s orbitals are called a_1 and the p orbitals t_1). In this case, a subindex defining their symmetry is generally added, of type g here (cf. Section 1.2). It is shown that the energy division among the

3.1. Crystal Field Theory

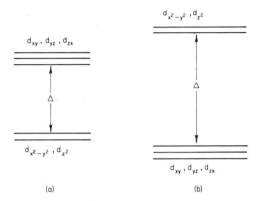

Fig. 3.2. Classification of d energy levels: (a) tetrahedral arrangement and (b) octahedral arrangement (based on Orgel [3.3]).

d electrons is less marked in the tetrahedral arrangement than in the octahedral arrangement. The difference Δ would in the first case be 44% of the difference obtained in the second case, all other things being equal.

Calculations based on the preceding model give satisfactory results although slightly inferior to experimental results. In 1960, Orgel [3.3] declared, "In general, I think that electrostatic theories provide an adequate approximation of size for the division, but do not offer a satisfactory quantitative correspondence with experiments. This can be obtained, however, by assuming a certain polarization of the anions by the central cation." Such polarization means taking the covalent aspect of the bond into account, completely left out by the image of point ionic charges. The author [3.6] suggested in 1966 that this aspect could be taken into account in the electrostatic theory, by replacing the ionic charges $(-n)$ on the anions by the effective charges $(-q)$ of the crystallochemical model. However, whereas the quantity n is by definition always positive, the charge q may be negative in certain compounds. The $(-q)$ charges in metalloid atoms are then positive, and an attraction replaces the repulsion exerted on the d orbital electrons. One therefore has to provide for a possible inversion of the t_{2g} and e_g levels in certain particular cases. The principle of such a phenomenon, incidentally, was admitted by Klixbull–Jørgensen [3.7] for p electrons.

The MO crystal field theory was developed by Van Vleck [3.8] from the approach mentioned in Section 1.2, which is used mainly for diatomic covalent molecules and aromatic hydrocarbons. It therefore applies to partially covalent bonds and no longer requires the point charge concept. It is known as the "ligand-field theory," the word ligand referring for example to a metalloid atom bonded to a transition metal. The MOs are formed by linear combinations of the AOs of a metal with the AOs of the ligands. The metal and metalloid AOs must have the same symmetry (cf. Section 1.4). The theory assumes that one obtains as many MOs as there are combined AOs, and that the energy range of the MOs is always wider than that of the AOs used. It is then shown, in the octahedral arrangement, that the t_{2g} AOs of the metal can form only π MOs, whereas the e_g AOs can form only σ MOs [3.2]. Here we again find the division shown by the electrostatic theory. Like it, the MO theory allows the energies of the d orbitals to be classified in relation to one another but not in relation to the energies of other orbitals.

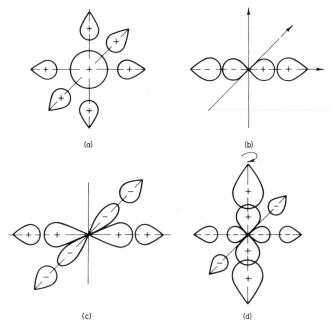

Fig. 3.3. Bonding σ MOs formed in the octahedral arrangement from the metal AOs: (a) s, (b) $p_x(p_y$ and $p_z)$, (c) $d_{x^2-y^2}$, and (d) d_{z^2} (based on Orgel [3.3])

Figure 3.3. shows the different σ MOs in the commonest case of octahedral arrangement, assuming the intervention of the s, p_x, p_y, p_z, $d_{x^2-y^2}$, and d_{z^2} AOs of the metal. One imagines instinctively the existence of a hybrid AO, $e_g^2 sp^3$, and the MOs it can form by superimposing the schemas above. The π bonds are harder to picture. They affect the p_x, p_y, p_z, d_{xy}, d_{yz}, and d_{zx} AOs of the metal. Figure 3.4 shows the π bonds established with the t_{2g} AOs of the metal by the AOs of the neighboring metalloids *or the t_{2g} AOs of other metals* (for it should not be forgotten that the MO method completely leaves aside the Lewis pair and can therefore cover the formation of MOs between two metal atoms). The intervention of the π bonds does not constitute an unnecessary com-

3.2. Magnetic Interactions

plication: it enables the increase or reduction in the energy difference Δ between the AOs of the metal to be explained, depending on whether the metalloid AOs are vacant ("acceptor") or occupied ("donor" atom).

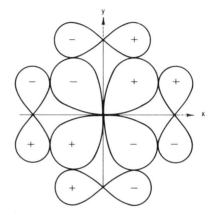

Fig. 3.4. Bonding π MO formed in the octahedral arrangement from the d_{xy} AO of the metal. Similar bonds are possible with the d_{yz} and d_{zx} AOs (based on Orgel [3.3]).

Table 3.1, adapted from Orgel [3.3], recapitulates the AOs of the metal and the LCOAs of its six metalloid neighbors capable of giving MOs in the octahedral arrangement. The group theory, which links the MOs with the symmetry of the crystal arrangement involved, imposes certain conditions of symmetry on the LCOAs from which they are constructed. Metalloid AOs are referred to by the symbol φ accompanied by the subindices 1, 2, 3, 4, 5, and 6 (corresponding to positions on axes x, x', y, y', z, and z' from the metal atom) and x, y and z (corresponding to the directions of the metalloid AOs).

Goodenough [3.9] has made wide use of the MO theory forms in interpreting the properties of certain transition metal compounds, notably the oxides of spinel and perovskite structures. Frequent reference will be made to this work throughout the rest of this book.

3.2. Magnetic Interactions

Only a fragmentary account will be given here, confined to the aspects of magnetism involved in interatomic bonds and to interactions in nonmetallic crystals. Further information can be obtained from general treatises, such as that of Rado and Suhl [3.10]. Among very recent works may be mentioned the small booklet by Michel [3.11] for the use of chemistry students.

Let us consider a system of G identical atoms (G = Avogadro's number), isolated from one another in such a way that there is no interaction. The manifestations of magnetism were described in 1905 by Langevin. The spins of the Z electrons contained in each atom may or may not be balanced. In the

TABLE 3.1

Symbol of the metal AOs	Symbol of the electrons occupying these AOs	LCAO of the metalloid neighbors	
		For σ MOs	For π MOs
a_{1g}	$4s$	$\varphi = (1/\sqrt{6})(\varphi_1 + \varphi_2 + \varphi_3 + \varphi_4 + \varphi_5 + \varphi_6)$	
e_g	$3d_{x^2-y^2}$	$\varphi = (1/2\sqrt{3})(2\varphi_5 + 2\varphi_6 - \varphi_1 - \varphi_2 - \varphi_3 - \varphi_4)$	
	$3d_{z^2}$	$\varphi = (1/2)(\varphi_1 + \varphi_2 - \varphi_3 - \varphi_4)$	
t_{1u}	$4p_x$	$\varphi = (1/\sqrt{2})(\varphi_1 - \varphi_2)$	$\varphi = (1/2)(\varphi_{3x} + \varphi_{4x} + \varphi_{5x} + \varphi_{6x})$
	$4p_y$	$\varphi = (1/\sqrt{2})(\varphi_3 - \varphi_4)$	$\varphi = (1/2)(\varphi_{1y} + \varphi_{2y} + \varphi_{5y} + \varphi_{6y})$
	$4p_z$	$\varphi = (1/\sqrt{2})(\varphi_5 - \varphi_6)$	$\varphi = (1/2)(\varphi_{1z} + \varphi_{2z} + \varphi_{3z} + \varphi_{4z})$
t_{2g}	$3d_{xy}$		$\varphi = (1/2)(\varphi_{1y} - \varphi_{2y} + \varphi_{3y} - \varphi_{4y})$
	$3d_{yz}$		$\varphi = (1/2)(\varphi_{1z} - \varphi_{2z} + \varphi_{5x} - \varphi_{6x})$
	$3d_{zx}$		$\varphi = (1/2)(\varphi_{3z} - \varphi_{4z} + \varphi_{5y} - \varphi_{6y})$

3.2. Magnetic Interactions

former case, a magnetic field H, perpendicular to an orbit, causes a variation in the speed of electrons. If the planes of the orbits are distributed at random and if $\bar{r}^2$ represents the average of the squares of the radii, one observes a variation in moment per field unit. This is the molar atomic susceptibility:

$$\chi_A = GM/H = -(GZe^2/6m) \sum \bar{r}^2 \tag{3.1}$$

This is *diamagnetism*. In the latter case, this underlying effect is covered by the orientation of the atomic moments in weak and medium fields, producing a positive susceptibility, much higher in absolute terms. This is the simple Curie law, in relation to temperature:

$$\chi_A' = GM^2/3kT \tag{3.2}$$

This is *paramagnetism* (k = Boltzmann constant).

Let us now consider a system of $2G$ atoms belonging to two different elements, a metal M and metalloid X, and let us assume that these atoms are bonded to form MX molecules or an MX crystal. One will find the two manifestations of magnetism already noted, but there will also be new phenomena, closely bound up with the nature of the bonds between M and X. The spins of the $(Z_M + Z_X)$ electrons contained by the atoms may be balanced (in the same atom or in a Lewis pair) or unbalanced (d and f bachelor electrons of M only). In the first case, Pascal showed in 1910 that the additivity rule of atomic susceptibilities had to be rectified by a positive constant C_p, which bears his name. Van Vleck [3.5] later showed that this slight paramagnetism depended basically on the change brought about in the symmetry of the electron cloud by the bonds. Sirota [3.12] has calculated it for A4 and B3 structures, from the covalent contribution to the wave function of the electrons. Finally, Bailly and Manca [3.13, 3.14], using the concepts described in Section 1.4 for this contribution, have obtained an expression for molar susceptibility (Fig. 3.5):

$$10^6 \chi_M = 10^6(-\sum \chi_A + C_p) = -(Z_M + Z_X) + 86.1(\delta/E_G)(1-\lambda) \quad \text{emu} \tag{3.3}$$

where δ is the asphericity of the purely covalent electron distribution defined by Sirota and E_G the height of the corresponding forbidden band in eV. The statistical distribution of the free electron spins, modified by the field, introduces another term, known as the Pauli term:

$$(8\pi/3)^{2/3} 2mh^{-2} D^{1/3} \mu_B^2 \tag{3.4}$$

where h is Planck's constant, D the electron density and μ_B Bohr's magneton. This term is negligible, however, in a conventional intrinsic semiconductor [3.15].

If the d or f electron spins are not balanced in a transition metal T or rare earth L, the susceptibility often follows the modified Curie–Weiss law:

$$\chi_M' = GM^2/3k(T-\theta) \tag{3.5}$$

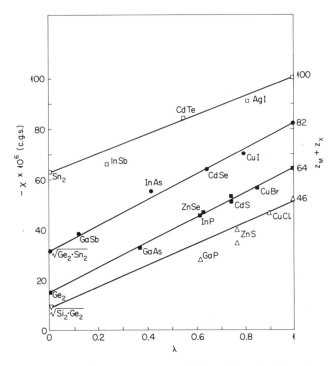

Fig. 3.5. Linear variation in the molar susceptibility with the ionicity λ for some series of isoelectronic semiconductor compounds within which (Z_M+Z_X) and (δ/E_G) remain constant. Theoretical straight lines and experimental points (in fact, $\lambda \neq 0$ for the hypothetical combination SiGe) (based on Bailly and Manca [3.14]).

If one assumes that the magnetic moments in the atom adopt the Russell–Saunders coupling, L and S being the resulting atomic vectors of the orbital and spin moments of each electron, respectively, and J their vectorial sum, $M = g[J(J+1)]^{1/2}$. In T metal compounds the orbital moments are frequently blocked, so that $L = 0$ and $J = S$. The author [3.16] has shown that if all the spin moments of the electrons in the atom also have parallel directions, the number of bachelor electrons is

$$2S \sim [1 + 8\chi_M(T-\theta)]^{1/2} - 1 \qquad (3.6)$$

Below the temperature of θ ($\theta > 0$) or a temperature of around $-\theta$ ($\theta < 0$) an entirely new phenomenon appears: the interaction of neighboring atom spins, in parallel or antiparallel directions. This is *ferromagnetism* ($\theta > 0$), which increases χ considerably, or *antiferromagnetism* ($\theta < 0$), which reduces it. The existence of two different categories of T atoms results in the complex phenomena of *ferrimagnetism* (Fig. 3.6).

3.2. Magnetic Interactions

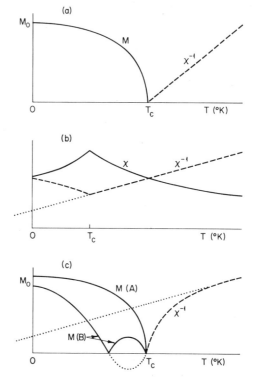

Fig. 3.6. Typical variation curves of the magnetization M, susceptibility χ and its inverse with temperature: (a) ferromagnetism, (b) antiferromagnetism, and (c) one of the cases of ferrimagnetism. [Based on Kouvel in "Intermetallic Compounds." (J. H. Westbrook, ed.), Fig. 1, p. 532. Wiley, New York, 1965. Used with permission of the copyright owner.]

The impression was given in the first chapter, in the condensed account of Heitler and London's work, that the energy of the bonding level (antiparallel spins, $S = 0$) of a Lewis pair was always less than that of the antibonding level (parallel spins, $S = 1$). A more exact analysis of the situation has been made by Heisenberg [3.18] in the case of the covalent bond ($\lambda = 0$). The energy of the bonding state may be put in the form

$$E = E_{\text{cov}} = E_0 + J_e \tag{3.7}$$

where E_0 regroups the coulombic interactions between the electrons 1 and 2 and between each of them and the nuclei i and j, while the second term corresponds to what is called the *exchange integral* between the two electrons

$$J_e = \int \varphi_i^*(1)\,\varphi_j^*(2)(e^2/r_{12})\,\varphi_i(2)\,\varphi_j(1)\,d\tau_{12} \tag{3.8}$$

φ^* is the conjugated function of φ, r_{12} the distance of the two electrons and $d\tau_{12} = d\tau_1\,d\tau_2$ the volume change associated with their coordinates. The energy of the antibonding state is then

$$E' = E'_{\text{cov}} = E_0 - J_e \tag{3.9}$$

It can be seen that the more stable state, E or E', depends on whether J_e is positive or negative and therefore no longer necessarily corresponds to the general wave function using the symmetrical space function (bonding state). It may be assumed that the ferromagnetic coupling occurs when $J_e > 0$, since the antibonding state $S = 1$ then has the lowest energy E'.

Heisenberg's work, confined to a covalent bond, obviously does not take the compounds of a transition metal T and a metalloid X into consideration. Can it be extended to such cases? Kramers [3.19] noted shortly afterwards that a T–X–T' interaction would be impossible in a purely ionic crystal, if in fact such a crystal can exist. In insulating materials or semiconductors with a low carrier density, ferromagnetism therefore seems to be bound up with the covalent aspect of the T—X bonds (superexchange), when these play a predominant role. The magnetic interaction energy can be estimated by the Curie temperature T_C ($\sim \theta$, ferromagnetic) or Neel temperature T_N (around $-\theta$, antiferromagnetic) at which the thermal agitation energy kT destroys the magnetic order. Its qualitative [3.20] and quantitative [3.21] relation to the ionicity of the T—X bond has been mentioned for numerous compounds with B1, B8$_1$, and B31 structures. The author [3.22], by an extension of the Weiss model, and then Bailly and Suchet [3.23], by an extension of the Heisenberg model, have justified this intervention of the parameter λ (Fig. 3.7). It is thus established that the magnetic coupling in one way or another involves the Lewis pair T—X.*

Bethe [3.24] has attempted to analyze the conditions governing the sign for J_e, in a simple way. The relative values of terms such as e^2/r_{12} or the product $\varphi_i \varphi_j$ are indeed open to qualitative reasoning. He has shown that J_e is most likely to be positive if (a) the bond length is great in relation to the orbit radii, and (b) the amplitude of the wave functions is fairly small near the nuclei. The first of these conditions is represented diagrammatically in "Bethe's curve," which clearly shows, for instance, the sign and value of J_e, estimated by the Curie temperature, in the series Mn, Fe, Co, Ni, Gd (Fig. 3.8), or MnAs, MnSb, and MnBi [3.25]. The interatomic distances depend closely on the nature of the bond, and this in no way contradicts the role of the parameter λ. Anyway, Bethe's curve concerns only metals and alloys, and even then is above all qualitative. The second condition is fulfilled when the azimuthal quantum number is high, which requires the intervention of d and f electrons of partly filled layers.

For the sign of J_e to determine a ferro- or antiferromagnetic coupling directly, the two Lewis pair electrons would have to be the same ones as are responsible for the atomic magnetic moments. This is impossible since it is

---

* The reader can find a more thorough analysis of the different types of magnetic coupling in Chapter 3, Section I,B of Goodenough's book [3.9].

3.2. Magnetic Interactions

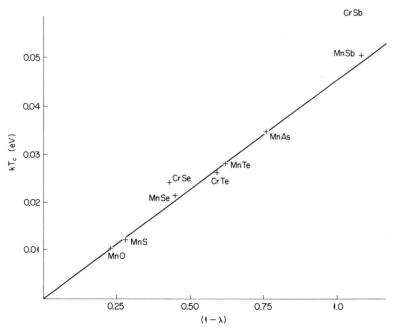

Fig. 3.7. Linear variation in the Curie (or Neel) temperature in relation to the term $(1-\lambda)$ for some compounds with an octahedral arrangement. $kT_c = 0.0456 \, (1-\lambda)$ eV (based on Suchet [3.16]).

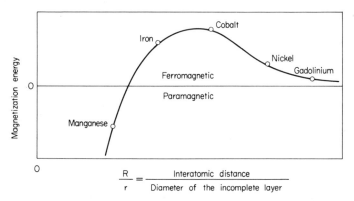

Fig. 3.8. Behavior of J_e in relation to the length of bond (based on Bethe [3.24]). Representative points for Mn, Fe, Co, Ni, and Gd have been added.

the d or f electrons that produce these moments, while the bond in the B1 and B3 structures and their derivatives is due to the s and p electrons (cf. Section 2.1). Is an indirect influence possible, as the result of an antiparallel coupling between d and s (or p) electrons in T atoms? It is unlikely in the tetrahedral arrangement where the successive pairs are strictly localized since the four electrons in the sp^3 hybrid AOs of X all have their spins parallel, and ferromagnetism will be obtained regardless of the sign of J_e (Fig. 3.9a). In the octahedral arrangement on the other hand, it is not absolutely certain that the spins of the three electrons of the AOs of X will retain their orientation during resonance (Fig. 3.9b) and it is possible to imagine their having an opposing orientation (Fig. 3.9c). Ferromagnetism would then be obtained only when $J_e > 0$. The author [3.26] suggested that case (c) could exist when the bonds were more ionic than covalent ($\lambda > 0.5$).

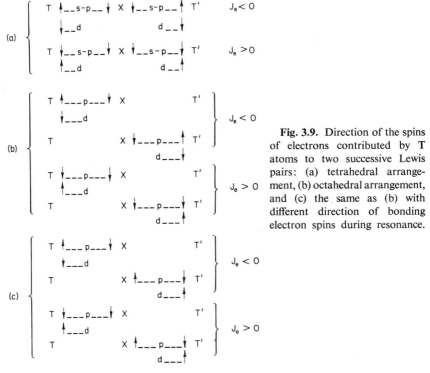

Fig. 3.9. Direction of the spins of electrons contributed by T atoms to two successive Lewis pairs: (a) tetrahedral arrangement, (b) octahedral arrangement, and (c) the same as (b) with different direction of bonding electron spins during resonance.

These difficulties would disappear if one assumed that, in addition to the previously known bond pairs, one had the formation of other T—T' Lewis pairs to which the Heisenberg model applied. The spatial distribution of the electron densities of the t_{2g} AOs could, for example, allow such a hypothesis in B1 and $B8_1$ structures (bonds directed along the bissectrixes of the angles defined by the p bonds). The number of bachelor t_{2g} electrons available to form such pairs, however, varies from 1 to 3, whereas one has 12 T′ neighbors. One would therefore be forced to envisage resonance mechanisms similar to those suggested by Pauling for the metal bond, or else bonds limited in space, compatible only with low conductibility. The T—T′ bond thus formed would be broken at the Curie (or Neel) temperature, so that kT_C (or kT_N) would represent its energy. The coupling between the $(1-\lambda)$ electrons which T contributes to the T—X pairs and those it contributes to the T—T′ pairs could explain the empirical relation expressed by Fig. 3.7. At low temperatures, each pair would give rise to two energy states, which one might call pseudobonding (PB) and pseudoantibonding (PAB), to distinguish them from the (B) and (AB) states defined in Sections 1.4 and 2.2 above.

If one leaves the VB approach and its Lewis pairs and moves on to the MO approach, there is no further obstacle to considering direct T—T′ bonds systematically, whenever the overlapping of the d AOs appears sufficient. The delocalization inherent in the electrons described by this method leads to a rather peculiar conception of the magnetic phenomena (theories under discussion on band magnetism). Readers interested in mathematical models can refer to [3.27–3.29].

3.3. Metal-to-Metal Transfers

In Section 2.4 there was a brief description of the transfer mechanism. A more detailed account of it must be given here. Landau [3.30], and then Gurney and Mott [3.31] showed that under certain conditions a carrier could be trapped in an ionic crystal by the polarization and distortion of the lattice induced by its own charge. This phenomenon is linked to the form in $1/r$ which the potential assumes (instead of $1/r^4$ in germanium). Let us take the example of the oxide FeO. The sum of ionic radii Fe^{2+} and O^{2-} is 2.20 Å. If a trivalent cation appears, the sum of the radii Fe^{3+} and O^{2-} would in theory be 1.98 Å. One can see that, within the limits permitted by the cohesive forces of the lattice, a very slight localized distortion is possible. Since this distortion requires a certain amount of time to reveal itself, it is vital for the electron or the hole to be localized on the metal atom for an amount of time that is long in relation to the vibration period of the lattice, which is 10^{-13} sec (Fig. 3.10).

Heikes and Johnston [3.32], following Verwey et al. [3.33], examined the case of NiO with increasing substitutions of lithium, $Li_xNi_{1-x}O$, up to $x = 0.35$. They noted that the variation of conductibility with temperature always remained exponential, whereas much lower concentrations would already have produced metal-type behavior in conventional semiconductors (impurity "band"). When $x = 0.10$, each Ni^{3+} has however at least one neighboring Li^+, so that the activation energy cannot be linked to the separation of an $Ni^{3+}Li^+$ pair. They conclude that it does not concern the number of carriers but their mobility. The reader will find a large number of experimental results involving NiO and other oxides, leading to the same conclusion, in Jonker and Van Houten's excellent digest [3.34]. The activation energy would therefore correspond to the work of shifting a hole (noted by the formula Ni^{3+}) at constant pressure and temperature, in other words the variation in Gibbs' free energy.

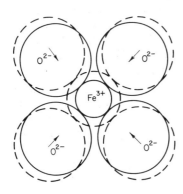

Fig. 3.10. Rough diagram of the local distortion brought about by a valency change in the cation if the anions remain in contact, in FeO for example.

Thermodynamics provide the expression of the conductibility [3.32]:

$$\sigma = Ne^2 D/kT \tag{3.10}$$

The hole diffusion coefficient is

$$D = \gamma(pa)^2 v_0 \exp(-E_A/RT) \tag{3.11}$$

where γ is a geometrical factor, pa the length of diffusion (p = integer, a = cell parameter), and v_0 the hopping frequency when T is infinite. Finally one obtains

$$\sigma = \sigma_0 T^{-1} \exp(-E_A/RT) \tag{3.12}$$

with

$$\sigma_0 = (4\gamma e^2 p^2 v_0 x/ak) \exp(\Delta S/RT) \tag{3.13}$$

where ΔS represents the entropy variation. These conclusions have been confirmed by Van Houten's research [3.35], which has in particular shown that the Seebeck coefficient is independent of the temperature and dependent on the density of Ni^{3+} ions. It may be noted that v_0 is not independent of the lattice vibration frequency since only energy fluctuations caused by phonons can explain the transfer of an electron taking place from atom A to atom B at a certain moment, and in the opposite direction at another moment [3.36].

The example of $Li_xNi_{1-x}O$ was not particularly well chosen since it later led to many controversies, when the very low Hall coefficient could be measured, enabling the mobility to be calculated, at around 10^{-3} cm^2 V^{-1} sec^{-1}. Ksendzov et al. [3.37], soon supported by other authors, showed that it was

3.3. Metal-to-Metal Transfers

not necessary to have recourse to the hopping mechanism between Ni atoms. Simultaneously, a large amount of theoretical work, such as that done by Sewell [3.38] and Klinger [3.39] applied the band model, not to the d electron (or hole), but to a much more complex entity called a "polaron," including the polarization that it causes in the surrounding medium. The polaron band has different properties from the electron band (a partly filled band no longer necessarily involves metallic conductibility), and so many of the previous difficulties were removed. Recent experimental results involving monocrystals of $Li_xNi_{1-x}O$ (cf. for instance [3.40]) reveal a complex situation difficult to interpret with certainty. It would seem however that in any case the simple Heikes and Johnston model is now outdated in this and similar cases (valency induction by an impurity whose presence complicates the phenomena).

Investigation of compounds in which the structure admits a variable number of vacancies provides a different viewpoint. Tannhauser's [3.41] work on wüstite $Fe_{1-x}O$ ($0.05 < x < 0.15$) offers a good example. The transfer mechanism provides for the number of carriers to be proportional to the probability of finding a divalent and a trivalent cation in adjacent octahedral sites. Since the electrical neutrality of the crystals requires a formula $\square_x Fe^{2+}_{1-3x} Fe^{3+}_{2x} O$, the number of carriers per cubic centimeter, and consequently the conductibility σ will be proportional to the product of the concentrations of the two types of cation. Experimental points confirm this law (Fig. 3.11). However, the passage from p-type to n-type should occur only for $x = 0.2$ ($Fe^{3+}/O = 0.4$), corresponding to an equivalent concentration of the two types of cation. Where $x < 0.2$, the minority group Fe^{3+} imposes a conductibility by holes (d^5 among d^6 cations) and where $x > 0.2$, the minority group Fe^{2+} imposes conductibility by electrons (d^6 among d^5 cations). The concentration of vacancies admitted by the B1 structure in which wüstite crystallizes is unfortunately very low at normal temperature. To find phases with a wider homogeneity domain, one has to move to the slightly distorted $B8_1$ structure, in which many transition metal compounds crystallize.

As regards this structure, the author [3.42] has shown first that the valency induction laws determining the fraction of T atoms changing valency will apply just as well to covalent crystals as to ionic crystals, and that the distribution of electrons in the Lewis pairs gives the same result as the condition of electrical neutrality. He has also shown that the order of the vacancies, which frequently affect one T atom plane in two perpendicularly to the hexagonal axis, can modify these laws. Let us consider the different planes of T atoms in which transfers take place. If there is no transfer possible in one of these planes, for a given composition, the density of transfers in the crystal, in other words of carriers, will be at a minimum level. This is what happens, for instance, perpendicularly to the hexagonal axis, when all the T(II) atoms are in full planes and all the T(III) atoms in planes with vacancies, or inversely. It is also shown that the density of carriers may reach a minimum at $x = 0, \frac{1}{3}$, and $\frac{1}{2}$ and that in the case of ordered vacancies, secondary minimum values may be observed at $x = \frac{1}{6}, \frac{1}{4}$, and $\frac{3}{8}$. Manca *et al.* [3.43] have investigated Fe_2Te_3 thoroughly and confirmed that the minimum is at $x = \frac{1}{3}$, by three different methods. The rapid increase in the number of transfers on each side of this value, and the existence of a strong Pauli paramagnetism, however, show that this compound is very close to the metallic state.

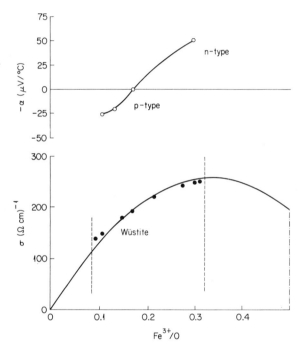

Fig. 3.11. Conductibility of wüstite at 1300°C in terms of the ratio Fe^{3+}/O. The points are experimental and the curve expresses the proportional value to the product $[Fe^{2+}][Fe^{3+}]$ (based on Tannhauser [3.41]).

The problems raised by this transition from a mechanism involving transfers like those of NiO or FeO, producing apparent semiconductor properties, to a mechanism involving transfers such as those of numerous antimonides and tellurides, and even a few oxides, producing apparent metallic properties, have been examined by Mott [3.44]. He has suggested that a crystal arrangement of atoms containing a partly filled layer might not necessarily have a metallic conductibility (persisting at absolute zero). If the parameter of a crystal such as this increased gradually, there would be *at absolute zero* an abrupt transition

3.3. Metal-to-Metal Transfers

from a metallic state, with above-zero conductibility (free electrons) to a nonmetallic state with zero conductibility (localized electrons). At higher temperatures, this second state would produce a hopping semiconductibility involving an activation energy. This effect of the crystal parameter was suggested by Morin [3.45], who used it as a basis for classifying oxides as metals or semiconductors. In the B1 structure, for instance, he postulated a varying amount of overlapping of the t_2 (or dε) cation MOs (Fig. 3.12). These theories were taken up and developed by Goodenough [3.46], who tried to assess the critical interatomic distance R_c in each structure, corresponding to the transition between these two types of behavior.

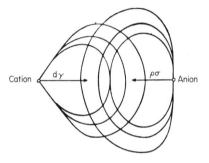

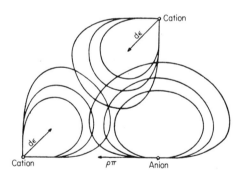

Fig. 3.12. Lines of equal amplitude of the MOs of a transition metal ion and an anion in the B1 structure. The arrows indicate the overlaps (dγ, pσ), (dε, dε), and (dε, pπ) (based on Morin [3.45]).

In fact, there is nothing to prove that this transition is necessarily abrupt at room temperature. Complete localization of the d electrons would prevent any transfer and the mechanism could not be observed. Such a case has never been met with, to the author's knowledge. Complete delocalization might with some reason be suspected in compounds presenting a strong Pauli paramagnetism. But partial localization, however transient, has to be taken into consideration in this case to explain the role played by the valency induction laws [3.42]. What is more, the concept of a d electron band generally has to be corrected by introducing a minimum state density for semioccupancy [3.43]. This shows that d electrons again in a way confirm Hund's rules and fill the states corresponding to a spin direction first, before filling those corresponding to the reverse direction (cf. Section 3.4).

The author [3.47] has suggested taking these points into account by assuming that the carrier is localized for an amount of time $(\tau - t)$ and free for an amount of time t (transfer period $\tau = 1/v$). The hopping time t would be negligible when the effective (positive) charge carried by the atom is high, and predominant when it is low (Fig. 3.13). The ratio $(\tau - t)/t$, which depends on the temperature through the intermediary of τ, varies between zero (delocalization)

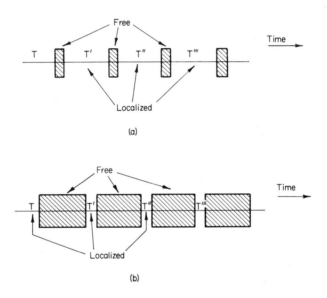

Fig. 3.13. Diagrammatic representation of the degree of localization of a d electron taking part in the transfer mechanism: (a) case of ionic compounds with high localization and (b) case of more covalent compounds with high localization (based on Suchet [3.47]).

and infinity (localization). This corresponds to the conversion of a free electron into a "polaron," trapped by its own polarization effect on the electronic distribution of the crystal or, to put it another way, by the modification imposed on the effective charge borne by the atom as a result of the accompanying local distortion. Delocalization is more likely when τ is low and, since the height of the "band" varies like hv, this in fact corresponds to a wide band. It should be noted that the effective mass varies in the same direction as the ratio just mentioned and within the same limits. In the series of Fe(VI) compounds, ionicity would vary as follows, for an ideal octahedral arrangement:

3.3. Metal-to-Metal Transfers

Fe—O	$\lambda = 0.76$ (Fe(II) VI)	and	0.62 ($\square$Fe$_2$(III) VI$_3$)
Fe—S	0.71		0.52
Fe—Se	0.47		0.20
Fe—Te	0.35		-0.12

It is clear that the d electrons of FeO and FeS will be strongly localized (polarons) and that those of FeSe will be less strongly attached. In telluride Fe$_2$Te$_3$, Fe(III) will have no tendency to trap an additional electron and the Fe(II) atom will not be stable; this is confirmed by experiment [3.48].

It will be gathered from this over-brief account that nonbonding d electrons (in the sense of the VB approach) give rise to an electrical conductibility that can take on many different forms. The term of transfer mechanism used in this book is a general one, referring to their single common characteristic, namely that their energies are grouped. The possible existence of a hopping activation energy E_A for the mobility indicates strong localization of these electrons, without the special model suggested by Heikes and Johnston necessarily applying to them. The possible absence of an activation energy for the mobility, on the other hand, does not necessarily indicate complete delocalization. If an optical transmission threshold exists and if the effective mass of the carriers is high, a second-type metal (or pseudometal), in which low localization can usually be shown indirectly, is involved.

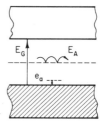

Fig. 3.14. Simultaneous appearance in a crystal of the mechanisms of conductibility by excitation and transfers (based on Suchet et al. [3.49]).

Finally, Suchet and Manca [3.49] have noted that the simultaneous intervention of excitation and transfer mechanisms is possible in certain compounds with a shallow forbidden band (Fig. 3.14) and have suggested, with certain reservations and as a simplifying hypothesis, that their contributions to conductibility are added together:

$$\sigma = N_1 e \mu_1 \text{ (excitation)} + N_2 e \mu_2 \text{ (transfers)} \tag{3.14}$$

Figure 3.15 shows the variations of N_1 and N_2, respectively, in relation to the temperature and density of impurities, in the event of the second mechanism becoming identical with hopping semiconduction.

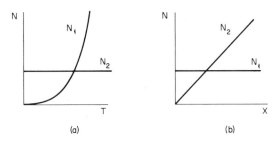

Fig. 3.15. Carrier density of the excitation (N_1) and transfer (N_2) mechanisms in terms of temperature (a) and density of defects (b) (based on Suchet et al. [3.49]).

3.4. MAGNETON–ELECTRON INTERACTIONS

Merely adding the conductibilities caused by the excitation and transfer mechanisms in fact implies that one ignores their interactions. In order to study these, and decide under which cases they can be ignored, one has to consider the *energy level diagram* of the crystal. No difficulty arises for the s and p levels, which generally provide the bonds (sp^3 in the tetrahedral arrangement, p^3 in the octahedral arrangement). The VB diagram (dielectronic) was given in the previous chapter (Fig. 2.11c), while Fig. 3.16 reproduces the Goodenough's [3.9] MO diagram (monoelectronic). The nonbonding d electron energies are inserted in the form of (VB) discrete levels or (MO) bands, the position of which it is in contrast extremely difficult to calculate with any accuracy. Ionization energies and electronic affinities, which apply to isolated ions, cannot very well be used in a crystal, even when it is fairly ionic, and it is difficult to calculate the polarization energies. The very nature of the transfer mechanism anyway makes it unlikely that any notable energy difference would be found, for instance between the levels corresponding to Ni^{2+} and Ni^{3+}. Otherwise, there would be excitation of an electron from one level to the other. Similarly, the concept of a donor or acceptor level, usual in the band model, is less obvious here. These problems are discussed in the digest [3.34].

The simplest attitude is to assume that the transfers result from momentary local disturbances that do not significantly affect the position of the t_{2g} and e_g levels. It was seen in Section 3.1 that the respective classification of these levels could be established in terms of the type of crystal arrangement and the plus or minus sign in front of the effective charge. A similar classification therefore remains to be made for the group of p (or s–p) bonding electron energies and of d electron energies. As a first example, let us take the simple case of manganese monochalcogenides. MnO, MnS, and MnSe crystallize in the B1 structure, where the metalloid octahedron is perfect; MnTe crystallizes in the $B8_1$ structure, where this octahedron is slightly deformed, but Pearson

[3.50] has shown that the same p^3 bonds applied there, at a first estimate anyway. Figure 3.17a shows the VB diagram for these, modified to take into account the existence of (B) and (AB) wide bands. The only difference in an MO diagram would be the substitution of t_{2g} and e_g narrow bands at the corresponding levels and the use of a different terminology. The transfer mechanism can be shown diagrammatically in Fig. 3.17b, which assumes the existence of manganese vacancies that have brought about the appearance of valency III atoms (cf. Chapter 7).

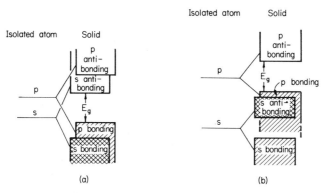

Fig. 3.16. Division of the energy bands in some typical semiconductors: (a) tetrahedral arrangement of the B3 structure, (b) octahedral arrangement of the B1 structure. [Based on Goodenough "Magnetism and the Chemical Bond," Fig. 39a,c, p. 158. Wiley (Interscience), New York, 1963. Used by permission of the copyright owner.]

The e_g level may be close to the bonding p^3 band. It has been seen that the spatial distribution of the electron density of the $d_{x^2-y^2}$ and d_{z^2} AOs is similar to that of the p_x, p_y, and p_z AOs. The existence of $e_g{}^2p$ and $e_g{}^2sp^3$ hybrid AOs, with the same octahedral symmetry as the p^3 AOs, is then possible, and can be shown by the *magnetic criterion*. Their presence alters the number of bachelor electrons, and consequently the magnetic properties of the crystal. The appearance of $e_g{}^2p$ AOs and $e_g{}^2p/p^3$ Lewis pairs between the atoms of a TX compound would free two electrons (Fig. 3.18a) and thus bring about metallic conductibility. In contrast, the appearance of $e_g{}^2sp^3$ AOs and $e_g{}^2sp^3/p^3$, $e_g{}^2sp^3/sp^2$, or $e_g{}^2sp^3/sp^3$ pairs between the atoms of TX_2 or $T(X-X)$ compounds is compatible with semiconductor properties (Fig. 3.18b) [3.52, 3.53]. As a second example let us consider the case of rutile TiO_2, crystallizing in the C4 structure. Figure 3.19a shows the modified VB diagram for this. Each O atom has a nonbonding p pair. The high value of E_G attaches t_{2g} in the forbidden band since Δ hardly rises above 1 eV. The t_{2g} level is empty in the stoichiometric oxide. The presence of interstitial Ti atoms leads a few atoms at normal sites to change to valency III (d^1), resulting in the transfer diagram

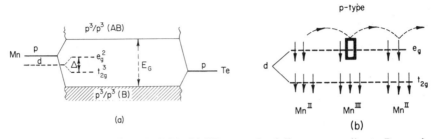

Fig. 3.17. Compound MnTe (B8$_1$): (a) VB energy level diagram according to Pearson's approximation and (b) scheme of the transfer mechanism in the e_g localized level in the presence of a slight excess of Te (based on Suchet [3.51]).

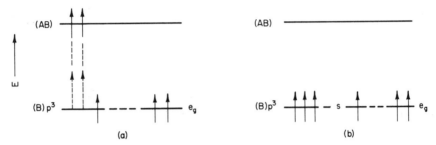

Fig. 3.18. Diagrammatic representation of the effect of an e_g—p hybridization: (a) TX compound and (b) TX$_2$ compound (based on Suchet [3.53]).

of Fig. 3.19b, Goodenough [3.54] has proposed an MO diagram, which is reproduced in Fig. 3.20. One will immediately note the taking into consideration of π bonds by overlapping of the p_z AOs of the anion with the d_{yz} and d_{zx} AOs of the cation, and their representation by bands. There remain the d_{xy} AOs, the varying amount of overlapping of which, from one cation to another, along axis c, can result in a band or a localized level. The partial filling of these bands gives isotropic metallic transfers for the π band, and anisotropic metallic or hopping transfers for d_{xy}.

The first interaction between the excitation and transfer mechanisms is the magnetic scattering of the carriers moving in the wide bands. It is caused by a disordered distribution of localized carriers, the spin quantum numbers of which have a certain atomic component S. Only transfers with a sufficient period of localization can produce it. Let us assume that the conductibility caused by transfers is negligible compared with that caused by excitations. It is shown that the resistivity of a crystal in which a magnetic order exists can be written [3.16, 3.51] as

$$\rho = 1/\sigma = (1/Ne)(1/\mu_1 + 1/\mu_1') \tag{3.15}$$

where μ_1 refers to the mobility of the carriers in the first mechanism, and the

3.4. Magneton–Electron Interactions

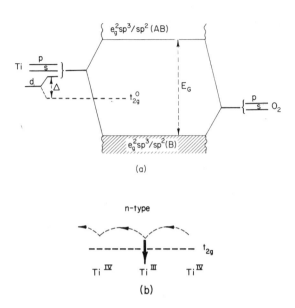

Fig. 3.19. Compound TiO_2 (C4): (a) VB energy level diagram and (b) schema of the transfer mechanism in the t_{2g} localized level in the presence of a slight excess of titanium (based on Suchet [3.51]).

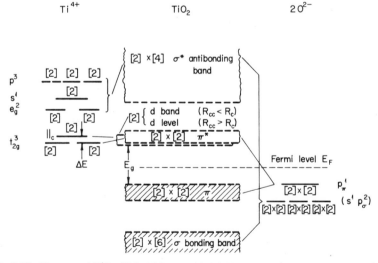

Fig. 3.20. Compound TiO_2 (C4): MO energy level diagram (Ti^{4+} AO on left and O^{2-} AO on right). The asterisk on σ and π signifies the "antibonding" bands in the MO terminology (based on Goodenough [3.54]).

second term, which is nil at absolute zero, expresses the magnetic scattering. It increases rapidly when the temperature nears the Curie (or Neel) point, and the magnetic order ends. In the paramagnetic domain, it has a constant value

$$1/\mu_1' \sim KS(S+1) \qquad (3.16)$$

where K summarizes the effect of factors independent of temperature. This term covers the positive temperature coefficient of the resistivity often observed below the Curie or Neel point, and the maximum resistivity in the neighborhood of this point. In Section 2.4, however, another possible cause of the maximum was mentioned, and a third one will be referred to at the end of this section. The magnetic scattering of d carriers moving in a narrow "band" has never, to the author's knowledge, been pointed out.

It was seen in Section 3.1 that the crystal electrostatic field produced a division of the energy levels, principally the d level. This is the energy of the space part of the total wave function [cf. Section 1.2, Eq. (1.18)]. The same applies to the inner magnetic field, known as the Weiss field, resulting from the establishment of a magnetic order in the crystal, for the spin part. It will be realized that the electron with its spin moment lying in the direction of the field (α) has a lower energy than one with its moment lying in the opposite direction (β). If localized electron levels are involved, there is a resulting additional division, which, in the simple case of MnO, for instance, will give the successive levels p(B), $t_{2g}\alpha$, $e_g\alpha$, $t_{2g}\beta$, $e_g\beta$, p(AB). If wide bands are involved, naturally no separation can occur, so that the energy lag between the α and β electron groups leads to a relative increase in the population of the α group. This is the origin of the Pauli paramagnetism mentioned in Section 3.2, which is proportional to the cube root of the electron density. Finally, if partly localized electron levels are involved, the situation is less straightforward. It can be assumed that partial separation could occur with the electron density reaching a minimum at the energy corresponding to the filling of half the widened level.

Experiments frequently show that these phenomena occur very slightly, even in the absence of magnetic order. In particular, nonionic compounds (in which the electrons do not behave like polarons) frequently present semiconductor properties when the α fraction of the level is filled and the β fraction empty, whereas they present apparent metallic properties (pseudometals) when one is partly filled [3.53]. Figure 3.21 shows the minimum electron density observed for a half-filled d band in $Fe_{1-x}Te$, where the disorder of the vacancies prevents the separate existence of the t_{2g} and e_g levels. Some authors call the α electrons "bonding" and the β electrons "antibonding," which makes the situation slightly more confusing, because of the different meaning of these words already met with in the VB and MO approaches.

Transitions from apparent metallic properties to apparent semiconductor

3.4. Magneton–Electron Interactions

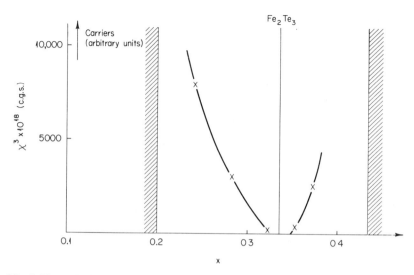

Fig. 3.21. Variation in the electron density (assumed to be proportional to the cube of the Pauli susceptibility, which is arguable here) in terms of x in $Fe_{1-x}Te$ (based on Manca et al. [3.43]).

properties, or vice versa, have been observed in many compounds. Several interpretations can be provided. The existence of T—T' bond pairs broken at the Neel point, referred to in Section 3.2, has been mentioned by Goodenough [3.55] in the case of VO, VO_2, and Ti_2O_3. If the position of the e_g level depends on the temperature, its hybridization with the p level might also occur at a given temperature. Morin [3.56] has also suggested that the semiconductor nature of VO below the Neel point comes from the division of the half-filled t_{2g}^3 level into $t_{2g}^3 \alpha$ (full) and $t_{2g}^0 \beta$ (empty). Finally, the author [3.51–3.53] has attributed the metallic character of CrTe, MnP, MnAs ($q > 0$), Cr_2Te_3, CrSb, MnSb ($q < 0$) below the Curie (or Neel) point to an interaction between the $p(AB)\alpha$ band and the upper d sublevel if it is partly filled (Fig. 3.22). It may be noted that a metal–semiconductor transition at rising temperatures is reflected by a resistivity maximum.

To complete the picture, it should be mentioned that the identification of a distorted structure with an ideal structure is only an approximation. Strictly speaking, the $B8_1$ structure, for instance, has two T—X—T' bond angles of $90° - \varepsilon$ and $90° + \varepsilon$, and its t_{2g} level is slightly divided into two sublevels containing 4 states and 2 states, respectively. The same applies to the C18 structure. In addition, some distortions are bound up with the number of bachelor d electrons of the T atom and result from the crystal field. This is the Jahn–Teller effect, which also causes a slight division in the t_{2g} or e_g levels. It takes place when an α or β sublevel is partly filled. In the tetrahedral arrangement, it occurs for d^1, d^3, d^4, d^6, d^8, and d^9, but not for d^2 ($e_g^2\alpha t_{2g}^0\alpha$), $d^5(e_g^2\alpha t_{2g}^3\alpha)$ or $d^7(e_g^2\beta t_{2g}^0\beta)$. In the octahedral arrangement, it occurs for d^1, d^2, d^4, d^6, d^7, and d^9, but not for d^3 ($t_{2g}^3\alpha\, e_g^0\alpha$), d^5 ($t_{2g}^3\alpha e_g^2\alpha$) or d^8 ($t_{2g}^3\beta e_g^0\beta$).

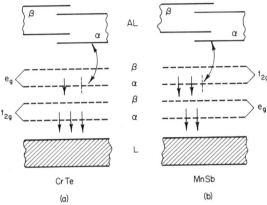

Fig. 3.22. Interaction between the antibonding band and the upper $d\alpha$ sublevel in the temperature range where a magnetic order exists: (a) with $e_g\alpha$ in CrTe and (b) with $t_{2g}\alpha$ in MnSb (based on Suchet [3.51]).

REFERENCES

[3.1] C. J. Ballhausen, "Introduction to Ligand Field Theory." McGraw-Hill, New York, 1964.
[3.2] F. A. Cotton, "Chemical Applications of Group Theory." Wiley (Interscience), New York, 1964.
[3.3] L. E. Orgel, "An Introduction to Transition-metal Chemistry." Methuen, London, 1960.
[3.4] H. Bethe, *Ann. Phys. (Leipzig)* **3**, 133 (1929).
[3.5] J. H. Van Vleck, "The Theory of Electric and Magnetic Susceptibilities." Oxford Univ. Press, London and New York, 1932.
[3.6] J. P. Suchet, *C. R. Acad. Sci. (Paris)* **262**, 689 (1966).
[3.7] C. Klixbüll-Jørgensen, "Inorganic Complexes." Academic Press, London and New York, 1963.
[3.8] J. H. Van Vleck, *J. Chem. Phys.* **3**, 803, 807 (1935).
[3.9] J. B. Goodenough, "Magnetism and the Chemical Bond." Wiley (Interscience), New York, 1963.
[3.10] G. T. Rado and H. Suhl, "Magnetism." Academic Press, New York, 1963.
[3.11] A. Michel, "Phénomènes magnétiques et structure." Masson, Paris, 1967.
[3.12] N. N. Sirota, "Khimicheskaya sviaz' v poluprovodnikakh i tverdykh telakh," p. 12 (N. N. Sirota, Ya. G. Dorfman, N. M. Olekhnovich and V. Z. Golodushko, eds.). Nauka i Tekhnika, Minsk, 1965.
[3.13] F. Bailly and P. Manca, *C. R. Acad. Sci. (Paris)* **262B**, 1075 (1966).
[3.14] F. Bailly and P. Manca, "Khimicheskaya sviaz' v kristallakh," p. 39 (N. N. Sirota, ed.). Nauka i Tekhnika, Minsk, 1969.
[3.15] M. Matyas, *Czech. J. Phys.* **9**, 257 (1959).
[3.16] J. P. Suchet, *Phys. Status Solidi* **2**, 167 (1962).

References

[3.17] J. S. Kouvel, "Intermetallic Compounds" (J. H. Westbrook, ed.). Wiley, New York, 1967.
[3.18] W. Heisenberg, *Z. Phys.* **49**, 619 (1928).
[3.19] H. A. Kramers, *Physica* **1**, 182 (1934).
[3.20] R. Benoit, *J. Chim. Phys.* **52**, 119 (1955).
[3.21] J. P. Suchet, *C. R. Acad. Sci. (Paris)* **253**, 2490 (1961).
[3.22] J. P. Suchet, *J. Phys. Radium* **23**, 497 (1962).
[3.23] F. Bailly and J. P. Suchet, "Séminaires de chimie de l'état solide" 1, 65 (J. P. Suchet, ed.). S.E.D.E.S., Paris, 1969.
[3.24] H. Bethe, "Handbuch der Physik," vol. 24, p. 595. Springer, Berlin, 1933.
[3.25] C. Guillaud, State thesis, Strasbourg, 1943.
[3.26] J. P. Suchet, *C. R. Acad. Sci. (Paris)* **256**, 2563 (1963).
[3.27] F. Bloch, *Z. Phys.* **61**, 206 (1931).
[3.28] J. C. Slater, *Phys. Rev.* **52**, 198 (1937).
[3.29] F. Seitz, "The Modern Theory of Solids." McGraw-Hill, New York, 1940.
[3.30] L. Landau, *Phys. Z. Sowietunion* **3**, 664 (1933).
[3.31] R. W. Gurney and N. F. Mott, *Proc. Phys. Soc. London, Suppl.* **49**, 32 (1937).
[3.32] R. R. Heikes and W. D. Johnston, *J. Chem. Phys.* **26**, 582 (1957).
[3.33] E. J. W. Verwey, P. W. Haayman, F. C. Romeyn and G. W. Van Oosterhout, *Philips Res. Rep.* **5**, 173 (1950).
[3.34] G. H. Jonker and S. Van Houten, *Halbleiterprobleme* **6**, 118 (1961).
[3.35] S. Van Houten, *J. Phys. Chem. Solids* **17**, 7 (1960).
[3.36] F. J. Morin, "Semiconductors," p. 600 (O. Hannay, ed.). Reinhold, New York, 1959.
[3.37] Ya. M. Ksendzov, L. A. Ansel'm, L. L. Vasil'eva and V. M. Latysheva, *Fiz. Tverd. Tela* **5**, 1537 (1963).
[3.38] G. Sewell, *Phil. Mag.* **3**, 1361 (1958); *Phys. Rev.* **129**, 597 (1963).
[3.39] M. I. Klinger, *Izv. Akad. Nauk SSSR, Ser. Fiz.* **25**, 1342 (1961).
[3.40] I. G. Austin, A. J. Springthorpe and B. A. Smith, *Phys. Lett.* **21**, 20 (1966).
[3.41] D. S. Tannhauser, *J. Phys. Chem. Solids* **23**, 25 (1962).
[3.42] J. P. Suchet, *Mater. Res. Bull.* **2**, 727 (1967).
[3.43] P. Manca, J. P. Suchet and G. A. Fatseas, *Ann. Phys. (Paris)* **1**, 621 (1966).
[3.44] N. F. Mott, *Phil. Mag.* **6**, 287 (1961).
[3.45] F. J. Morin, *Proc. Int. Conf. Semicond. Phys. (Prague 1960), Czech. J. Phys.* 858 (1961).
[3.46] J. B. Goodenough, "Magnetism and the Chemical Bond." Wiley (Interscience), New York, 1963.
[3.47] J. P. Suchet, "Khimìcheskaya svìaz' v kristallakh," p. 29 (N. N. Sirota, ed.), Nauka i Tekhnika, Minsk, 1969.
[3.48] J. P. Suchet and P. Imbert, *C. R. Acad. Sci. (Paris)* **260**, 5239 (1965).
[3.49] J. P. Suchet and P. Manca, Conf. El. Proc. Low Mobility Solids (Sheffield 1966), Rend. Sem. Fac. Sc. Univ. Cagliari (Italie) 37, 217 (1967).
[3.50] W. B. Pearson, *Can. J. Phys.* **35**, 886 (1957).
[3.51] J. P. Suchet, "Séminaires de chimie de l'état solide" (J. P. Suchet, ed.), vol. 1, p. 37. S.E.D.E.S., Paris, 1969.
[3.52] J. P. Suchet, *Coll. Int. Prop. Dérivés Semimétalliques (Orsay, 1965)* p. 293. C.N.R.S., Paris, 1967.
[3.53] J. P. Suchet, *Phys. Status Solidi* **15**, 639 (1966).
[3.54] J. B. Goodenough, *Coll. Int. Comp. Ox. Elém. Trans. (Bordeaux 1964)* p. 162. C.N.R.S., Paris 1965.
[3.55] J. B. Goodenough, *Phys. Rev.* **117**, 1442 (1960).
[3.56] F. J. Morin, *Phys. Rev. Lett.* **3**, 34 (1959).

Part Two BIBLIOGRAPHICAL DIGEST
(1947–1967)

Chapter 4 | IIIB, IVB, and VB Metalloid Compounds

4.1. Main Structures

The main crystallographic structures encountered in this class of compounds have the symbols B1 (rock salt), $B8_1$ (NiAs), $B8_2$ (Ni_2In), B31 (MnP), C2 (pyrite), C18 (marcasite), DO_2 (skutterudite) and C40 ($CrSi_2$). The B1 structure is sufficiently familiar for there to be no need to return to what was said in Chapter 2. Here we shall deal mainly with the standard $B8_1$ and C2 structures. Readers wishing for more detailed information may refer to the general books by Wells [4.1] and Wyckoff [4.2] as well as a small specialized work by Aronsson et al. [4.3].

Figure 4.1a represents the primeval solid of the NiAs structure: a right prism with a centered hexagon as its base. Figure 4.1b represents the corresponding simple cell: an orthorhombic prism with an angle of 120°. The metalloid atoms would form an ideal compact hexagonal sublattice for $c/a = 1.633$. Experimental results in fact range from 1.2 to 2, and these divergencies are often bound up with deviations in the stoichiometry. Such a sublattice contains one octahedral cavity and two adjoining tetrahedral cavities (in other words, a trigonal bipyramidal cavity) for each crystallographic site (Fig. 4.2). The metal atoms fill all the octahedral cavities in the TX stoichiometric composition, but this can generally be obtained only at high temperatures. The case of structures with metal vacancies $\square_x T_{1-x} X$, very frequent for chalcogenides, will be examined in Chapter 7. Metalloid vacancies are met with in interstitial compounds. An excess of metal TXT_x, occupying the bipyramidal cavities, characterizes the compounds of metalloids in columns V, IV, and III of the table and, when these are all filled, produces the formula

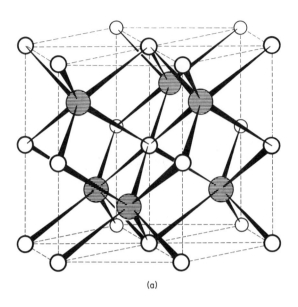

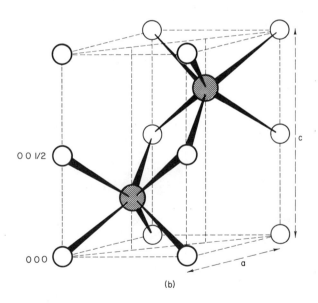

Fig. 4.1. B8₁ structure of nickel arsenide NiAs. The large hatched circles represent the As atoms: (a) hexagonal primeval solid with six molecules and (b) bimolecular single cell (based on Wells [4.1]).

4.1. Main Structures

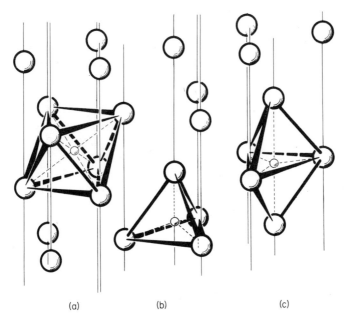

Fig. 4.2. Cavities left by a compact hexagonal sublattice of metalloid atoms: (a) octahedral, (b) tetrahedral and (c) trigonal bipyramidal. [Reprinted with permission from Kjekshus and Pearson, *Progress in Solid State Chemistry* **1**, 83 (1964), Pergamon Press.]

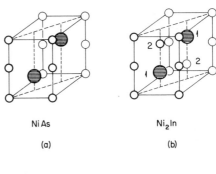

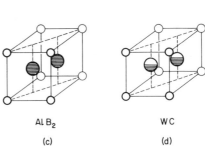

Fig. 4.3. Structures derived from the (a) NiAs structure: (b) Ni_2In, (c) AlB_2, and (d) WC (based on Wells [4.1]).

TX$^{(1)}$ T$^{(2)}$ (B8$_2$ structure of Ni$_2$In, Fig. 4.3b). If the two atoms 1 and 2 were identical, the height of the elementary cell could be reduced by half: this is what happens in a TX$_2$ formula (C32 structure of AlB$_2$, Fig. 4.3c) or even in a TX formula, when the metalloid is distributed at random among the normal sites of the compact sublattice and the bipyramidal cavities (WC structure, Fig. 4.3d). It might be added that slight distortions in the symmetry of the octahedral cavities are frequently encountered at normal temperatures. Finally, in the B31 structure of MnP, the T atoms are systematically offset in relation to the centers of these cavities, so as to produce fairly close T—T' pairs.

In order to illustrate the essential role always played by the crystallographic structure in interpreting experimental data, let us mention two important points here, in connection with the B8$_1$ structure: irregularities in properties close to the Curie temperature, and the special part played by the (001) planes in the transfers. Reexamining and putting earlier observations into a systematic form, Willis and Rooksby [4.5] have shown the existence, in several compounds of manganese, notably MnAs and MnSb, of abrupt changes in the derivative of the cell parameters in relation to temperature, bound up with the magnetic transitions of the second order (Curie temperatures). In MnBi, a neutron-diffraction investigation has shown the movement of manganese atoms from normal sites to the bipyramidal cavities when the paramagnetic region is reached [4.6]. This interaction between the crystallographic order and the magnetic order frequently takes the form of an hysteresis of the electrical properties (cf. Section 4.4). As regards the role of the (001) planes in the transfers, Fig. 4.4 shows that the t_{2g} AOs of a metal atom are directed towards its 12 metal neighbors in the B1 structure, while in the B8$_1$ distorted structure only the three t_{2g} AOs of the (001) plane are directed exactly towards the six neighbors on this plane. Despite the longer T—T interatomic distance in these planes, the density of electron transfers will not be negligible, compared with what occurs in the direction of axis c (cf. Chapter 7).

Kjekshus and Pearson [4.4] recently stated: "The choice of the MnP structure, or the NiAs structure with c/a fairly close to the ideal value of 1.63, is undoubtedly due to a pronounced directional character of the chemical bonds." In other words, the B8$_1$ and B31 structures are adopted by fairly covalent compounds. The octahedral coordination of the metal atom, with X—T—X angles close to 90°, can correspond only to the use of p AOs, as in the B1 structure, or a d–p or d–s–p hybrid AOs. It was also noted in Section 3.4 that hybridization can occur only between AOs with their electron density maxima in the same directions, in other words e_g and p, and that the magnetic criterion generally allows it to be identified with certainty. For instance Mn, with the electronic formula 3d^{5}4s^2, and Sb, with the formula 5s^{2}p^3, become 3d^{4}4s^{0}p^3 and 5s^{2}p^3, respectively, giving Mn$^{(p^3)}$Sb$^{(p^3)}$ with $2S = 4$ bachelor

4.1. Main Structures

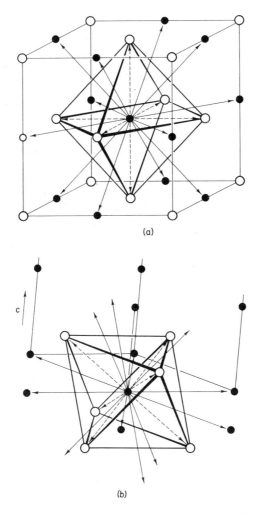

Fig. 4.4. Directions of maximum electron densities of the AOs of a metal atom: (a) B1 structure and (b) $B8_1$ structure. [Reprinted with permission from Kjekshus and Pearson, *Progress in Solid State Chemistry* **1**, 83 (1964), Pergamon Press.] (⟶) t_{2g} orbitals and (---→) p or e_g orbitals.

electrons. The $e_g{}^2$p hybridization would reduce the number by 2, while two p electrons would be excited at the AB level and would move freely (Fig. 3.17a). Equation (3.6) allows one, from experimental data, to calculate, for clearly metallic compounds such as VAs or CoAs, a $2S$ number reduced by 2 units in comparison with the expected value, while the scale is more or less respected for those such as CrSb or MnSb, which have less clear-cut properties [4.7].

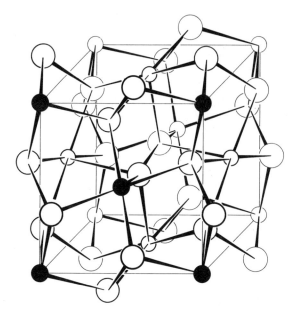

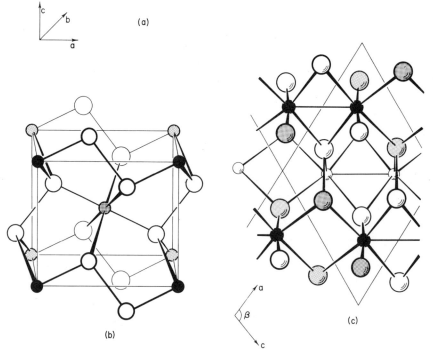

Fig. 4.5. (a) C2 cubic structure of pyrite, (b) C18 orthorhombic structure of marcasite, and (c) EO$_7$ monoclinic structure of arsenopyrite. The small shaded circles represent T atoms, and the large white circles the X atoms. [Reprinted with permission from Hulliger and Mooser, *Progress in Solid State Chemistry* **2**, 330 (1965), Pergamon Press.]

The $e_g{}^2sp^3$ AOs, which do not have any resonance, could not be associated to form the MO with metalloid p AOs, which have to include a resonance.

Figure 4.5 shows the arrangement of atoms in structure C2 of pyrite (a) and C18 of marcasite (b), in which the two varieties of the compound FeS_2 crystallize, and in the EO_7 structure of arsenopyrite FeAsS (c). All three contain X—X' bonds: each T atom is surrounded by an X_6 octahedron and each X atom by an XT_3 tetrahedron, these polyhedra being distorted to varying degrees. The C2 structure can be considered as linked to the B1 structure of rock salt, where the X_2 doublet would occupy the metalloid sites. Whereas in pyrite the X_6 octahedra are joined only by one corner, they have an edge in common in marcasite. Arsenopyrite involve an alternation of short and long distances between T atoms, the short distances probably corresponding to a T—T' bond. Strictly speaking, the distortions of the X_6 octahedron rule out the division of the d level of the T atom into only two sublevels, particularly in the case of marcasite (Fig. 4.6).

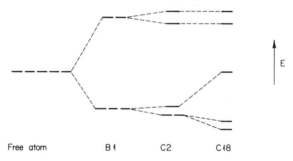

Fig. 4.6. Division by the crystal field of the d level of a T atom in different structures. [Reprinted with permission from Hulliger and Mooser, *Progress in Solid State Chemistry* **2**, 330 (1965), Pergamon Press.]

The e_g–s–p hybridization of the metal AOs is widespread in these structures, and Table 4.1 summarizes the known compounds and classifies them in terms of the number of d electrons, according to Hulliger and Mooser [4.9]. The electronic formula of the T atom has been added, with parentheses around the electrons contributing to the T—X Lewis pairs, the spins of which are thus balanced. The number $2S$ of bachelor nonbonding d electrons that could be expected in each case has also been mentioned. In fact, while experiments confirm this figure for $CrSb_2$ and the homologues of $NiAs_2$, the same is not true for other series in which no magnetic moment is apparent. For FeP_2 homologues, it has been seen that the t_{2g} level of the C18 structure is divided into two sublevels with four and two quantum states, separated by an appreciable energy. It must be assumed that the lower sublevel is here completely filled (2α electrons and 2β), the upper being empty. This particular separation causes, through the Jahn–Teller effect, a distortion of the cell, and the ratio of

TABLE 4.1

Known compounds	Structure	2S	T Electronic formula	
			Level before higher level	Higher level
CrSb$_2$	C18	2	$3d^2(d^2$ $\cdots$	$4sp^3)d^0$
		3		
FeP$_2$ FeAs$_2$ FeSb$_2$			$3d^4(d^2$ $\cdots$	$4sp^3)d^0$
RuP$_2$ RuAs$_2$ RuSb$_2$	C18	2	$4d^4(d^2$ $\cdots$	$5sp^3)d^0$
OsP$_2$ OsAs$_2$ OsSb$_2$			$5d^4(d^2$ $\cdots$	$6sp^3)d^0$
CoAs$_2$ CoSb$_2$			$3d^5(d^2$ $\cdots$	$4sp^3)d^0$
RhP$_2$ RhAs$_2$ RhSb$_2$ RhBi$_2$	EO$_7$	1	$4d^5(d^2$ $\cdots$	$5sp^3)d^0$
IrP$_2$ IrAs$_2$ IrSb$_2$			$5d^5(d^2$ $\cdots$	$6sp^3)d^0$
NiAs$_2$ NiSb$_2$	C18		$3d^6(d^2$ $\cdots$	$4sp^3)d^0$
PdAs$_2$ PdSb$_2$		0	$4d^6(d^2$ $\cdots$	$5sp^3)d^0$
PtP$_2$ PtAs$_2$ PtSb$_2$ PtBi$_2$	C2		$5d^6(d^2$ $\cdots$	$6sp^3)d^0$
AuSb$_2$	C2	1	$5d^6(d^2$ $\cdots$	$6sp^3)d^1$

the sides of the X_6 octahedron, parallel and perpendicular to the smaller axis, drops to about 0.75. For CoAs$_2$ homologues, there is one bachelor d electron on each T atom, and one therefore has to imagine the balancing of their spin within the T—T' Lewis pairs localized on the shortest distances. Finally, the metallic character of AuSb$_2$ must result from the freeing of the electron occupying level 6d.

4.1. Main Structures

Figure 4.7 shows the plane hexagonal arrangement of atoms constituting the basic pattern for several disilicide structures. Let A be the layer of atoms in question. The crystal lattice is obtained by superimposing other similar layers on it, but, contrary to what happens in the compact hexagonal arrangement,

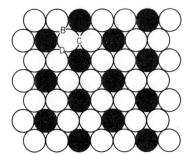

Fig. 4.7. Hexagonal arrangement of the atoms in the single atomic layers which make up most disilicides (based on Wallbaum [4.10]). (●) T atom, (○) Si atom.

the atoms in the following layer do not take up position above the cavity left by three atoms in A, preferring the positions B, C, or D. Different structures are thus obtained, depending on the alternation of the various layers of atoms and their situation: structure C11b of $MoSi_2$ [4.11] for the alternation ABAB... of layers situated in the (110) planes (Fig. 4.8), structure C40 of $CrSi_2$ for the alternation ABCABC... of layers situated in the (001) planes, distorted $TiSi_2$ structure for the alternation ABCDABCD... of layers situated in the (001) planes. The coordinance of T is 10, less than that of a compact arrangement, and it is difficult to determine the bonding orbitals involved.

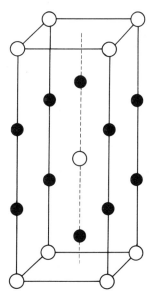

Fig. 4.8. C11b structure of $MoSi_2$. (●) Si atom, (○) Mo atom. [Based on Pascal, "Nouveau Traité de Chimie Minérale." Masson, Paris, 1956–1963.]

Estimation of the electrical properties is of great importance in this field, and we shall return to it frequently in the following chapters. Let it be stated clearly that no simple theoretical criterion can be applied generally. The best example is Heikes' rule [4.12] according to which the ferromagnetic compounds are metallic and the antiferromagnetic ones semiconducting. This rule, although frequently confirmed, has too many exceptions for one to be able to rely on it. On the other hand, in a group of different structures, empirical rules are often valuable. Dudkin [4.13], for instance, predicts semiconductor properties when the ratio between the metallic diameter of T and the shortest distance T—T' is inferior or equal to a certain figure. If one uses Bokii's [4.14] metallic diameters, this figure is 0.855. If one refers to Pauling's [4.15] metallic radii, it is apparently around 0.82 [4.4]. Finally, as seen in Section 2.2, the crystallochemical model provides an approximate criterion, in terms of the sign of λ.

4.2. Interstitial Compounds

Although it is extended to other bodies, the expression "interstitial compounds" can really be applied with certainty only to hydrides, borides, carbides, and nitrides. In this case, the distances between T atoms are very little different from those in pure metal, and the physical properties, with the exception of a few substances which we shall consider in particular detail, are close to those of the corresponding metals. The original work on their electrical properties was mainly done by Soviet research workers, and L'vov et al. [4.16] summarized the first results. Comparison of the Hall coefficients R_H is particularly interesting. Figure 4.9 shows that they always increase in the direction boride–carbide–nitride–metal and in the direction IVA–VA–VIA. In an attempt to interpret the behavior of interstitial compounds, Bilz [4.17] has proposed a model in which the T—X bonds form two bands, bonding (B) and antibonding (AB), corresponding to the p orbitals in the B1 structure (Fig. 4.10). Between these two there is a d band corresponding to the T—T' metallic bonds, and slightly overlapping with the two preceding bands. An isolated band, with low energies, corresponds to the nonbonding pair of s electrons of X. It is estimated that there is a minimum electron density for compounds with approximately eight valency electrons per T atom, and inversion of the sign of R_H would result from the start of filling of the d band before that of the (B) band is completed. If the crystal field intervened, the d band referred to would be a dε or t_{2g} band, while the dγ^2p or e_g^2p hybrid orbitals might replace the p^3 orbitals in the T—X bond pairs. Ern's [4.18] calculations using the MO method show that the metallic character increases in the oxide–nitride–carbide–boride–hydride series of titanium, as the ionization potential of X drops. A purely metallic model, in which the metalloid intervenes only by contributing electrons, can even be used for the final terms [4.19].

4.2. Interstitial Compounds

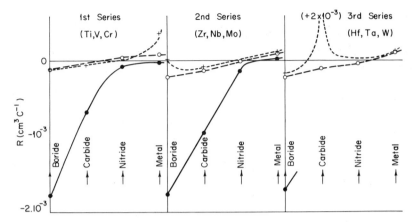

Fig. 4.9. Variation of the Hall coefficient in borides, carbides, nitrides, and transition metals (based on L'vov et al. [4.16]). The value of WC is abnormal. (—●—) T = IVA, (—O—) T = VA, (—+—) T = VIA.

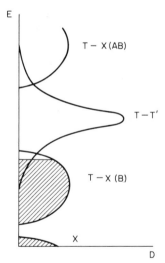

Fig. 4.10. Band structure of interstitial compounds (based on Bilz [4.17]).

There are numerous transition metal borides, and 5 or 6 different phases have been identified in some diagrams [4.20]. The commonest and most familiar have the formula TB_2 (Ti, Zr, Hf, V, Nb, Ta, Cr, Mo, C32 hexagonal structure of AlB_2), TB (V, Nb, Ta, Cr, Mo, W or Mn, Fe, Co, Ni, orthorhombic structures of CrB or B27 structure of FeB), T_2B (Ta, Cr, Mo, W, Mn, Fe, Co, Ni, C16 quadratic structure of $CuAl_2$). T—B bonds are particularly important in refractory diborides (3000°C for T = IVA or VA). Resistivities at room temperatures are around 10^{-5} Ω cm, with a positive temperature coefficient, and supraconductivity around 1°K is frequently observed [4.21]. The Seebeck effect has been studied in this connection by Samsonov et al. [4.22]. The Hall effect and the magnetoresistance of ZrB_2 monocrystals have recently been studied by Piper [4.23] between 1 and 300°K, with fields ranging up to 13 kOe. Some conclusions are given on the band structure. One suggestion to explain the low number (0.06) of free electrons per Zr atom is that there is a minimum electron density in the middle of the d band, resulting from the existence of the Zr—B bonds.

The phases in which carbides crystallize are fewer in number. The commonest have the formula TC (Ti, Zr, Hf, V, Nb, Ta, B1 structure of rock salt or Mo, W, Os, Ru, hexagonal structure of WC). One might also mention T_3C (Fe, Co, Ni, DO_{11} orthorhombic structure) and various T_2C structures [4.20]. Compounds of B1 structure, where the paramagnetism is very low, are among the most refractory substances known (3800°C for HfC, NbC, and TaC). The homogeneity range of the phase generally varies between 40 and 50% C atoms. Resistivities at room temperature are around 10^{-4} Ω cm, and supraconductivity is frequently observed at low temperatures (6°K for NbC) [4.21]. The Seebeck effect has been studied by Samsonov and Stryel'nikova [4.22] and Neshpor and Samsonov [4.24]. A negative temperature coefficient of the resistivity has been observed for TiC by Moers [4.25] on sintered samples and by Munster et al. [4.26, 4.27] on thin layers, whereas the weak Seebeck effect suggested metallic properties [4.28, 4.29]. On monocrystals in the [100] direction, Hollander [4.30] has measured a resistivity of 2×10^{-4} Ω cm at room temperature, with a positive temperature coefficient, and a Seebeck effect of -8 μV deg^{-1}, while Tsuchida et al. [4.31] and Piper [4.32] have, respectively, found a roughly linear increase and decrease in the Hall coefficient as a function of temperature. The origin of these contradictions probably lies in the possible variations in the stoichiometric ratio. Golikova et al. [4.33, 4.34] have systematically studied the resistivity variation, Seebeck effect, Hall effect, mobility and effective mass of electrons in terms of x in TiC_x. Figure 4.11 shows how the density of free electrons varies in the crystal. It suggests a semiconductibility for the extreme composition $x = 1$, which is claimed to be confirmed by the variation in relation to temperature of its resistivity and Seebeck effect [4.34]. The electronic formulas for Ti and C would then probably be $3d^04s^0(p^3)$ and $3s^2(p^3)$, the electrons between parentheses taking part in the bond pairs (displaced towards the carbon). The effect of neutron irradiation has been studied [4.35]. Umanski [4.36] was the first to suggest that for ZrC there was a transfer of electrons by the carbon to the lattice, a hypothesis subsequently generalized in the "metallic" model of interstitial compounds. The resistivity of this compound is indeed metallic [4.37]. The resistivities of TiC_x, ZrC_x, and HfC_x have been compared by Golikova et al. [4.38]. The Seebeck effect has been measured for VC_x [4.28], and the magnetic suscepti-

4.2. Interstitial Compounds

bility varies in inverse ratio to x [4.39]. For TaC_x, Cooper and Hansler [4.40] have shown that the resistivity varies in inverse ratio to x, and the metallic character has been established by Steinitz and Resnick [4.41] (see also [4.42, 4.43]). Chromium carbides are metallic [4.44, 4.45].

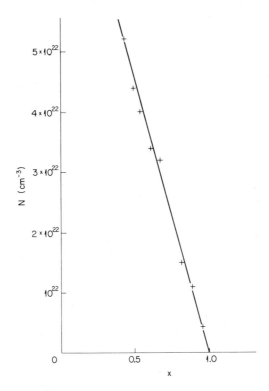

Fig. 4.11. Electron density variation in TiC_x (based on Golikova et al. [4.33]).

Many nitrides crystallize with the formula TN (Ti, Zr, Hf, V, Nb, Cr, W, B1 structure of rock salt, or Ta, Mo, W, Mn, unidentified complex structures occasionally of WC type). The structures of the subnitrides T_4N, T_3N, or T_2N are generally obscure [4.20]. Compounds with B1 structure present certain analogies with carbides of the same structure: refractoriness (3000°C for TiN, ZrN, TaN, 3300°C for HfN), a wide homogeneity range (30 to 50% N atoms for TiN, 40 to 50% for VN), resistivities at room temperature of around $10^{-4}\,\Omega$ cm, and supraconductivity at low temperatures (9°K for ZrN and 15°K for NbN) [4.21]. The Seebeck effect has been studied here by Neshpor and Samsonov [4.24]. For TiN, a negative temperature coefficient for the resistivity has been observed by Munster et al. [4.26, 4.27] on thin layers, and by Samsonov and Verkhoglyadova [4.45] and L'vov et al. [4.46] on sintered samples (Fig. 4.12). The latter give an activation energy of around 0.5 eV for 40% N atoms, dropping to zero for 50%. There has been controversy about the ionic or metallic nature of the Ti—N bonds [4.45, 4.47–4.49]. ZrN and HfN have also been studied [4.45]. The former is said to be semiconducting, with an activation energy of 0.8 eV for 40% N atoms, dropping to zero for 50% [4.46]. For VN, the resistivity decreases above 450°C, possibly because of oxidation [4.50]. It has been measured for NbN [4.51], TaN [4.21] and CrN [4.52]. Up to 1500°C, TaN is said to present a very slight decrease in resistivity; CrN contains covalent bonds [4.53] and is said to be semiconducting [4.46].

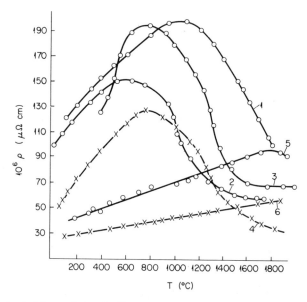

Fig. 4.12. Resistivity variation in TiN_x in relation to temperature (based on Samsonov and Verkhoglyadova [4.45]): (1) $x = 0.53$; (2) $x = 0.63$; (3) $x = 0.69$; (4) $x = 0.77$; (5) $x = 0.94$; (6) $x = 0.99$.

Interesting information can also be supplied by the properties of solid solutions containing these different compounds. TiC_xN_{1-x} solutions have been studied by L'vov et al. [4.54], who note the increased importance of 3d electrons in nitride-rich solutions, and then by Itoh et al. [4.55], who find a regular variation from 0 to 1 free electron per Ti atom from carbide to nitride, agreeing with Bilz' model. The variation is more complex for $Ti_xV_{1-x}C$ (Fig. 4.13). TiN_xO_y solutions are said to be semiconducting [4.46] (see also [4.47, 4.49]).

4.2. Interstitial Compounds

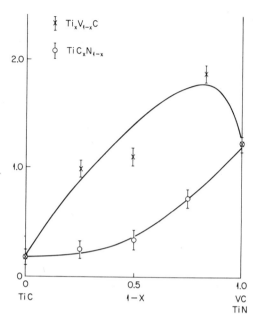

Fig. 4.13. Variation in the number of conduction electrons per Ti atom in terms of x (calculated from the Hall coefficient) in TiC_xN_{1-x} and $Ti_xV_{1-x}C$ solid solutions (based on Itoh et al. [4.55]).

An attempt can be made to interpret all these experimental facts, in the straightforward case of titanium compounds, using the diagrams in Fig. 4.14, which apply solely to stoichiometric compositions. If they reflected the situation exactly, TiC would be a semiconductor by excitation (t_{2g} empty) and TiN and TiO would be second category or pseudometals (t_{2g} partly filled). In fact, as Bilz' general model suggests (Fig. 4.10), there is probably partial overlapping between the d band and the (B) band, so that the former is never empty and the transfer mechanism is always present. For nonstoichiometric compositions, the study of TiO_x done by Denker [4.57] can be extended to TiN_x. The metalloid vacancies, reducing the number of valency electrons per Ti atom, lower the Fermi level and consequently tend to empty the d band. Let us consider the vacancy as an atom with zero valency: the electronic formula $3d^1 4s^0(p^3)$ of the titanium in $Ti^{(p^3)}N^{(p^3)}$ will be reduced to $3d^0 4s(p^3)$ in $Ti^{(p^3)}N_{2/3}^{(p^3)} \square_{1/3}^{(p^3)}$, namely for 40% N atoms. It is in fact around $x = \frac{2}{3}$ that the semiconductor properties are most marked in Fig. 4.12. This argument does not seem to apply to VC_x [4.39].

It should be noted that the ionicity parameter of B1 structure carbides and nitrides is usually positive (TiC 0.76, TiN 0.71), so that their semiconductibility can be presumed. But since this appears to be bound up with the absence of electrons in the d level, it obviously cannot be accompanied by magnetic properties.

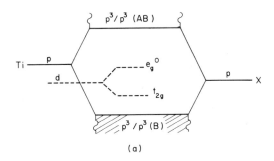

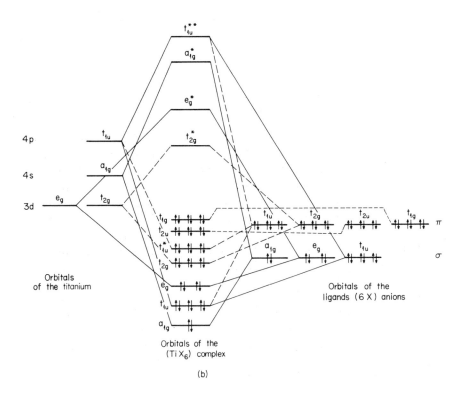

Fig. 4.14. Energy level diagrams for TiX: (a) VB, t_{2g}^0 for X = C, t_{2g}^1 for X = N, t_{2g}^2 for X = O; (b) MO, as such for X = C, add 1 electron for X = N and 2 for X = O to the t_{2g}^* level. [case (b) based on Nelson Tech. Rep. 179, Laboratory for Insulation Research, May, 1963]. (——) σ bonding and antibonding, (———) π bonding and antibonding, ($\cdots$) π nonbonding [4.56].

4.3. Silicides, Germanides, Stannides

Silicides crystallize in a large number of phases. The main ones have the formulas T_5Si_3 (except for Co, Ru, Ni, Pd, Pt: quadratic structure of Cr_5Si_3, which impurities frequently replace by the $D8_8$ structure of Mn_5Si_3, with roughly the same stability), TSi (Ti, Zr, Hf, B27 orthorhombic structure of FeB; Cr, Mn, Re, Fe, Ru, Os, Co, Rh, B20 cubic structure of FeSi; Rh, Ir, Ni, Pd, Pt, B31 orthorhombic structure of MnP) and TSi_2 (Ti, Zr, Hf, C49 or C54 orthorhombic structures; V, Nb, Ta, Cr, C40 hexagonal structure of $CrSi_2$; Mo, W, Re, C11b quadratic structure of $MoSi_2$; Co, Ni, C1 cubic structure of fluorite) [4.20]. Ferromagnetism exists notably in the B20 and $D8_8$ structures. The $D8_8$ structure contains two separate T atom sites. Resistivities of disilicides at room temperature are around 10^{-5} (C11b, C40) or 10^{-4} Ω cm (C49, C54). Supraconductivity has been observed in the C1 structure [4.21]. Early work was done on the Seebeck effect in silicides of Fe, Cr, and Mn [4.58]. We shall return to the electrical properties of monosilicides. Among disilicides, $TiSi_2$ has been studied by Neshpor et al. [4.29]; $MoSi_2$, known for its excellent resistance to oxidation up to 1500°C, has a clearly established metallic character [4.59–4.61]; $ReSi_2$, on the other hand, with the same C11b structure, is, according to Neshpor et al. [4.62], a semiconductor with an 0.13 eV forbidden band gap (Fig. 4.15). It is interesting because of its high melting point (1980°C), and its low vapor pressure. We shall consider the important case of disilicides with C40 structure later.

Only two compounds have a silicon/metal atomic ratio of more than 2: Fe_2Si_5 and $IrSi_3$. The former, which was for a long time written $FeSi_2$, has an unfamiliar quadratic structure. Its semiconductor properties had been pointed out by Abrikosov [4.63, 4.64] and Neshpor and Jupko [4.65]. According to Ware and McNeill [4.66], the activation energy is 0.88 eV and the n- and p-types are obtained by substituting Co and Al, respectively, in the metal and metalloid sites. It is hoped to use it in thermoelectric generators (see below). As for $IrSi_3$, neither its structure (hexagonal) nor its electrical properties have yet been studied, but its high silicon content makes it a probable semiconductor.

The interest aroused by silicides may be explained in particular by their possible application in thermoelectric generators. Such a use, which does not involve their magnetic properties in any systematic way, will not be mentioned in part three. Readers interested in this subject can refer, among others, to the classic works by Ioffe [4.67] and Heikes [4.68]. The principle can be summarized here very briefly. Let us assume that electrons on the one hand, and holes on the other, form two perfect gases in two cylinders joining a cold wall to a warm wall (Fig. 4.16). The gases contract when close to the cold wall, where the carrier concentration is thus higher, producing the direction of the current. Since the two cylinders are traversed by this current, their electrical conductivity σ must be high to avoid an unnecessary energy loss. On the other hand, their thermal conductivity k must be low, to avoid the temperatures of the walls being equalized. Let us assume that each cylinder consists of a block 0.5 cm high and with a cross-sectional area of 1 cm², and that the resistance of the utilization circuit is equal to that of the generator [4.69]. The electrical power is $V^2\sigma = (\alpha\Delta T)^2\sigma$ (α = Seebeck's coefficient) and the thermal power $k\Delta T$, giving a thermoelectrical conversion efficiency of $\alpha^2\sigma\Delta T/k = Cf$, where $C = \Delta T/T$ is Carnot's efficiency and $f = \alpha^2\sigma T/k$ is the factor of merit of the semiconductor material. This increases with T, whence the value of materials

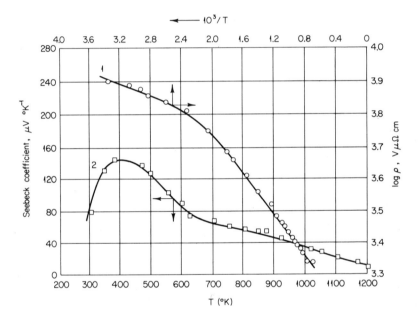

Fig. 4.15. Variation in resistivity and Seebeck coefficient of ReSi₂ in terms of temperature (based on Neshpor and Samsonov [4.62]).

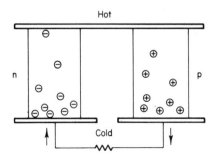

Fig. 4.16. Diagrammatic representation of a thermoelectric generator (based on Suchet [4.69]).

4.3. Silicides, Germanides, Stannides

functioning at high temperatures. For Fe_2Si_5, mentioned above, f/T exceeds 2.10^{-4} between 150 and 650°C [4.66]. An account of the use of the various silicides in this field has been provided by Mayer and Mlavsky [4.70].

Although the behavior of monosilicides seems to be metallic, research into their electrical properties is still of some interest. Andreyeva et al. [4.71] have recently compared CrSi, MnSi, FeSi, and CoSi. The resistivity of FeSi decreases exponentially between 20 and 300°K, possibly because of a change in the valency of the iron. TiSi [4.29], CrSi [4.72, 4.73], MnSi [4.74, 4.75], CoSi [4.74, 4.76, 4.77] had already been studied, mainly for their Seebeck effect; CoSi seems to be a possible thermoelectrical material at high temperatures (n branch). Asanabe et al. [4.78] have made a thorough investigation of $Co_{1-x}Fe_xSi$ solid solutions between 4 and 800°K. Resistivity decreases with temperature (Fig. 4.17). The type changes from n to p for increasing values of x

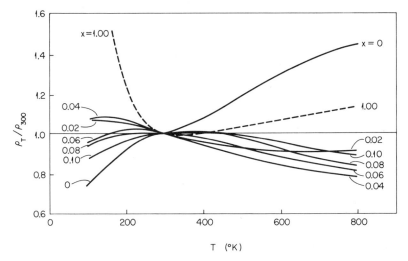

Fig. 4.17. Variation in relative resistivity ρ_T/ρ_{300} in relation to temperature, for different $Co_{1-x}Fe_xSi$ solid solutions (based on Asanabe et al. [4.78]).

(Fig. 4.18). Everything takes place as if the replacement of a Co atom by an Fe atom produced a hole. For $x = 0$ to 4°K, one finds $\mu_n = 800$ cm^2 V^{-1} sec^{-1} and $m_n^*/m_0 = 2$, $\mu_p = 160$ cm^2 V^{-1} sec^{-1} and $m_p^*/m_0 = 4$. The authors concluded that these solutions were *semimetals*. It was seen in Section 2.3 that the minimum for E_2 (conduction band) and the maximum for E_1 (valency band) occur in Si and Ge for different directions of propagation (indirect transitions, cf. Fig. 2.9b and c). The same applies here, but in addition $E_2 < E_1$. They are therefore semiconductors of a very special type. This behavior has been confirmed by the study between 20 and 600°C done by McNeill and Ware [4.79]. Subsequently, Asanabe [4.80] showed, when studying $Co_{1-x}Mn_xSi$,

$Co_{1-x}Cr_xSi$ and $Co_{1-x}Ni_xSi$ solutions, that the substitution of Mn and Cr produced holes, while that of Ni produced electrons. We believe that these results can be explained by assuming that the d band presents a minimum density for semioccupancy (d^5). One need only assume that the metal has valency 4, corresponding to that of silicon. One then obtains the formulas d^5 for Co, d^4 for Fe, d^3 for Mn and d^2 for Cr, but d^6 for Ni, in accordance with the magnetic properties.

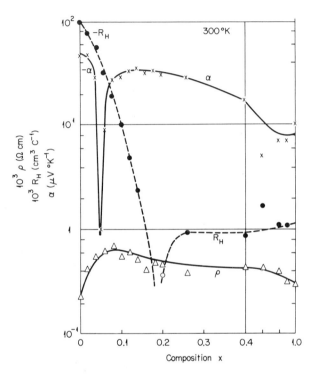

Fig. 4.18. Resistivity ρ, Hall coefficient R_H, and Seebeck coefficient α in relation to x in $Co_{1-x}Fe_xSi$ (based on Asanabe et al. [4.78]).

Among disilicides with the C40 structure, the electrical [4.81] and magnetic [4.82] properties of VSi_2 have been measured. Paramagnetism corresponds to $2S = 0.75$, namely to valency 4 for V. But work has been concentrated on $CrSi_2$ and $MnSi_2$. After preliminary measurements of the resistivity and Seebeck effect in $CrSi_2$ [4.74, 4.83], Nikitin [4.72] has studied its use in thermoelectric generators, within the framework of general work on the Cr–Si system (Fig. 4.19), mentioning its doping with B and Ag. Others [4.84] suggest a metallic character. Sakata and Tokushima [4.77] have repeated the measurements, including the Hall effect between -200 and $+1200°C$, and doping

4.3. Silicides, Germanides, Stannides

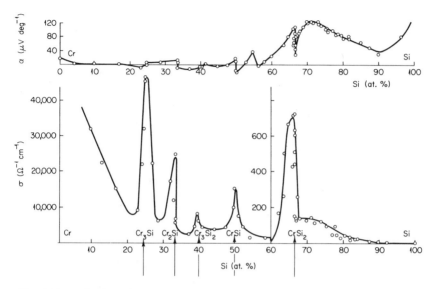

Fig. 4.19. Seebeck coefficient and electrical conductivity of the various phases of the chromium–silicon system, in relation to their composition (based on Nikitin [4.72]).

with V and Mn. They suggest the use of $CrSi_2$ for the p branch of a generator and CoSi for the n branch. The conductibility of liquid chromium silicides has been measured by Baum et al. [4.73], who have revealed the formation of quasi-molecular configurations. Finally, Shinoda et al. [4.85] have carried out a very full study of different stoichiometric ratios, as well as the addition of Mn on monocrystals obtained by the Czochralski method. Figures 4.20a, b, and c show the variation of ρ, R_H, and α in relation to temperature. Mobility follows a law in $T^{-3/2}$, characteristic of scattering caused mainly by the vibrations of the atoms. Radovskii et al. [4.82] have found that $CrSi_2$ is diamagnetic, and have discussed, without solving it, the problem of bonds in the C40 structure. The conductibility and Seebeck effect in $MnSi_2$ have been measured by Nikitin [4.72, 4.74, 4.75], who have pointed out the usefulness of $CrSi_2$–$MnSi_2$ alloys in thermoelectric generators. Korshunov and Gel'd [4.86, 4.87] have shown that the composition corresponding to minimum conductivity is $MnSi_{0.7 \pm 0.03}$ (Fig. 4.21). In the compositions $MnSi_{1.73}$ and $MnSi_{1.75}$ (Mn_4Si_7), the number of holes remains around 10^{-20} cm^{-3}, while their mobility, which is 40 cm^2 V^{-1} sec^{-1} at room temperature, varies exponentially. However, the transfer mechanism gives way above 500°C to an intrinsic excitation mechanism, with an 0.6 eV forbidden band. This compound can be used in the p branch of a thermoelectric generator, with $f/T = 5 \times 10^{-4}$ at 550°C (Fig. 4.22). According to Bienert and Skrabek [4.88], it is a separate quadratic phase of C40 that is involved here, though this is stabilized by the presence of Al [4.89].

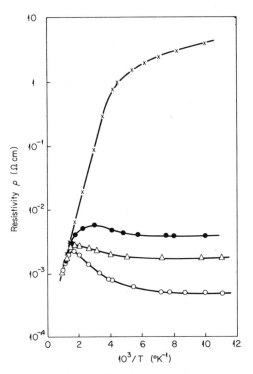

Fig. 4.20a

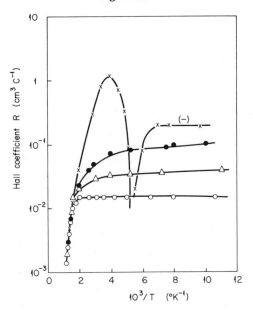

Fig. 4.20b

4.3. Silicides, Germanides, Stannides

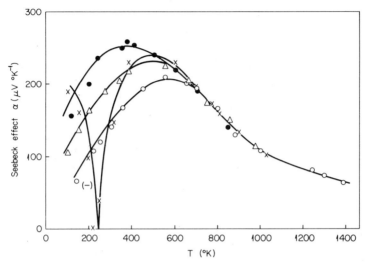

Fig. 4.20c

Fig. 4.20. Variation in the resistivity (a), and Hall (b) and Seebeck (c) coefficients, in relation to temperature, for monocrystals with formulas (—O—) $CrSi_2$, (—△—) $CrSi_{2.01}$, (—●—) $CrSi_{2.02}$, and (—X—) $Cr_{0.88}Mn_{0.12}Si_2$ (based on Shinoda et al. [4.85]).

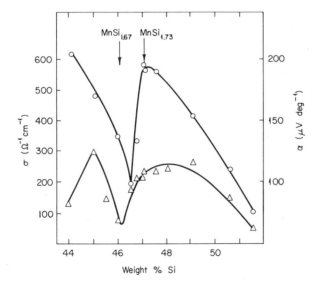

Fig. 4.21. Variation in the conductivity and Seebeck coefficient of manganese silicides at room temperature, between 44 and 52% in weight of silicon (based on Korshunov and Gel'd [4.86]).

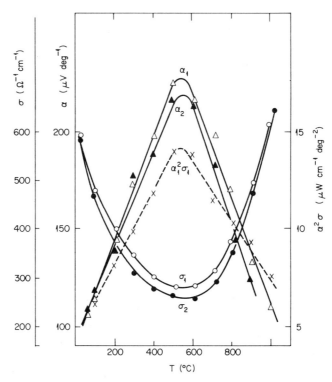

Fig. 4.22. Variation in the conductivity, Seebeck coefficient, and product $\alpha^2\sigma$ in terms of temperature (based on Korshunov and Gel'd [4.86]). (1) 47% and (2) 47.5% in weight of Si.

Germanides and stannides crystallize in a wide variety of compositions and structures, the main ones being T_3X (V, Nb, Ta, Cr, Mo, A15 structure of β-W; Ti, Fe, Mn, Ni, DO_{19} structure of Fe_3Sn; Ni, Pt, $L1_2$ structure of Cu_3Au), T_2X (Pd, Pt, C22 structure of Fe_2P; Hf, C16 structure of $CuAl_2$), T_5X_3 (Ti, Zr, Hf, V, Nb, Ta, Mn, $D8_8$ structure of Mn_5Si_3), $T_{1+x}X$ (Ti, Mn, Fe, Co, Rh, Ir, Ni, Pd, Pt, $B8_1$ structure of NiAs to $B8_2$ structure of Ni_2In), TX (Cr, Rh, B20 structure of FeSi; Rh, Ni, Pd, Pt, B31 structure of MnP; Fe, Co, Ni, B35 structure of PtTl), TX_2 (Ir, Pt, C1 structure of CaF_2; Mn, Fe, Co, Rh, C16 structure of $CuAl_2$; Hf, Nb, Ta, C40 structure of $CrSi_2$; Ti, Zr, Hf, C49 and C54 structures) [4.20]. Ferromagnetism is present notably in the DO_{19} and $B8_1$–$B8_2$ structures. Their electrical properties, which always seem to be metallic, have had little work done on them [4.21]. The metallic character of the bonds is less marked, however, for the germanides. Thus, the iron atoms in germanide with the $B8_2$ structure give two separate Mössbauer resonances, while those of stannide give only one (cf. Chapter 6) [4.90].

Dudkin's criterion [4.13] applied to disilicides with structures C11b, C40, C49, and C54 indicate $CrSi_2$ as the only semiconductor (the cases of Fe_2Si_5 and Mn_4Si_7 were not considered). Applied to digermanides and distannides with C16 structure, it rules out any semiconductors. As for the ionicity λ, it can really be calculated only for simple structures such as C1 and $B8_1$, and possibly B31. In the octahedral arrangement of $B8_1$ and B31, the equilibrium ionicity λ_0 equals $-\frac{1}{3}$, so that λ is usually negative or nil. It is obvious that there is little hope of finding magnetic semiconductors here.

4.4. Phosphides, Arsenides, Antimonides

Phosphides crystallize with formulas T_3P (V, Cr, Mo, Mn, Fe, Ni, quadratic structure of Fe_3P; Pd, Pt, DO_{11} orthorhombic structure of Fe_3C), T_2P (Mn, Fe, Ni, C22 hexagonal structure of Fe_2P; Re, Ru, Co, C23 orthorhombic structure of $PbCl_2$; Rh, Ir, C1 cubic structure of antifluorite), TP (Ti, Zr, V, $B8_1$ structure with or without superstructure; Cr, W, Mn, Fe, Re, Co, B31 structure; Mo, structure of WC, etc), TP_2 (Fe, Ru, Pt, C18 and C22 structures; V, Nb, Ta, Mo, W, Rh, Ir, Ni, Pd, complex structures) and TP_3 (Co, Rh, Ir, Ni, Pd, cubic structure, probably DO_2 of skutterudite $CoAs_3$) [4.20]. In the Zr–P system, there exists a phase containing phosphorus vacancies, with B1 structure and the formula $ZrP_{0.9}$, where the Zr–Zr distances are 15% greater than those of metal. Iron phosphides are ferromagnetic (except FeP_2) and $2S$ varies between 0.4 and 1.9 [4.21]; RhP_2 and IrP_2 apparently have the EO_7 structure of arsenopyrite FeAsS [4.8], but many TP_2 compounds still remain to be studied. Figure 4.23 represents the DO_2 structure and its characteristic groups of four metalloid atoms linked directly with one another. Each

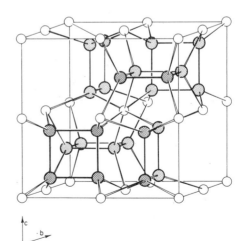

Fig. 4.23. DO_2 cubic structure corresponding to the formula TX_3 (the shaded circles represent X). [Reprinted with permission from Hulliger and Mooser, *Progress in Solid State Chemistry* **2**, 330 (1965), Pergamon Press.]

T atom is surrounded with a distorted X_6 octahedron, and each X atom by a deformed T_2X_2 tetrahedron ($e_g{}^2\text{sp}^3/\text{sp}^3$ MOs).

Fakidov and Krasovskii [4.91, 4.92] have studied the electrical properties of Mn_2P and MnP. They have the properties of ferromagnetic conductors with 2×10^{22} holes cm^{-3}. The Seebeck effect in MnP is minimum at $-50°\text{C}$, and abnormal below this temperature. The magnetoresistance effects, also measured by Suzuki et al. [4.93], will be described in Part Three. Let us consider the case of MnP: if one ignores the distortion of the B31 structure, one can use the VB energy diagram mentioned in Section 4.2 (Fig. 4.14a) as a rough approximation, except that the occupancy of the d sublevels is $t_{2g}^3\alpha$ and $e_g{}^1\alpha$. The second is therefore partly filled and the transfer density is equal to that of the Mn atoms in the crystal, around 10^{22} cm^{-3} (pseudometal). In an MO energy diagram, one will naturally talk of d bands but, in contrast with Bilz' model mentioned in Section 4.1, perhaps there is no further need to allow for overlapping with the wide bands. Figure 4.24 reproduces the diagram proposed by

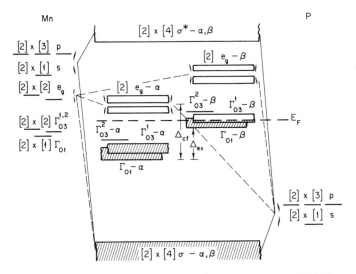

Fig. 4.24. MO energy diagram of MnP (based on Goodenough et al. [4.94]).

Goodenough et al. [4.94], which takes into account the difference caused by the crystal field (Δ_{cf}), the magnetic order (Δ_{ex}) and the serious distortion of the P_6 octahedron in the B31 structure. The relative position of the α and β subbands is arbitrary. The number $2S$, provided by the paramagnetism formula (3.6), is 3.1, whereas the bonds per p^3 AO would give 4 for strictly localized electrons and zero spin–orbit coupling. Below the Curie temperature $T_C = 22°\text{C}$, investigation of the saturation magnetization gives an even lower value of 1.2, which can be explained by a change in the structure of the d subbands

4.4. Phosphides, Arsenides, Antimonides

(Fig. 4.24), or an e_g–p hybridization reducing $2S$ by 2 (cf. Section 4.1). The resistivity of MnP monocrystals has been measured by Suzuki et al. [4.95, 4.96] in the direction of the three axes, between 4 and 70°K (Fig. 4.25). It confirms metamagnetism below 50°K, with a helicoidal spin structure presenting a periodicity along axis a. The resulting forbidden energy band disappears when a magnetic field is applied. The existence of a magnetic order for FeP seems to be ruled out, and paramagnetic susceptibility gives $2S = 2.5$ instead of 5 [4.8] perhaps because of an e_g–p hybridization. Flechon and Ormancey [4.97] have studied the resistivity of thin layers of nickel phosphides, and observed a negative temperature coefficient for phosphorus-rich compositions.

A classification of semiconducting diphosphides was given in Section 4.1; RhP_2 and IrP_2, in the EO_7 structure of $CoSb_2$, are diamagnetic and could be semiconductors [4.98]. The resistivity of VP_2, NbP_2, TaP_2, and α-WP_2, isomorphs of $NbAs_2$, has a positive temperature coefficient, but fuller investigation is needed, as well as for TiP_2, ZrP_2, and HfP_2, in the C23 structure. A P—P bond exists in these two structures, and MoP_2 and β-WP_2 are metallic

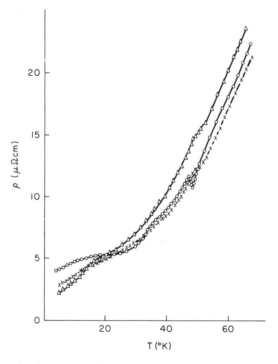

Fig. 4.25. Variation in resistivity in relation to temperature in a monocrystal of MnP (based on Suzuki et al. [4.95]). Current parallel to (—O—) a-axis, (—△—) b-axis, and (—X—) c-axis.

[4.99]. In triphosphides with DO_2 structure, Hulliger [4.100] has shown that CoP_3, RhP_3, and IrP_3, with t_{2g}^6 configuration (nonmagnetic), are semiconductors, and NiP_3 and PdP_3 are metallic.

Arsenides crystallize with the formulas T_3As (V, Mn, Pd), T_2As (Cr, Mn, Fe, C38 structure of Cu_2Sb; Co, C22 structure; Rh, C1 structure; V, Ir), TAs (V, Cr, Mn, Fe, Ru, Co, Rh, B31 structure; Mn, Ni, $B8_1$ structure; Ti, Zr) and TAs_2 (Fe, Ru, Os, Ni, C18 structure; Pd, Pt, C2 structure; Zr, C23 structure; Nb, Ta, Mo, W, Re, Co, Rh); $CoAs_3$ and $RhAs_3$ with DO_2 also exist, and also $IrAs_3$. Many structures are complex and unfamiliar [4.20]. Most of these compounds are slightly paramagnetic; Cr_3As_2, a high-temperature phase with an unfamiliar quadratic structure, and MnAs are magnetic. MnAs presents irregularities at $T = T_C$ (sharp specific heat peak, maximum magnetic induction, thermal hysteresis, 2.5% contraction of the lattice). The compound $PdAs_2$ becomes supraconducting at $1.6°K$ [4.21]. Antiferromagnetism appears to be the rule for arsenides of C38 structure. The Neel temperature T_N is $393°K$ for Cr_2As, $573°K$ for Mn_2As [4.101] and $323°K$ for Fe_2As [4.21]. The resistivity of a Co_5As_2 (Co_2As?) composition has been compared by Kochnev [4.102] with its dissociation pressure at various temperatures. The nonlinearity of the curve $1/\chi(T)$ reveals ferrimagnetism in the case of Cr_3As_2, with $T_C = 213°K$ [4.101]. The existence of a magnetic compound Mn_3As_2 has not been confirmed [4.21].

Among the monoarsenides, properties appear that sometimes recall those of semiconductors. Guillaud [4.103] has studied the electrical and magnetic properties of MnAs. Figure 4.26 shows, at $47°C$, the existence of a singular point corresponding to T_C, above which the resistivity decreases exponentially. Figure 4.27 gives some idea of the complex magnetic properties, which have been interpreted in several different ways. The directions of the spin moments are contained in the (001) planes. Investigation of the saturation magnetization for Mn gives $2S = 3.4$, while one finds 3.7 (instead of 4) in the paramagnetic region [4.3, 4.101]. According to Fischer and Pearson [4.104], the resistivity/temperature curve presents an hysteresis, and depends on the previous history of the sample. They claim that the exponential decrease corresponds to an activation energy of several tenths of an electron volt. Recent research shows that the $B8_1$ structure changes into B31 at $315°K$, and back into $B8_1$ at $400°K$ [4.105]. This transformation of the first order explains the magnetic irregularities. For MnAs, Goodenough et al. [4.94] have proposed the MO energy diagram in Fig. 4.28 (to be compared with the diagram for MnP), and have backed it up with the magnetic study of $MnAs_{1-x}P_x$ and $MnAs_{1-y}Sb_y$ solid solutions (Fig. 4.29). Busch and Hulliger [4.106] have found semiconductor properties in CrAs and FeAs, with an activation energy of 0.1 eV for the latter. Semiconductor diarsenides were classified in Section 4.1. The case of $PdAs_2$ remains uncertain however [4.8]. Activation energies are 0.2 ($FeAs_2$), 0.15

4.4. Phosphides, Arsenides, Antimonides

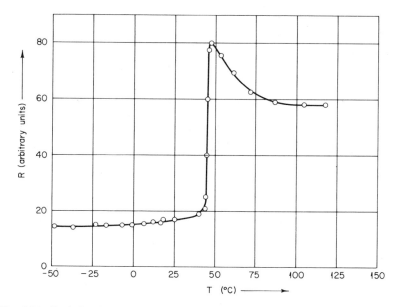

Fig. 4.26. Variation in the resistance of MnAs in relation to temperature (based on Guillaud [4.103]).

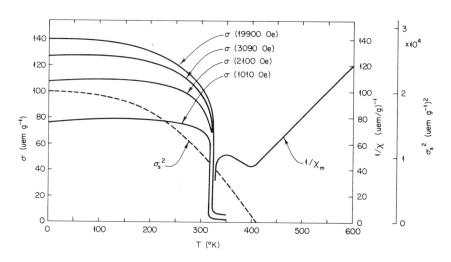

Fig. 4.27. Variation in the saturation magnetization σ and the reciprocal of susceptibility $1/\chi$ in relation to temperature, for MnAs. [Based on Kouvel, in "Intermetallic Compounds." (J. H. Westbrook, ed.), Fig. 10, p. 546. Wiley, New York, 1965. Used with permission of the copyright owner.]

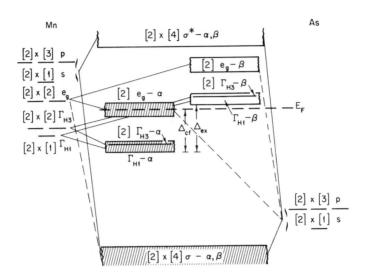

Fig. 4.28. MO energy diagram of MnAs (based on Goodenough et al. [4.94]).

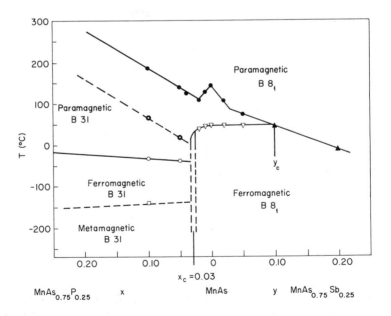

Fig. 4.29. Magnetic and crystallographic phase diagram of MnP–MnAs–MnSb systems (based on Goodenough et al. [4.94]).

4.4. Phosphides, Arsenides, Antimonides

($CoAs_2$), and 0.05 eV ($NiAs_2$) [4.107]. In triarsenides with DO_2 structure, $CoAs_3$, $RhAs_3$, and $IrAs_3$, with t_{2g}^6 configuration (nonmagnetic), are semiconducting, and $NiAs_3$ is metallic [4.100].

Antimonides crystallize with formulas T_3Sb (Ti, V, Nb, Ta, A15 structure), Mn_2Sb and Ti_5Sb_2 (C38), Fe_3Sb_2 and Pd_5Sb_3 ($B8_2$), TSb (Ti, V, Cr, Mn, Co, Ir, Ni, Pd, Pt, $B8_1$ structure; Rh, B31 structure, TSb_2 (Ti, V, C16 structure; Cr, Fe, Os, Co, Ni, C18 structure; Co, Rh, Ir, complex monoclinic structure; Pd, Pt, C2 structure) and TSb_3 (Co, Rh, Ir, DO_2 structure) [4.20]. Compounds with a low antimony content frequently have fairly wide homogeneity ranges between the $B8_2$ and $B8_1$ structures (notably for Mn and Fe). The only magnetic compounds are Mn_2Sb and MnSb; Fe_3Sb_2 is reported to show slight ferromagnetism below 220°C. Supraconductivity has been observed in CrSb at 1°K and in PdSb at 1.5°K [4.21]. The first suggestion of a ferrimagnetic structure was made in connection with Mn_2Sb [4.108]. The solution $Mn_{2-x}Cr_xSb$ ($0.01 < x < 0.2$) has been studied by Bierstedt [4.109]. Its resistivity is independent of x, but depends closely on the magnetic structure, showing an abrupt change at the transition temperature (cf. Part Three). The Hall effect shows that there is one hole per molecule. The composition $x = 0.12$ has been studied for monocrystals by Grazhdankina [4.110].

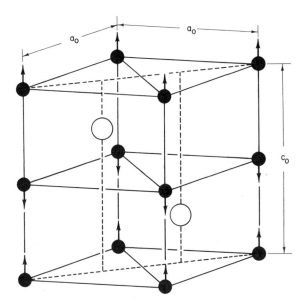

Fig. 4.30. Orientation of spin moments in the cell of CrSb. (●) Cr and (○) Sb. [Based on Kouvel in "Intermetallic Compounds." (J. H. Westbrook, ed.), Fig. 4, p. 538. Wiley, New York, 1965. Used with permission of the copyright owner.]

Among monoantimonides, CrSb and MnSb are of particular interest. Their magnetic properties have been recalled by Lotgering and Gorter [4.111]: CrSb is antiferromagnetic with $T_N = 720°K$ (Fig. 4.30), while MnSb is ferromagnetic with $T_C = 587°K$ (spin moments parallel to axis c, in contrast to MnAs). The number $2S$ is 2.7 (neutrons, $T < T_N$) or 3.4 ($T > T_N$) instead of 3 for chromium and 3.5 ($T < T_C$) or 4.1 ($T > T_C$) instead of 4 for Mn. Resistivity variations in CrSb, MnSb, and $Cr_xMn_{1-x}Sb$ solutions, in relation to temperature, have been studied by Suzuoka [4.112], using sintered samples. Figure 4.31 shows that a negative temperature coefficient exists above T_C ($0 \leqslant x \leqslant 0.6$) or T_N ($0.6 \leqslant x \leqslant 1$). The ferrimagnetic Curie point of solutions $0.7 \leqslant x \leqslant 0.8$ is reflected by a small irregularity. Later work by Fakidov and Afans'ev [4.113] confirms the resistivity variation mentioned above for CrSb. The author

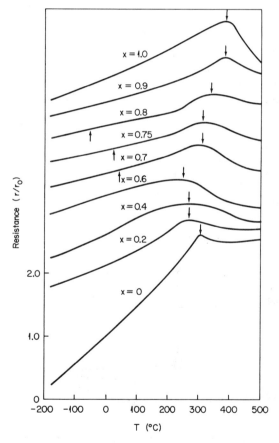

Fig. 4.31. Variation in the resistivity in relation to temperature, in $Cr_xMn_{1-x}Sb$ solid solutions (based on Suzuoka [4.112]). The arrows indicate the Curie or Neel points.

4.4. Phosphides, Arsenides, Antimonides

[4.114] has also found an exponential law (0.2 eV) following minimum conductivity at 690°K. The work by Fischer and Pearson [4.104] on melted samples of MnSb shows a resistivity of around 10^{-3} Ω cm, which is very high for a metal alloy. Figure 4.32 shows very high electrical hysteresis around T_C, probably due to the changes in lattice mentioned in Section 4.1. The galvanomagnetic effects of MnSb have been considered by Fakidov and Grazhdankina [4.115], Kikoin *et al.* [4.116], and later by Nogami *et al.* [4.117], Nogami [4.118] (cf. Part Three). The ordinary Hall coefficient, measured by Nogami *et al.* in the ferromagnetic range, indicates a reduction in the number of holes per cubic centimeter when the temperature rises, thus confirming metallic properties below T_C. Finally, some recent work has been devoted to the transport properties of NiSb [4.119].

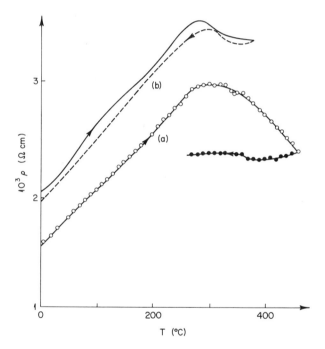

Fig. 4.32. Typical variations in resistivity in relation to temperature during the heating and cooling of polycrystal samples of MnSb. [Reproduced by permission of the National Research Council of Canada from Fischer and Pearson, *Can. J. Phys.* **36**, 1014 (1958).]

A classification of semiconductor diantimonides was given in Section 4.1. The case of $PdSb_2$ is uncertain however [4.8]. The Seebeck effect in $CrSb_2$ has been studied [4.64, 4.120]. It presents two successive activation energies at 0.16 and 0.32 eV. For $FeSb_2$, see [4.121]. For $CoSb_2$, see [4.64, 4.122]. Very comprehensive research on the properties of PtSb$_2$ has been done by Damon *et al.* [4.123], using monocrystals obtained by the Czochralski method. Figure 4.33 summarizes the main results obtained on diamagnetic samples containing 3×10^{18} holes cm^{-3}, or doped with Te to introduce electrons. Mobility varies from 200 (electrons) to 1500 cm^2 V^{-1}sec^{-1} (holes), and drops when the temperature rises (scattering by the lattice vibrations and impurities). It is concluded that there is a classic semiconductibility mechanism by excitation across a forbidden band of 0.07 eV. The triantimonide CoSb$_3$ is a semiconductor suitable for use in thermoelectric generators [4.64, 4.122]. Dudkin [4.124] has given the scheme of bonds for it (Fig. 4.34). The effect of different impurities on the Seebeck effect and conductivity have been studied [4.125], 4.126]. The other compounds with DO$_2$ structures are also semiconducting [4.100].

Dudkin's criterion [4.13], applied to phosphides, arsenides, and antimonides with B8$_1$ structure, indicates VP as the only semiconductor, together with CrSb above 400°C. It suggests metallic properties for TiSb$_2$ and VSb$_2$ in the C16 structure. Applied to structures C2 and C18, it suggests semiconducting properties, with the exception of FeP$_2$. The nonstoichiometric compound NiSb$_{2.2}$ is excluded from this classification [4.127]. The ionicity parameter λ is more or less zero for monoantimonides with structures B8$_1$ and B31, and reaches -0.1 for FeSb. It is positive for the other compounds. Interaction between the (AB) α band and the upper d sublevel, if it is partly filled, may however explain metallic behavior below T_C or T_N (cf. Section 3.4). Finally, Goodenough's criterion sets the critical distance T—T' at around 3 Å. If one

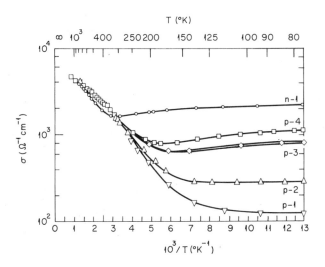

Fig. 4.33a

4.4. Phosphides, Arsenides, Antimonides

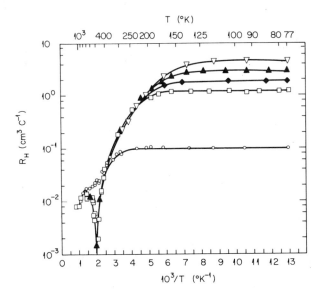

Fig. 4.33b

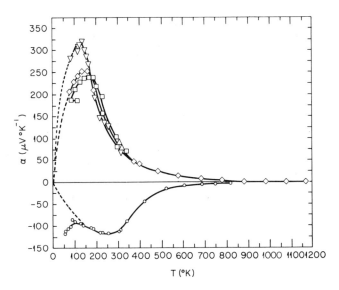

Fig. 4.33c

Fig. 4.33. Variation with temperature of the properties of $PtSb_2$: (a) electrical conductivity, (b) Hall coefficient, and (c) Seebeck coefficient (based on Damon et al. [4.123]).

accepts that this is equal to the half-height $c/2$ of the cell, semiconducting properties seem likely for

compound:	VAs	TiSb	VP	CrAs	FeAs
$c/2$:	3.16	3.15	3.11	3.11	3.01 Å

while doubt will remain for

compound:	MnP	MnAs $(T > T_c)$	MnSb	FeP
$c/2$:	2.96	2.90	2.90	2.90 Å

None of these criteria can be relied on entirely. It was pointed out in Section 4.1 in particular that the distance T—T′ along a c-axis is not the only one involved, conduction in the (001) planes probably occupying a preferential position.

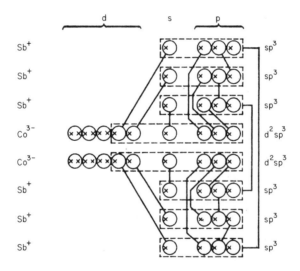

Fig. 4.34. Diagram of the bonds in CoSb$_3$ (based on Dudkin [4.124]).

To sum up, the semiconductor character increases when one moves from metalloids in column III to those in columns IV and particularly V, but in this chapter only magnetic pseudometals have been found, in which the d electrons are highly delocalized It is among the metalloids in column VI (oxides and chalcogenides) that in the following chapters we shall meet with the first magnetic semiconductors.

REFERENCES

[4.1] A. F. Wells, "Structural Inorganic Chemistry," 3rd ed. Oxford Univ. Press (Clarendon), London and New York, 1962.
[4.2] R. W. G. Wyckoff, "Crystal Structures," 2nd ed., Vol. 1. Wiley (Interscience), New York, 1963.
[4.3] B. Aronsson, T. Lundström and S. Rundqvist, "Borides, Silicides and Phosphides." Methuen, London 1965.
[4.4] A. J. Kjekshus and W. B. Pearson, *Progr. Solid State Chem.* **1**, 83 (1964).
[4.5] B. T. M. Willis and H. P. Rooksby, *Proc. Phys. Soc. London* **B67**, 290 (1954).
[4.6] B. W. Roberts, *Phys. Rev.* **104**, 607 (1956).
[4.7] J. P. Suchet, *Phys. Status Solidi* **15**, 639 (1966).
[4.8] F. Hulliger and E. Mooser, *Progr. Solid State Chem.* **2**, 330 (1965).
[4.9] F. Hulliger and E. Mooser, *J. Phys. Chem. Solids* **26**, 429 (1965).
[4.10] H. J. Wallbaum, *Z. Metallk.* **33**, 378 (1941).
[4.11] R. Kieffer and E. Cerwenka, *Z. Metallk.* **43**, 101 (1952).
[4.12] R. R. Heikes, *Phys. Rev.* **99**, 1232 (1955).
[4.13] L. D. Dudkin, *Dokl. Akad. Nauk SSSR* **127**, 1203 (1959).
[4.14] G. B. Bokii, "Vvedenie v Kristallokhimiyu." Izd. Mosk. Gosudarstv. Univ., 1954.
[4.15] L. Pauling, "The Nature of the Chemical Bond." Cornell Univ. Press, Ithaca, New York, 1950.
[4.16] S. N. L'vov, V. F. Nemchenko and G. V. Samsonov, *Dokl. Akad. Nauk SSSR* **135**, 577 (1960).
[4.17] H. Bilz, *Z. Phys.* **153**, 338 (1958).
[4.18] V. Ern and A. C. Switendick, *Phys. Rev.* **137A**, 1927 (1965).
[4.19] E. Dempsey, *Phil. Mag.* **8**, 285 (1963).
[4.20] M. Hansen, "Constitution of Binary Alloys." McGraw-Hill, New York, 1958; R. P. Elliott, idem (1961).
[4.21] P. Pascal, "Nouveau traité de chimie minérale." Masson, Paris, 1956–1963.
[4.22] G. V. Samsonov and N. S. Stryel'nikova, *Ukr. Fiz. Zh.* **3**, 135 (1958).
[4.23] J. Piper, *J. Phys. Chem. Solids* **27**, 1907 (1966).
[4.24] V. S. Neshpor and G. V. Samsonov, *Dokl. Akad. Nauk SSSR* **157**, 834 (1964).
[4.25] K. Moers, *Z. Anorg. Chem.* **198**, 243 (1931).
[4.26] A. Münster, K. Sagel and G. Schlamp, *Nature* **174**, 1154 (1954).
[4.27] A. Münster and K. Sagel, *Z. Phys.* **144**, 139 (1956).
[4.28] S. Noguchi and T. Sato, *J. Phys. Soc. Japan* **15**, 2359 (1960).
[4.29] V. S. Neshpor, V. F. Nemchenko, S. N. L'vov and G. V. Samsonov, *Ukr. Fiz. Zh.* **5**, 839 (1960).
[4.30] L. E. Hollander, *J. Appl. Phys.* **32**, 996 (1961).
[4.31] T. Tsuchida, Y. Nakamura, M. Mekata, H. Sakurai and H. Takaki, *J. Phys. Soc. Japan* **16**, 2453 (1961).
[4.32] J. Piper, *J. Appl. Phys.* **33**, 2394 (1962).
[4.33] O. A. Golikova, A. I. Avgustinnik, G. M. Klimashin and L. V. Kozlovskii, *Fiz. Tverd. Tela* **7**, 2860 (1965).
[4.34] O. A. Golikova, F. L. Feigel'man, A. I. Avgustinnik and G. M. Klimashin, *Fiz. Tekh. Poluprov.* **1**, 293 (1967).
[4.35] M. S. Koval'chenko and V. V. Ogorodnikov, *At. Energ.* **21**, 302 (1966).
[4.36] J. S. Umanski, *Ann. Sect. Anal. Phys. Chim. Inst. Chim. Gen. URSS* **16**, 127 (1943).
[4.37] L. N. Grossman, *J. Amer. Ceram. Soc.* **46**, 457 (1963).
[4.38] O. A. Golikova, A. I. Avgustinnik, G. M. Klimashin, L. V. Kozlovskii, S. S. Ordan'yan and V. A. Snetkova, *Fiz. Tverd. Tela* **7**, 3698 (1965).

[4.39] H. Bittner and H. Goretzki, *Monatsh. Chem.* **93**, 1000 (1962).
[4.40] J. R. Cooper and R. L. Hansler, *J. Chem. Phys.* **39**, 248 (1963).
[4.41] R. Steinitz and R. Resnick, *J. Appl. Phys.* **37**, 3463 (1966).
[4.42] G. Santoro, *Trans. Met. Soc. AIME* **227**, 1361 (1963).
[4.43] L. B. Dubrovskaya, I. I. Matveyenko, and P. V. Gel'd, *Fiz. Metal. Metalloved.* **20**, 243 (1965).
[4.44] S. N. L'vov, V. F. Nemchenko, T. Ya. Kosolapov, and G. V. Samsonov, *Fiz. Metal. Metalloved.* **11**, 143 (1961).
[4.45] G. V. Samsonov and T. S. Verkhoglyadova, *Dokl. Akad. Nauk SSSR* **138**, 342 (1961).
[4.46] S. N. L'vov, V. F. Nemchenko, G. V. Samsonov and T. S. Verkhoglyadova, *Ukr. Fiz. Zh.* **8**, 1372 (1963).
[4.47] W. H. Philipp, *Acta Met.* **10**, 583 (1962) and **12**, 740 (1964).
[4.48] W. B. Pearson, *Acta Met.* **10**, 1123 (1962).
[4.49] W. J. James and M. E. Straumanis, *Acta Met.* **12**, 739 (1964).
[4.50] L. Glasser and J. Hoy, *J. Phys. Chem.* **70**, 281 (1966).
[4.51] W. Dürrschnabel and G. Hörz, *Naturwissenschaften* **50**, 687 (1963).
[4.52] T. S. Verkhoglyadova, S. N. L'vov, V. F. Nemchenko, and G. V. Samsonov, *Fiz. Metal. Metalloved.* **12** 622 (1961).
[4.53] S. A. Nemnonov, *Zh. Tekh. Fiz.* **18**, 247 (1948).
[4.54] S. N. L'vov, V. F. Nemchenko, and G. V. Samsonov, *Ukr. Fiz. Zh.* **7**, 331 (1962).
[4.55] F. Itoh, T. Tsuchida, and H. Takaki, *J. Phys. Soc. Japan* **19**, 136 (1964).
[4.56] C. W. Nelson, Tech. Rep. 179, M.I.T. Lab. Insulation Res. (1963).
[4.57] S. P. Denker, *J. Phys. Chem. Solids* **25**, 1397 (1964).
[4.58] P. V. Gel'd, *Zh. Tekh. Fiz.* **27**, 113 (1957).
[4.59] F. W. Glaser, *J. Appl. Phys.* **22**, 103 (1951).
[4.60] M. J. Arvin, *J. Appl. Phys.* **24**, 498 (1953).
[4.61] M. J. Arvin and R. F. Tipsord, *J. Phys. Chem. Solids* **9**, 336 (1959).
[4.62] V. S. Neshpor and G. V. Samsonov, *Fiz. Metal. Metalloved.* **11**, 638 (1961).
[4.63] N. Kh. Abrikosov, *Izv. Akad. Nauk SSSR, Ser. Fiz. Khim. Anal.* **27**, 157 (1956).
[4.64] N. Kh. Abrikosov, *Izv. Akad. Nauk SSSR, Ser. Fiz.* **2**, 141 (1957).
[4.65] V. S. Neshpor and V. L. Yupko, *Porosh. Met.* **32**, 55 (1963).
[4.66] R. M. Ware and D. J. McNeill, *Proc. IEE* **111**, 178 (1964).
[4.67] A. F. Ioffe, "Semiconductor Thermoelements and Thermoelectric Cooling." Infosearch, London, 1957.
[4.68] R. R. Heikes and R. W. Ure, "Thermoelectricity." Wiley (Interscience), New York, 1961.
[4.69] J. P. Suchet, Cours de composés chimiques semiconducteurs, CTA, Sao Jose dos Campos (SP) (Brazil) (1963).
[4.70] S. E. Mayer and A. I. Mlavsky, "Properties of Elemental and Compound Semiconductors" (H. C. Gatos, ed.). Wiley (Interscience), New York, 1960.
[4.71] L. P. Andreeva, F. A. Sidorenko, and P. V. Gel'd, *Fiz. Metal. Metalloved.* **19**, 784 (1965).
[4.72] E. N. Nikitin, *Fiz. Tverd. Tela* **2**, 2685 (1960).
[4.73] B. A. Baum, P. V. Gel'd, and S. I. Suchil'nikov, *Fiz. Metal. Metalloved.* **16**, 939 (1963).
[4.74] E. N. Nikitin, *Zh. Tekh. Fiz.* **28**, 23 (1958).
[4.75] E. N. Nikitin, *Fiz. Tverd. Tela* **1**, 340 (1959).
[4.76] E. N. Nikitin, *Fiz. Tverd. Tela* **2**, 633 (1960)
[4.77] T. Sakata and T. Tokushima, *Trans. Nat. Res. Inst. Metals (Japan)* **5**, 34 (196ɔ).
[4.78] S. Asanabe, D. Shinoda, and Y. Sasaki, *Phys. Rev.* **134**, A774 (1964).

[4.79] D. J. McNeill and R. M. Ware, *Brit. J. Appl. Phys.* **15**, 1517 (1964).
[4.80] S. Asanabe, *J. Phys. Soc. Japan* **20**, 933 (1965).
[4.81] E. A. Zhurakovskii and V. P. Dzeganovskii, *Fiz. Metal. Metalloved.* **22**, 193 (1966).
[4.82] I. Zh. Radovskii, F. A. Sidorenko, and P. V. Gel'd, *Fiz. Metal. Metalloved.* **19**, 915 (1965).
[4.83] L. N. Guseva and B. I. Ovechkin, *Dokl. Akad. Nauk SSSR* **112**, 681 (1957).
[4.84] V. P. Trusova, V. S. Kutsev, and B. F. Ormont, *Zh. Neorg. Khim.* **5**, 538 (1960).
[4.85] D. Shinoda, S. Asanabe, and Y. Sasaki, *J. Phys. Soc. Japan* **19**, 269 (1964).
[4.86] V. A. Korshunov and P. V. Gel'd, *Fiz. Metal. Metalloved.* **11**, 945 (1961).
[4.87] V. A. Korshunov and P. V. Gel'd, "Thermoelectric Properties of Semiconductors," p. 54. Consultants Bureau, New York, 1964.
[4.88] W. B. Bienert and E. A. Skrabek, *Proc. IEEE (AIAA Thermoelectric Conf. Washington D.C. 1966)*, pp. 10.1, 10.12. IEEE, New York, 1966.
[4.89] J. B. Kusma and H. Nowotny, *Monatsh. Chem.* **95**, 1266 (1964).
[4.90] G. A. Fatseas and P. Lecocq, *C.R. Acad. Sci. (Paris)* **262**, 408 (1966).
[4.91] I. G. Fakidov and V. P. Krasovskii, *Fiz. Metal. Metalloved.* **7**, 302 (1959).
[4.92] V. P. Krasovskii and I. G. Fakidov, *Fiz. Metal. Metalloved.* **7**, 477 (1959).
[4.93] T. Suzuki, Y. Matsumura, and E. Hirahara, *J. Phys. Soc. Japan* **21**, 1446 (1966).
[4.94] J. B. Goodenough, D. H. Ridgley, and W. A. Newman, *Proc. Int. Conf. Magnetism, Nottingham, 1964* Inst. Phys. Phys. Soc., London, 1964.
[4.95] T. Suzuki, Y. Matsumura, and E. Hirahara, *J. Phys. Soc. Japan* **21**, 1621 (1966).
[4.96] T. Suzuki, Y. Matsumura, and E. Hirahara, *Proc. Int. Conf. Phys. Semicond., Kyoto, 1966 J. Phys. Soc. Japan suppl.* **21**, 615 (1966).
[4.97] J. Flechon and G. Ormancey, *J. Phys. Radium* **23**, 128 (1962).
[4.98] F. Hulliger, *Phys. Lett.* **4**, 282 (1963).
[4.99] F. Hulliger, *Nature* **204**, 775 (1964).
[4.100] F. Hulliger, *Helv. Phys. Acta* **34**, 782 (1961).
[4.101] J. S. Kouvel, "Intermetallic Compounds" (J. H. Westbrook, ed.). Wiley, New York, 1967.
[4.102] M. I. Kochnev, *Dokl. Akad. Nauk SSSR* **105**, 1236 (1955).
[4.103] C. Guillaud, *J. Phys. Radium* **12**, 223 (1951).
[4.104] G. Fischer and W. B. Pearson, *Can. J. Phys.* **36**, 1010 (1958).
[4.105] C. P. Bean and D. S. Rodbell, *Phys. Rev.* **126**, 104 (1962).
[4.106] G. Busch and F. Hulliger, *Helv. Phys. Acta* **31**, 301 (1958).
[4.107] F. Hulliger, *Helv. Phys. Acta* **32**, 615 (1959).
[4.108] C. Guillaud, State thesis, Strasbourg (1943).
[4.109] P. E. Bierstedt, *Phys. Rev.* **132**, 669 (1963).
[4.110] N. P. Grazhdankina, *Zh. Eksperim. Teor. Fiz.* **47**, 2027 (1964).
[4.111] F. K. Lotgering and E. W. Gorter, *J. Phys. Chem. Solids* **3**, 238 (1957).
[4.112] T. Suzuoka, *J. Phys. Soc. Japan* **12**, 1344 (1957).
[4.113] I. G. Fakidov and A. Ya Afans'ev, *Fiz. Metal. Metalloved.* **6**, 176 (1958).
[4.114] J. P. Suchet, *Ann. Phys. (Paris)* **8**, 285 (1963).
[4.115] I. G. Fakidov and N. P. Grazhdankina, *Dokl. Akad. Nauk SSSR* **66**, 847 (1949).
[4.116] I. K. Kikoin, N. A. Babushkina, and T. N. Igosheva, *Fiz. Metal. Metalloved.* **10**, 488 (1960).
[4.117] M. Nogami, M. Sekinobu, and H. Doi, *Jap. J. Appl. Phys.* **3**, 572 (1964).
[4.118] M. Nogami, *Rep. Fac. Eng. Shizuoka Univ.* **27** (1965).
[4.119] H. Wagini, *Z. Naturforsch.* **21a**, 362 (1966).
[4.120] N. Kh. Abrikosov and V. F. Bankina, *Dokl. Akad. Nauk SSSR* **108**, 627 (1956).
[4.121] L. D. Dudkin and V. I. Vaidanich, "Voprosy Metallurgii i Fiziki Poluprovodnikov IV" (N. Kh. Abrikosov, ed.). Izd. Akad. Nauk, Moscow, 1961.

[4.122] L. D. Dudkin and N. Kh. Abrikosov, *Zh. Neorg. Khim.* **1**, 2096 (1956).
[4.123] D. H. Damon, R. C. Miller, and A. Sagar, *Phys. Rev.* **138**, A636 (1965).
[4.124] L. D. Dudkin, *Zh. Tekh. Fiz.* **28**, 240 (1958).
[4.125] L. D. Dudkin and N. Kh. Abrikosov, *Fiz. Tverd. Tela* **1**, 142 (1959).
[4.126] B. N. Zobina and L. D. Dudkin, *Fiz. Tverd. Tela* **1**, 1821 (1959).
[4.127] L. D. Dudkin and V. I. Vaidanich, *Fiz. Tverd. Tela* **2**, 1526 (1960).

Chapter 5 Oxides of the Metals Ti, V, Cr, Mn, and Homologues

5.1. TO, T_3O_4, AND T_2O_3 OXIDES

A digest by Morin [5.1] has attempted to give a general view of transport phenomena in the oxides of transition metals and the vital role played by d electrons. These are frequently separated from the (B) and (AB) bands by forbidden energy bands. Ionicity of the T—O bonds is high, and the positive effective charge carried by the T atom in its upper valency is sufficient to attract an extra electron, which is thus localized. There are exceptions to this general schema, however, and some of these will be met with immediately among monoxides. The only ones known to exist are those of Ti, V, Nb, and Mn. They crystallize in the B1 structure, but all (except MnO) admit vacancies for the two components [5.2]. Their composition $\square_x T_{1-x} O_{1-y} \square_y$ varies from $y = 0.3$ ($x = 0$) to $x = 0.2$ ($y = 0$), including $x = y = 0.15$ for titanium, and from $y = 0.1$ ($x = 0$) to $x = 0.2$ ($y = 0$) including $x = y = 0.14$ for V, with disordered vacancies. In the case of Nb, the homogeneity domain is fairly limited around $x = y = 0.25$, and the order of the vacancies rather curiously involves plane tetragonal coordination for each type of atom.

Melted TiO has been studied by Pearson [5.3]. The variation in resistivity in relation to temperature (Fig. 5.1), and the thermoelectric effect indicate metallic character. The pure equiatomic composition, annealed below 950°C, takes on a more complex structure than B1. Ariya and Bogdanova [5.4], working on conglomerated powders, confirmed the metallic character and showed the absence of compounds between TiO and Ti_2O_3. It was also using conglomerates that Samokhvalov and Rustamov [5.5] studied the variations in numerous properties, notably the Hall effect. Magnetic susceptibility is constant, at

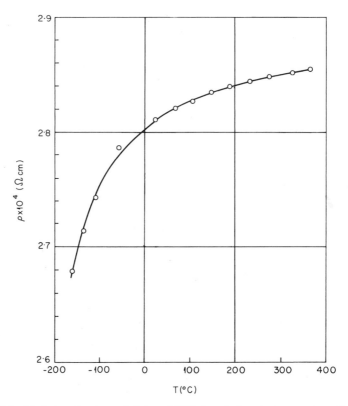

Fig. 5.1. Variation in the resistivity of TiO in relation to temperature (based on Pearson [5.3]).

around 2×10^6 emu gm^{-1}. Figure 5.2 shows that conduction is due mainly to the d electrons with a mobility of 1 cm^2 V^{-1}sec^{-1} and a high effective mass. In (a), below 200°K, major drifting of carriers through the vibrations of the lattice has been noted. Denker [5.6] repeats these measurements, adding optical reflection. L'vov et al. [5.7] observed semiconducting properties in TiN/TiO solutions. As regards the oxide VO, constant paramagnetism (50×10^{-6}) and high conductivity were referred to at a very early stage [5.8]. A semiconductor–metal transition, which Morin [5.9] believed he had observed around 120°K, has not been confirmed by Kawano et al. [5.10] and might be due to the presence of V_2O_3 (see p. 126). These authors have varied the composition of the conglomerates between $VO_{0.9}$ and $VO_{1.25}$, and have shown that the susceptibility follows the Curie–Weiss law, with $2S$ varying between 0.6 and 0.8. A Neel point appears at low temperatures for O/V > 1.1. Resistivity follows the law of semiconductors, with an activation energy varying between 0.004 and 0.16 eV (Fig. 5.3).

5.1. TO, T$_3$O$_4$, and T$_2$O$_3$ Oxides

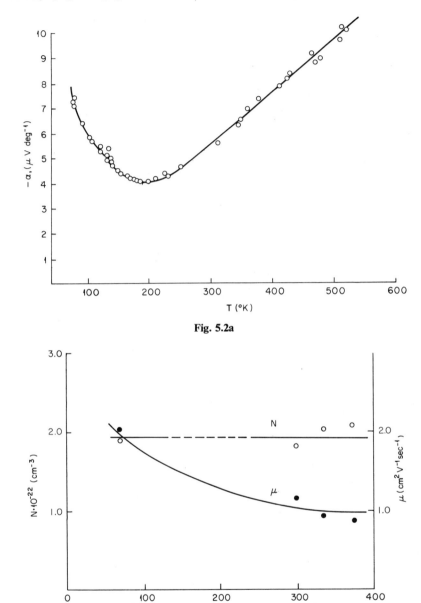

Fig. 5.2. Other electrical properties of TiO in relation to temperature: (a) Seebeck coefficient and (b) carrier density and mobility (based on Samokhvalov and Rustamov [5.5]).

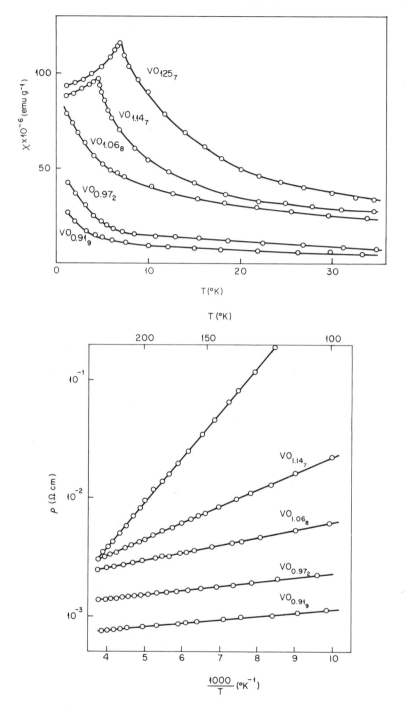

Fig. 5.3. Variation with temperature of the (a) (top) magnetic susceptibility and (b) (bottom) resistivity of various oxides, $VO_{0.9}$ to $VO_{1.25}$ (based on Kawano et al. [5.10]).

5.1. TO, T_3O_4, and T_2O_3 Oxides

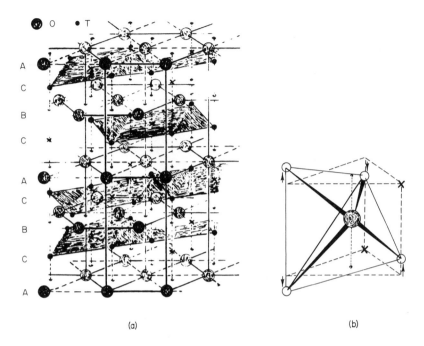

Fig. 5.4. Structure of corundum T_2O_3: (a) succession of atomic planes O—T—O—T···, (b) T_4 tetrahedron (based on Krebs "Grundzuge der anorganischen Kristallchemie," Enke, Stuttgart, 1968, Figs. 158 and 159, p. 238). The crosses indicate vacant sites [5.17].

The difference in behavior between TiO and VO had been explained by Morin on the assumption that the t_{2g} AOs of each metal atom overlapped with those of the 12 neighbors of the same kind in the B1 structure, and that the corresponding bands were partly filled (t_{2g}^2 and t_{2g}^3), but that the existence of a magnetic order at low temperatures in VO separated the subbands $t_{2g}^3 \alpha$ (full) and $t_{2g}^0 \beta$ (empty). The questioning of this result by Kawano et al. does not alter this reasoning: it was indicated in Section 3.4 that it appears often as if the α and β subbands were separate, even in the absence of magnetic order. Denker [5.11] has recently tried to explain the presence of vacancies by stabilization of the lattice caused by lowering of the Fermi level (cf. Section 4.2 and Fig. 4.14). The oxide MnO has a high magnetic susceptibility, presents a Neel point at 120°K and is insulating [5.8]. It is generally accepted that the t_{2g} AOs of Mn do not overlap there. The arrangement of the atomic moments at low temperatures is known [5.12], and produces a slight rhombohedral distortion. Miller and Heikes [5.13] have studied transport by holes in $Li_xMn_{1-x}O$. We shall return to this in Section 6.2. Haussmannite Mn_3O_4 has a structure derived from the $H1_1$ structure of spinel, to which we shall return in Section 6.3, in connection with magnetite. We shall then consider its

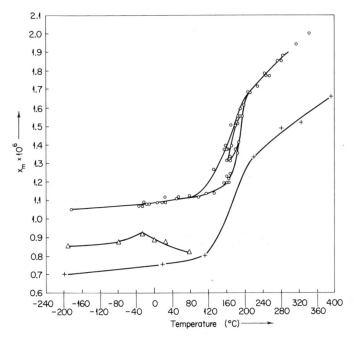

Fig. 5.5. Variation in the magnetic susceptibility of Ti_2O_3 in relation to temperature (based on (△) Adler and Selwood [5.20], (+) Foëx and Wucher [5.21], and (O) Pearson [5.3]).

conductibility, by comparison with Fe_3O_4 and the $Fe_xMn_{3-x}O_4$ solutions. Its semiconductor character is known [5.14–5.16].

Among the sesquioxides T_2O_3, those of Ti, V, and Cr crystallize in the $D5_1$ rhombohedral structure of corundum Al_2O_3, while α-Mn_2O_3 is the prototype of a different structure. There also exists a form of haussmannite containing vacancies (quadratic γ-Mn_2O_3). In the structure of Al_2O_3, the oxygen planes follow one another as in the compact hexagonal assembly (ABABA, Fig. 5.4). It is close to it anyway, corresponding to an ionic radii ratio $r_T/r_O < 0.6$. A C-plane of metal occurs between each two consecutive oxygen planes. If all the sites on this plane were occupied, one would obtain an equiatomic formula TO. In reality, every third site is unoccupied, involving the existence of distorted O_6 octahedrons and T_4 tetrahedrons (p^3/sp^3 bonds). The distortion is more marked for α-Mn_2O_3, which crystallizes in the cubic structure of bixbyite $(Fe, Mn)_2O_3$, corresponding to a radius ratio $0.6 < r_T/r_O < 0.87$; Ti, V, Cr, and Mn have 1, 2, 3, and 4 d electrons, respectively.

The electrical properties of Ti_2O_3 have been studied by Foëx and Loriers [5.18], who have revealed the semiconducting character (0.1 eV) of samples sintered in vacuum at 1400°C. A slight contraction of the cell and a sudden fall in resistivity take place above 200°C. This temperature corresponds to a

5.1. TO, T_3O_4, and T_2O_3 Oxides

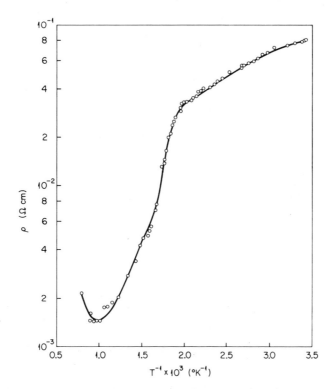

Fig. 5.6. Logarithmic variation in the resistivity of a monocrystal of Ti_2O_3 in terms of the reciprocal of the absolute temperature (based on Abrahams [5.25]).

specific heat peak [5.19]. Pearson [5.3], also using conglomerate samples, has found a p-type and an activation energy varying from 0.1 to 0.2 eV between 50 and 400°C. The diluted powder method (cf. Section 2.4) gives an absorption threshold corresponding to a forbidden band gap of 0.15 eV. Figure 5.5 shows the complex variation in the magnetic susceptibility, compared with the earlier results [5.20, 5.21]. Morin [5.9], and Yahia and Frederikse [5.22] believed that monocrystals behave like metals above 450°K. Yahia et al. measured the Hall and Seebeck effects. Kawakubo et al. [5.23] found no resistivity irregularity in the $Ti_{1.8}V_{0.2}O_3$ solution. Al'Shin and Astrov [5.24] observed in Ti_2O_3 the magnetoelectric effect discovered in Cr_2O_3 (see p. 128). Semiconductibility is confirmed for monocrystals by Abrahams [5.25] (Fig. 5.6) and neutron diffraction indicates an antiferromagnetic order of moments of around 0.2 μ_B with $T_N = 660°K$. It is assumed that T_N varies from 450 to 650°K depending on the samples, and corresponds to the irregularities in other properties.* Keys and Mulay [5.26] suggest a noncooperative transition.

---

* Recent work has shown the absence of any magnetic order.

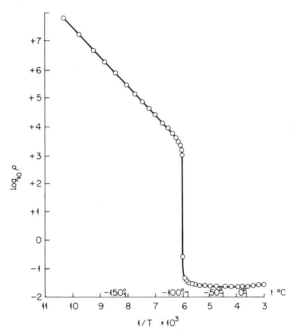

Fig. 5.7. Variation in the resistivity of V_2O_3 in relation to temperature (based on Foëx [5.27]).

The low resistivity of V_2O_3 (55×10^{-4} Ω cm) was pointed at an early stage, as well as its dilatometric transformation around $-100°C$, with a drop in magnetic susceptibility at low temperatures [5.8]. Foëx [5.27] has discovered the simultaneous existence of a semiconductor (0.2 eV)–metal transformation with an increase in conductivity by a factor of 10^5 (Fig. 5.7) and a slight contraction in the cell. He used samples sintered at 1200°C. A comprehensive study of physical properties was published shortly afterwards by Foëx et al. [5.28]. The structure below the transformation point has a lower symmetry. A maximum of resistivity exists around 250°C, followed at higher temperatures by a strictly exponential decrease (0.04 eV). Variation in magnetic susceptibility is complex (Fig. 5.8). The conclusion is that there is an antiferromagnetism–paramagnetism transformation at $-100°C$, perhaps followed by a return to antiferromagnetism, with the moments directed differently, between $+110$ and $+250°C$ (specific heat maxima). Teranishi and Tarama [5.29] have continued investigation of the susceptibility, concluding that there is a Neel point between 260 and 380°C, with $2S \sim 2$. Morin [5.9], followed by Goodman [5.30], have confirmed the electrical transition in monocrystals. Ariya and Bogdanova [5.4]. followed by Kachi et al. [5.31], have measured the conductivity of different compositions. Acket and Volger [5.32] have measured

5.1. TO, T_3O_4, and T_2O_3 Oxides

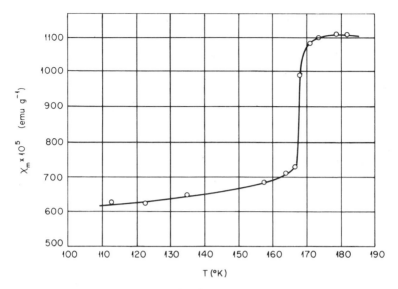

Fig. 5.8a

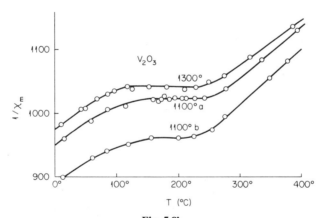

Fig. 5.8b

Fig. 5.8. Variation, in relation to temperature, of (a) the susceptibility and its (b) reciprocal for samples of V_2O_3 sintered at different temperatures (based on Foëx et al. [5.28]).

the Seebeck and Hall effects, concluding that there is a metallic character between -100 and $+25°C$, with the hopping mechanism occurring outside this range. Austin [5.33], followed by Minomura and Nagasaki [5.34], have studied the effect of pressure. Finally, vital information has been provided by Shinjo et al. [5.35, 5.36], who have diffused ^{57}Fe and observed the Mössbauer

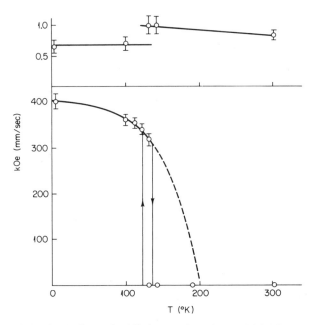

Fig. 5.9. Variation in the isomeric shift (top) and the internal field (bottom) in relation to temperature, in $V_{1.98}Fe^{57}_{0.02}O_3$ (based on Shinjo and Kosuge [5.36]).

effect (see Section 6.1). The Zeeman effect up to 130°K proves the existence of antiferromagnetism, while its disappearance at 140°K, after the resistivity drop, confirms the subsequent paramagnetism. Figure 5.9 also shows that the extrapolated T_N would be around 200°K. The transformation of V_2O_3, which seems to take place between 140 and 170°K, depending on the samples, is thus not magnetic in origin.

The oxide Cr_2O_3 is a semiconductor that is highly resistant, refractory (2400°C) and antiferromagnetic ($T_N = 33°C$) with $2S = 3$. It can be mixed in any proportion with α-Fe_2O_3 [5.8]. Bevan et al. [5.37] have measured the resistivity of conglomerate samples, and found a p-type, interpreted as the passage of a few Cr atoms to valency 4 (excess oxygen). In a neutral or reducing atmosphere, what is more, the resistivity increases sharply, and the activation energy rises from 0.6 to approximately 1.5 eV. The effect of defects has been discussed by Hauffe et al. [5.38], using Schottky and Wagner's model, along with the mixed valency theory. Brockhouse [5.39] has measured the magnetic structure by neutron diffraction. Roche and Jaffray [5.40] have observed a break in the resistivity/temperature slope at 33°C. Astrov [5 41] has discovered a major magnetoelectric effect, which has led to a great deal of work (cf Part Three). Foner and Hanabusa [5.42] have observed it in Cr_2O_3–Al_2O_3 solutions. Schultz and Bober [5.43] mention the absence of photoconductivity in Cr_2O_3.

5.1. TO, T₃O₄, and T₂O₃ Oxides

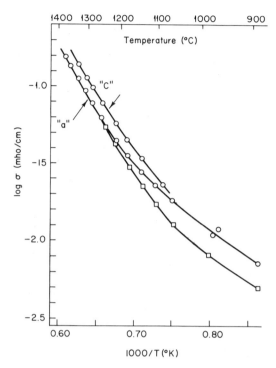

Fig. 5.10. Logarithmic variation in conductivity in terms of the reciprocal of the temperature, for monocrystals lying along the a and c-axes (based on Crawford and Vest [5.45]). Pressures P_{O_2} are at (○) 1 atm and at (□) 10^{-6} atm.

This indicates the predominance of d electron transfers. Fischer and Dietrich [5.44], using bars of melted material, followed by Crawford and Vest [5.45], using monocrystals (Fig. 5.10) and De La Banda et al. [5.46], in a comprehensive study using conglomerate samples, have continued the work of Bevan et al. According to Crawford et al., the very low mobility found in the intrinsic region implies high localization of the carriers. The substitution of lithium has been investigated by Hagel [5.47].

Many authors have tried to give a general interpretation of the properties of Ti_2O_3, V_2O_3, and Cr_2O_3. The explanation suggested by Morin for a possible semiconductor–metal transition in VO cannot be extended to V_2O_3 (2d electrons, partly-filled $t_{2g}\alpha$ subband) except to the extent that distortion of the O_6 octahedron introduces additional divisions. In any case, the work of Shinjo et al. does not fall into line with this. For Goodenough [5.48], the Cr—Cr distance is higher than the critical distance, and the d electrons are therefore localized. On the other hand, the Ti—Ti and V—V distances are lower, and the strong interactions among d electrons result at low temperatures in the formation of T—T' pairs (cf. Section 3.2), explaining the noncooperative transition, shift of T atoms in relation to the symmetry center, absence of magnetic moment and semiconductibility. This model probably needs to be modified, as far as magnetism is concerned, in the light of the results provided by Abrahams and Shinjo et al. Other explanations have been suggested for the transition of V_2O_3 [5.49–5.51].

5.2. Rutile TiO_2

Titanium forms three TO_2 phases with different structures: quadratic rutile (C4), quadratic anatase (C5) and orthorhombic brookite (C21). Only the first interests us here, and Fig. 5.11a shows the arrangement of the atoms in the centered quadratic cell. Each titanium atom is in the center of a very slightly elongated O_6 octahedron, the distances to the oxygen atoms 1 and 2 being 1.99 Å, compared with 1.95 for the others. In addition, the oxygen atoms 3, 4, 5, and 6 form a rectangle within which the O—Ti—O angles are 81 and 99° alternately. Each oxygen atom is in the center of an almost equilateral Ti_3 triangle [5.52]. In view of the absence of magnetism, these facts leave no other choice than d^2sp^3–sp^2 bonds, where the only two d electrons of the titanium are involved in d–s–p hybrid AOs. These overlap with the s–p hybrid AOs of the oxygen, and form six Lewis pairs per titanium atom, greatly displaced towards the oxygen because of the high ionicity of the bonds (0.78). No bachelor electron thus remains (formula t_{2g}^0). Figure 5.11b shows, in two consecutive cells, the arrangement of the O_6 octahedrons, which have a shared edge within the same chain and a shared corner between two chains. Figure 5.11c shows the arrangement of these chains. Rutile is a semiconductor with high resistivity (10^{10} Ω cm) and low paramagnetism (0.08×10^{-6}). Its dielectric constant is 180 along axis c, and 90 in a perpendicular direction. Approximately 115 is found for polycrystals (78 for brookite and 48 for anatase). It melts at 1855°C [5.8].

The forbidden band gap in stoichiometric rutile monocrystals has been estimated at 3 eV by Cronemeyer et al. [5.54]. It is deduced simultaneously from the fundamental absorption threshold at low temperature at 0.4 μm, from the photoconductivity peak on the same wavelength, and from the law of decreasing in resistivity between 350 and 850°C. An optical study (transmission, reflection, refraction, photoconductivity) was completed shortly afterwards, and the resistivity/temperature curve suggests successive values of 3.05 and 3.67 eV for E_G (Fig. 5.12) [5.55]. Various magnetic susceptibility values have been mentioned. In his digest, Grant [5.52] refers to a range of -0.3 to $+0.3 \times 10^{-6}$ emu gm^{-1}. The pure stoichiometric oxide is probably diamagnetic. Parker and Wasilik [5.56] have shown that the abnormal dielectric constant ε sometimes observed comes from barrier layers at the electrodes. Soffer

5.2. Rutile TiO$_2$

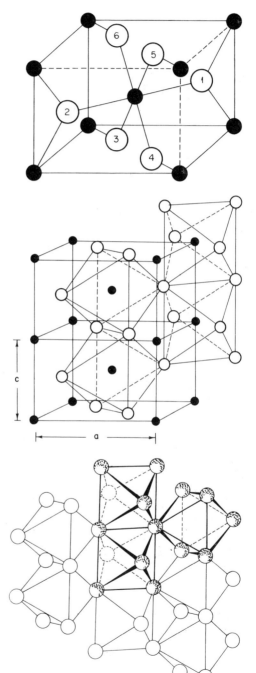

Fig. 5.11. Structure of rutile TiO$_2$: (a) (top) Ti—O bonds (based on Grant [5.52]), (b) (middle) O$_6$ octahedra (based on Breckenridge and Hosler [5.53]), (c) (bottom) Octahedra chains along the c-axis (based on Cronemeyer [5.55]). Key: (●) Ti^{4+} and (○) O^{2-}.

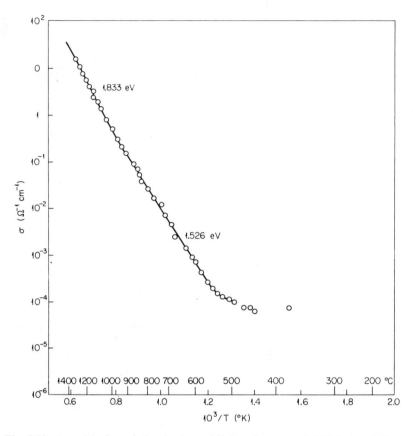

Fig. 5.12. Logarithmic variation in the resistivity of monocrystal TiO_2 in relation to temperature (based on Cronemeyer [5.55]).

[5.57] has found a temperature coefficient of -7 to -9×10^{-4} eV deg^{-1} for E_G, and an absorption band due to OH grouping at 0.4 eV. Von Hippel et al. [5.58], in a long survey, have linked the optical properties with the vibrations of the lattice, and measured the complex constant $\varepsilon = \varepsilon' - i\varepsilon''$ (Figs. 5.13 and 5.14) and the dielectric relaxation. These polarization phenomena (field emission, currents limited by the space charge) have also been studied by [5.59–5.62]. In his digest, Morin [5.1] repeats the suggestion of Von Hippel et al.: 3d level at 3.05 eV of the valency band (O^{2-}) and 0.6 of the following band.

After reduction TiO_2 becomes a conductor. If one refers to a TiO_x formula, conductivity varies as follows [5.8]:

x	2.0000	1.9995	1.995	1.75	
σ	10^{-10}	10^{-1}	0.8	10^2	mhos cm^{-1}

5.2. Rutile TiO_2

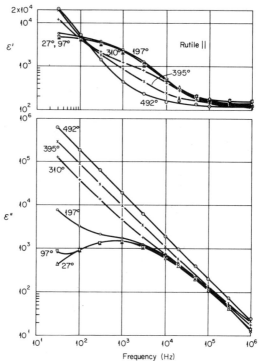

Fig. 5.13. Variation in the real and imaginary dielectric constants, in relation to the frequency in rutile, parallel to axis c (based on Von Hippel et al. [5.58]). Key: (▲) 27°C, (▽) 97°C, (●) 197°C, (+) 310°C, (×) 395°C, (○) 492°C.

Contrary to the first hypotheses, the discrepancy in relation to stoichiometry is not due to oxygen vacancies. The homogeneity domain of the actual C4 structure is narrow, limited to $TiO_{1.98}$ [5.2]. Beyond this exists a certain number of Ti_nO_{2n-1} phases, where blocks of n octahedrons have the C4 structure, but where these octahedrons have a face in common at the edges of two adjoining blocks [5.63–5.65]. The values $n = 4$ to 10 have been isolated, and infinity corresponds to rutile. The interstitial titanium atoms are thus ordered along certain planes [5.2]. Hurlen [5.66], Frederikse [5.67], Von Hippel et al. [5.58], and Tannhauser [5.68] have confirmed this point of view in turn. The blue color is caused by an absorption band centered on 1.8 μm and extending from 0.7 to 5 μm [5.54]. It probably results from the Ti^{3+} ions.

The first work on electrical properties of reduced rutile were done by Earle [5.69], Boltaks et al. [5.70], Gorelik [5.71] on sintered samples and Cronemeyer [5.55] on monocrystals. Earle showed that ionic transport was negligible, and he established a law for conductivity and oxygen pressure (confirmed by Hauffe [5.72]):

$$\log \sigma = A - B \log P \tag{5.1}$$

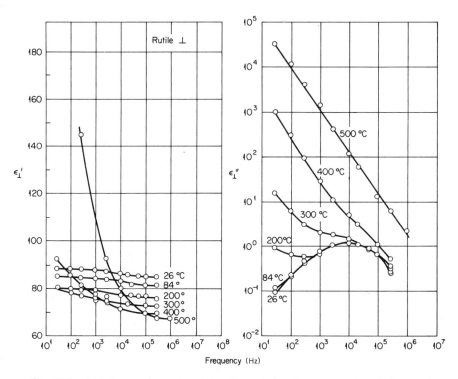

Fig. 5.14. Variation in the real and imaginary dielectric constants in relation to the frequency in rutile, at right angles to axis c (based on Von Hippel et al. [5.58]).

Boltaks et al. used products with compositions $TiO_{1.80}$ to $TiO_{1.95}$, sintered in H_2 or CO with conductivity

$$\sigma = CT^{-3/4} \exp(-e_d/kT) \tag{5.2}$$

with $C \sim 10^5$ and $e_d \sim 0.1$ eV. The Seebeck effect increases slightly with temperature ($TiO_{1.8} \sim 60$, $TiO_{1.95} \sim 300$ μV deg^{-1} at room temperature). The number of electrons is higher than 10^{20} cm^{-3}, while their effective mass is approximately $16m$ and their mobility around 0.1 cm^2 V^{-1} sec^{-1}. Cronemeyer [5.55] has pointed out an abrupt transition to nonohmic behavior for a certain potential (Fig. 5.15) (cf. also [5.60, 5.73, 5.74]). Breckenridge and Hosler [5.53] have found comparable results for sintered samples and monocrystals between -190 and $+500$°C (Fig. 5.16). The electron density varies little, and even stay constant if one takes account of the negative power of T associated with their dispersal by impurities (mobility variation, cf. Section 2.4) [5.52]. Kataoka and Suzuki [5.75] have shown that the Seebeck coefficient can reach 1 mV deg^{-1}, but decreases rapidly as reduction and conductivity increase. These results are summarized by Grant [5.52].

5.2. Rutile TiO_2

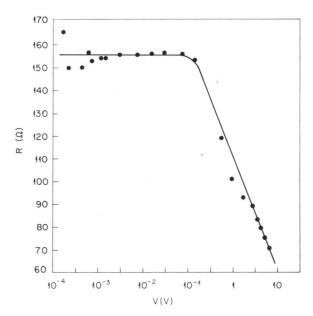

Fig. 5.15. Variation in the resistance of a highly reduced ceramic TiO_2, in relation to the tension applied (based on Cronemeyer [5.55]).

Research became more detailed after 1960. Kleber *et al.* [5.76] confirmed the diminution of e_d and the independence of conductivity from temperature in highly reduced crystals. Hollander *et al.* [5.77] studied the effect of pressure, then [5.78] found a high resistivity anisotropy (Fig. 5.17) for a certain degree of reduction, attributing it to the change from the classic mechanism to conduction by impurities (interstitials). Frederikse *et al.* [5.79] have studied the EPR of the electron bonded to the titanium, and have concluded that there is a polaron of 0.01 to 0.07 eV below 5°K and a narrow d band above. Sakata [5.80] has nevertheless found an intrinsic region above 1150°C. Frederikse [5.67] has reviewed these results and estimated the effective mass of the electrons at $25m$. Hasiguti *et al.* [5.81, 5.82] have defined the behavior at low temperatures. Hollander and Castro [5.83], working on slightly reduced monocrystals (10^4–10^{13} Ω cm) have observed quasi-ferroelectric phenomena at field gradients higher than 500 V cm^{-1}, and have proposed a model of conducting needles in an insulating medium. Acket and Volger [5.84] have criticized the work [5.78]. Yahia [5.85], followed by Tannhauser [5.68] prefer, instead of Eq. (5.1)

$$\sigma \sim P^{-1/5} \quad (<10 \text{ mm}) \quad \text{or} \quad P^{-1/6} \quad (>10 \text{ mm}) \tag{5.3}$$

Keezer (5.86), followed by Keezer *et al.* [5.87], have studied the role of the surface-chemisorbed oxygen layer, which depends on P, on photoconductivity,

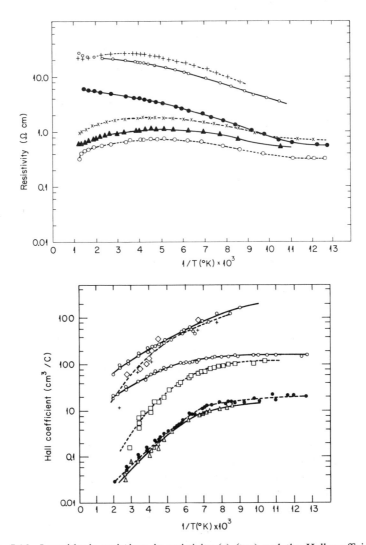

Fig. 5.16. Logarithmic variations in resistivity (a) (top) and the Hall coefficient (b) (bottom) of a monocrystal of TiO_2, in terms of the reciprocal of the temperature, for various temperatures and lengths of reduction (based on Breckenridge and Hosler [5.53]). Key: (O - - O - - O) 800°C, reduced ∞ min; (●—●—●) 700°C, reduced 5 min; (O—O—O) 650°C, reduced 15 min; (▲—▲—▲) 800°C, reduced 15 min; (X - - X - - X) 700°C, reduced 5 min; (+ - - + - - +) 650°C, reduced 15 min.

and on response time. Greener [5.88] has found a law in $P^{-1/4}$ and $E_G = 3.6$ eV, above 1200°C. Blumenthal et al. [5.89], studying monocrystals between 1000 and 1500°C, have found $E_G = 3.8$ eV, and an equation:

5.2. Rutile TiO_2

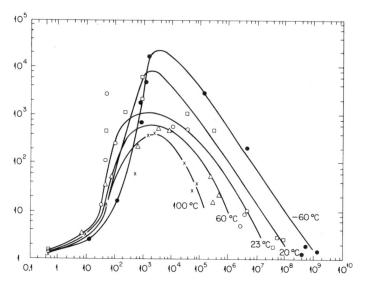

Fig. 5.17. Variation in the anisotropy of the resistivity (axis a/axis c) in terms of resistivity along axis c (based on Hollander and Castro [5.78]).

$$(A\sigma + B)/P + C\sigma^4 - \sigma^5 = 0 \tag{5.4}$$

Application of the mass-action law to the reactions

$$Ti + 2O = O_2(gas) + Ti^{3+}(interst.) + 3 \text{ electrons} \tag{5.5}$$

$$Ti + 2O = O_2(gas) + Ti^{4+}(interst.) + 4 \text{ electrons} \tag{5.6}$$

$$Ti^{3+}(interst.) = Ti^{4+}(interst.) + 1 \text{ electron} \tag{5.7}$$

then enables them to calculate the formation enthalpies: (9.24, 10.67, and 1.43 eV).

Investigation of the carrier mobility is essential. In Section 2.4, this measurement was linked to the Hall effect because, when electrical conduction is due simultaneously to carriers with widely differing mobilities, the "Hall" mobility μ corresponds to that of the majority carriers. When, as is frequently the case, the transfer mechanism predominates, μ is thus the electron mobility at the d level. Photoconductivity, which creates excess carriers by an excitation mechanism, can be used for more direct measuring of a "drift" mobility μ_E. The method is to send out a pinpoint light impulse and measure the speed of drift (in the direction of the electrical field applied) of the group of carriers created by the increase in conductivity it causes at a certain distance; μ_E depends on the different mobilities of the electrons and holes created simultaneously. In the case of excitation of the electron from the (wide) valency band to a (narrow) d band, it is therefore higher than that of the d electrons.

Breckenridge and Hosler [5.53] have measured the variation $\mu(T)$ and suggested scattering caused by lattice vibrations at low temperatures, and by impurities above 100°K. Bogomolov et al. [5.90, 5.91] have shown that the anisotropy mentioned earlier [5.78] is mainly due to that of the mobilities μ and μ_E. Figure 5.18 shows average variation in $T^{-5/2}$, independent of the carrier density and therefore not corresponding to scattering by impurities. Acket and Volger [5.92] have confirmed these results. Yahia [5.93], using a third method, has found a mobility that increases with temperature. Becker and Hosler [5.94] attribute the variation in the Hall coefficient anisotropy with temperature to two competing conduction mechanisms. A digest by Acket [5.95] has been followed by other speculations [5.96–5.98]. Bogomolov and Zhuze [5.99] refer to the theory of small polarons. In conclusion, it may be said simply that stoichiometric rutile has an excitation mechanism but that, by reduction, the titanium t_{2g} sublevel begins to fill up when certain atoms change to valency 3 (Fig. 3.18b) and is subject to transfers. A growing reduction first brings about delocalization along the octahedron chain parallel to the c-axis (maximum anisotropy), then in all directions. Goodenough's viewpoint was given in Section 3.4 (Fig. 3.19).

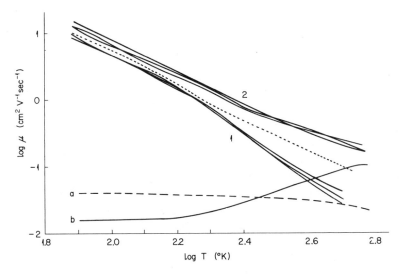

Fig. 5.18. Logarithmic variation in the Hall mobility in terms of temperature in reduced TiO_2: (1) at right angles to axis c, (2) along axis c (the different curves of each group concern samples with different carrier densities). Below, anisotropies of μ (a) and μ_E (b) ratio of directions (1) on (2). (Based on Bogomolov and Zhuze [5.91]).

Impurities play an important role [5.100, 5.101]. In accordance with the valency induction laws, the addition of an oxide of lower valency reduces

5.3. Other Dioxides

conductivity, while one of higher valency increases it [5.102]. Gorelik [5.71] has studied alkaline–earth oxides, and Acket and Volger [5.84], Yahia [5.85] and Zijlstra et al. [5.103] alumina. The latter have obtained μ_E by noise measurements. Cauville [5.104, 5.105] had studied the properties and reduction of $TiO_2 + Al_2O_3$ mixtures. Kzendzov [5.106], Ishikura and Sakata [5.107], Ivukina and Panova [5.108], and Bogoroditskii et al. [5.109] have studied Nb_2O_5 and Ta_2O_5. The oxide V_2O_5 seems to be an exception [5.107]. Finally, lithium diffusion takes place in the interstitial sites, reaching a limit of 2.5×10^{19} atoms cm^{-3} [5.110]. Acket and Volger [5.97] have mentioned its donor properties. None of these additions reveals any new phenomenon compared with reduced rutile.

5.3. OTHER DIOXIDES

All the other dioxides have the C4 quadratic structure (distorted for some of them), except for ZrO_2 and HfO_2, which crystallize in the C1 cubic structure (distorted in C43 at room temperature). This is because the high r_T/r_O ratios favor this structure in a compact TO_2 arrangement. They decrease as the d layer fills up in each transition series. They are less than 0.5 in the first series, approaching 0.6 for the first terms in the others, and even exceeding 0.7 for rare earths (cf. Chapter 8) [5.8, 5.111]. The low-temperature phase of VO_2 presents a slight monoclinic distortion, accentuated in MoO_2, WO_2, TcO_2, and ReO_2, where the T atoms move away a little from the center of symmetry of the O_6 octahedrons with alternately short and long T—T' distances [5.111]. Lewis pairs would therefore be possible [5.48]. All have a homogeneity domain and application of the usual formula TO_x would give as a limit $x = 1.7$ (Ti), 1.8 (V), and 1.93 (Mn). There is some argument about the domain of CrO_2. A series of V_nO_{2n-1} oxides has been observed, similar to that of Ti [5.2].

Vanadium dioxide has the C4 structure only above 70°C. Susceptibility, which is low and constant below this temperature, corresponds to $2S \sim 0.7$ for the rutile phase [5.112]. The sudden resistivity drop accompanying this transformation has been mentioned by Jaffray and Dumas [5.113], and studied using monocrystals by Morin [5.9], who attributes the metallic properties of the C4 phase to the existence of a partly-filled t_{2g} narrow band, the divisions of which at low temperatures (distortion, possible magnetic order), he suggests, bring about semiconductor behavior. Kosuge et al. [5.114] have identified nine different phases between V_2O_3 and V_2O_5, paramagnetic or antiferromagnetic. Kachi et al. [5.31] have found forbidden bands of around 0.3 eV at low temperatures, and they have shown other resistivity irregularities (Fig. 5.19). Minomura et al. [5.34], followed by Neuman et al. [5.115], have studied the effect of pressure on VO_2. Kosuge et al. [5.116] have studied the phase change in V_6O_{13} at 155°K (T_N) and concluded that there is a second-order

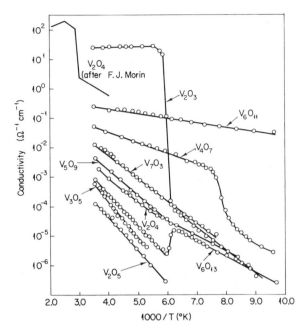

Fig. 5.19. Logarithmic variation in conductivity in terms of the temperature reciprocal for various compositions between V_2O_3 and V_2O_5 (based on Kachi et al. [5.31]).

transition between two semiconducting phases. Kawakubo [5.117] has speculated on the transition of VO_2. Bongers [5.118] has shown that the resistivity anisotropy is not affected by the transition, contradicting Morin's hypothesis. It is suggested that the transition is of the first order [5.119]. Umeda et al. [5.120], using NMR, have found a new paramagnetic phase immediately below 68°C. Further measurements of resistivity, Hall effect (on powder), and Seebeck effect have been done [5.121–5.123]. Barker et al. [5.124] have confirmed, by an optical investigation, the forbidden band gap of 0.3 eV and the absence of antiferromagnetism below 68°C. Measurements of the Hall effect give 3×10^{21} electrons cm^{-3} in the "metallic" phase. They have calculated $m^* \sim 0.5m$ and $\mu_E \sim 2$ cm^2 V^{-1}sec^{-1}, compared with $\sim 4m$ and ~ 20 cm^2 V^{-1}sec^{-1} in the semiconducting phase. Finally, Kosuge [5.125] has diffused ^{57}Fe and observed the Mössbauer effect (see Section 6.1). The absence of magnetic order invalidates Morin's interpretation and strengthens Goodenough's, according to which the transition is caused by a change in interatomic bonds, as in the case of V_2O_3.

Chromium dioxide is metastable at atmospheric pressure, and decomposes at about 250°C. It is known to be ferromagnetic ($T_C \sim 120°C$, $2S = 2$) [5.126]. It has recently aroused fresh interest [5.127]. Investigation of the

5.3. Other Dioxides

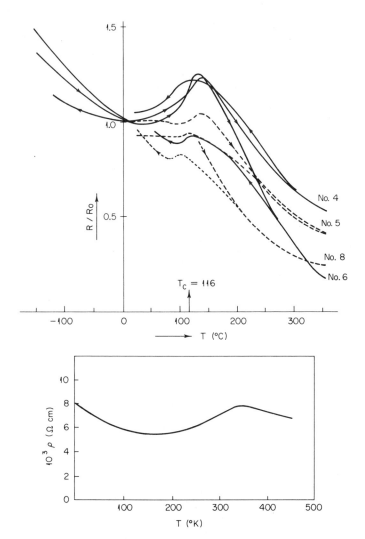

Fig. 5.20. Variation in the resistivity of CrO_2 in terms of temperature: (a) (top) based on Kubota and Hirota [5.129], (b) (bottom) based on Chapin et al., *J. Phys. Chem.* **69**, 402 (1965), Fig. 1, p. 403. Copyright 1965 by the American Chemical Society. By permission of the copyright owner.

resistivity, done with conglomerates by Hulliger [5.128] and Kubota and Hirota [5.129] (Fig. 5.20a), seemed to indicate semiconductor behavior with an activation energy of 0.2 eV. Siratori and Iida [5.130] have pointed out an abnormal variation in the cell parameter depending on the temperature. Goodenough [5.48] believes that the negative temperature coefficient results solely from the disappearance of magnetic order. Reflectivity in the infrared range gives $E_G \leqslant 0.25$ eV [5.131]. Chapin et al. [5.132] have found a similar variation in resistivity (Fig. 5.20b), but they reach the conclusion of metallic properties, for theoretical reasons. Goodenough has proposed the energy diagram in Fig. 5.21 (compare with Fig. 3.19). Druilhe and Bonnerot [5.133]

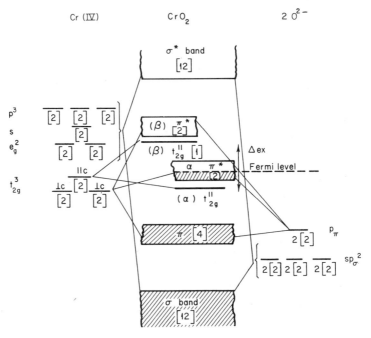

Fig. 5.21. MO energy level diagram for CrO_2. [Based on Goodenough, quoted by Chapin et al., *J. Phys. Chem.* **69**, 402 (1965), Fig. 4, p. 405. Copyright 1965 by the American Chemical Society. By permission of the copyright owner.]

have found an electronic specific heat of above zero, which confirms this hypothesis. On the other hand, they claim to have observed an absorption threshold in the infrared range using the diluted powder method, but Chrenko and Rodbell [5.134] have shown that it is a parasitic thermal effect that is involved. Rodbell et al. [5.135] have published the resistivity/temperature curve for a monocrystal, obtained by epitaxy on rutile, which up to 250°C presents a positive coefficient along the directions [001] and [010]. Druilhe and Suchet [5.136] have emphasized the analogy in the Seebeck effect variation between CrO_2 (Fig. 5.22) and the semiconductor MnTe, where a particular phenomenon has been pointed out (cf. Chapter 7). They attribute the high positive coefficient of the resistivity between 260 and 360 K to the magnetic scattering of carriers in a wide band. To conclude, one might say that it is probably a

5.3. Other Dioxides

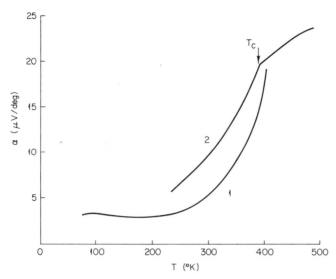

Fig. 5.22. Variation in the Seebeck effect in two samples of CrO$_2$, in terms of temperature. [Based on Chapin et al., J. Phys. Chem. **69**, 402 (1965), Fig. 3 and 5, pp. 404–407. Copyright 1965 by the American Chemical Society. By permission of the copyright owner.] Also based on Druilhe and Suchet [5.136].

pseudometal that is involved, but that its behavior requires further research, which is unfortunately hindered by the difficulties of preparation.

Manganese dioxide has several structures [5.137]. The β phase (pyrolusite, polianite) has that of rutile. It is known to be antiferromagnetic ($T_N \sim 84°K$) [5.138]. Neutron diffraction has led Erickson [5.139, 5.140] and Yoshimori [5.141] to imagine a complex helicoidal magnetic structure. Certain activation treatments endow it with depolarizing properties, used in Leclanché cells [5.142], increased paramagnetism and, perhaps, two Mn—O bonds of different ionicities [5.143]. The variation in resistivity in relation to temperature, investigated by Bizette [5.144], using conglomerates, seems to indicate a strong magnetic scattering between 50 and 100°K. The semiconductor behavior indicated by Fig. 5.23 has been studied by Das [5.145], who gives an activation energy of around 0.3 eV, with 4×10^{19} electrons cm^{-3}, with a mobility of 1 cm^2 V^{-1}sec^{-1} and a Seebeck effect of 200 to 500 μV deg^{-1}. Rectifying effects have been recorded. Chevillot and Brenet [5.146], using a very pure sample (115 Ω cm), have found $E_G = 0.26$ eV. Bhide et al. [5.147, 5.148], using natural monocrystals, have found a resistivity irregularity at 50°C, accompanied by a maximum dielectric constant (10^6) and minimum losses. They have concluded that there are ferroelectric properties below this temperature (cf. Section 5.4), and have confirmed this hypothesis by studying the dielectric hysteresis (Fig. 5.24). They claim that saturation polarization is 10^4 esu, and the coercive

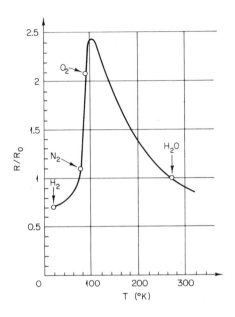

Fig. 5.23. Variation in the resistivity of β-MnO$_2$ in relation to temperature (based on Bizette [5.144]).

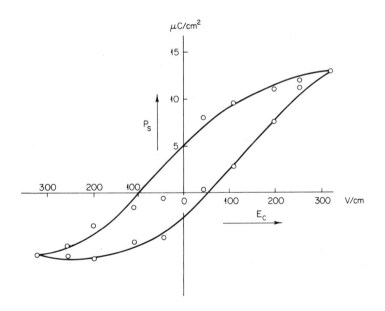

Fig. 5.24. Dielectric hysteresis of β-MnO$_2$ at $-60°$C. [Based on Bhide and Damle, *Physica* **28**, 513 (1960), Fig. 5, p. 575. By permission of North-Holland Publishing Company.]

5.3. Other Dioxides

field 100 V cm^{-1}. A shift of 0.045 Å by the Mn atoms in their O_6 octahedron would be enough to explain the phenomenon. A similar resistivity irregularity exists in $BaTiO_3$ [5.149]. However, the existence of ferroelectricity is repudiated by Yousef and Farag [5.150]. Wiley and Knight [5.151] have continued Das' research. Zeilmaker and Drotschmann [5.152] compare β- and γ-MnO_2 during the discharge of a Leclanché cell. Druilhe [5.153], and then Druilhe and Suchet [5.136], working with conglomerates, have found an (extrinsic) activation energy of 0.08 eV up to 250°K, an irregularity around 50°C, and $E_G = 0.28$ eV above 170°C. The effect of impurities has been studied by Chevillot and Brenet [5.146, 5.154]. All these properties are those of a semiconductor and not a pseudometal, as ferroelectric phenomena prove. The $t_{2g}^3 \alpha$ and $t_{2g}^0 \beta$ sublevels are thus separate. An MO diagram of energy levels has been proposed [5.155].

Among the different solid solutions containing dioxides with C4 structure, $Mn_xCr_{1-x}O_2$ is of particular interest. Siratori and Iida [5.156] have studied compositions with $x < 0.75$, and shown that the magnetic properties are midway between those that would result from parallel orientation of the Mn and Cr moments, and a simple dilution of CrO_2 in MnO_2. Druilhe [5.153], then Druilhe and Suchet [5.136] have studied their resistivity and mentioned their semiconducting character (Fig. 5.25). The maximum resistivity, around T_C,

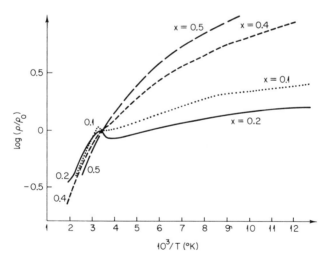

Fig. 5.25. Logarithmic variation in the resistivity of $Mn_xCr_{1-x}O_2$ solutions in terms of the temperature reciprocal (based on Druilhe [5.153]).

disappears at $x > 0.2$, probably because magnetic scattering of carriers through the disorder of the Mn and Cr moments, which is independent of temperature, becomes more important than that of the Cr moments among themselves at T_C. The (intrinsic) activation energy is around 0.3 eV, as for VO_2

and β-MnO$_2$. Using the diluted powder method, one can observe an absorption threshold that tallies with this value. Villers and Druilhe [5.157], continuing the magnetic study done by Siratori *et al.*, have mentioned, where $x > 0.2$, saturation magnetization with the form

$$\sigma = \sigma_s + \chi H \tag{5.8}$$

σ_s, thus defined, tends at 0°K towards a value σ_0. The variation of which in relation to x, as well as that of T_C, is given in Fig. 5.26. The direction of the Mn and Cr moments seems to be antiparallel where $x < 0.1$. A noncollinear structure is then likely, because of the absence of saturation at 0°K. Druilhe [5.158] has summarized these results and defined the optical spectrum. From the value $x = 0.1$, a threshold appears clearly on Fig. 5.27. Chromium dioxide therefore seems to be an extreme case.

Zirconia ZrO$_2$ crystallizes at high temperatures in the C1 cubic structure (slight quadratic distortion), but below 1000°C it changes to the C43 monoclinic structure with cell dilation. Additions of di- and trivalent oxides stabilize C1 at low temperatures, with a high concentration of O vacancies. The first

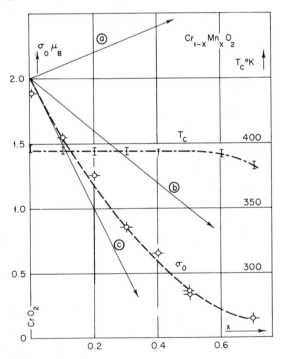

Fig. 5.26. Variation of $\sigma_0 = \sigma_s(0°K)$ and T_C in terms of x in Mn$_x$Cr$_{1-x}$O$_2$ solutions. The straight lines a, b, and c would correspond to parallel Cr and Mn moments, the dilution of CrO$_2$ in MnO$_2$, and antiparallel Cr and Mn (based on Villers and Druilhe [5.157]).

5.3. Other Dioxides

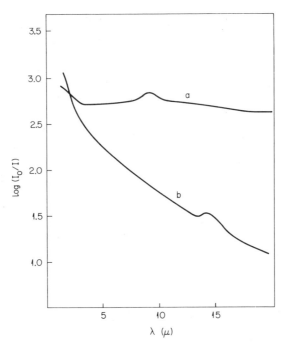

Fig. 5.27. Infrared absorption spectra of CrO_2 (a) and $Mn_{0.1}Cr_{0.9}O_2$ (b), obtained by the diluted powder method. The parasitical thermal effect has been eliminated (based on Druilhe [5.158]).

researches concern the ionic conductivity of cubic solutions ZrO_2–Y_2O_3 and ZrO_2–CaO. The list will be found in [5.159] (see also [5.160]). Anthony [5.161], and then Anthony et al. [5.162, 5.163], have found an electronic conductivity in ZrO_2–La_2O_3 solutions. Vest et al. [5.164, 5.165] have studied the effect of the oxygen pressure P on the electronic conductivity of (pure) monoclinic zirconia. The Zr vacancies create holes with very low mobility ($P > 10^{-16}$ atm), and the O vacancies create electrons ($P < 10^{-16}$ atm). Cubic zirconia has partly ionic conductivity. The activation energy rises from 1 (C1) to 2 eV (C43). Tallan et al. [5.166] discussed the conditions under which the two types of conductivity appear under the influence of lattice defects and impurities. Kröger [5.167] discusses the electronic conductivity of ZrO_2–CaO solutions. Anthony et al. [5.168] have studied the (total) conduction of monocrystals of cubic zirconia. The oxide HfO_2 presents the same structures as zirconia, but the phase change at 1850°C raises fewer problems. Curtiss et al. [5.169] have observed an exponential decrease in resistivity along with temperature.

Niobium dioxide is semiconducting [5.170, 5.171]. Janninck and Whitmore [5.172], using sintered samples, have found an n-type and an activation energy (assumed to be intrinsic) of 1.2 eV. The Seebeck coefficient drops from more than 1 mV deg^{-1} at room temperature to 100 μV deg^{-1} at 1000°K. MoO$_2$ has a catalytic effect [5.173] and is probably semiconducting [5.174]. Vickery and Hipp [5.175] have confirmed this result for n-type monocrystals of 0.5 Ω cm. WO$_2$ has susceptibility of 0.4×10^{-6} and a resistivity of 10^{-3} Ω cm [5.8]. Choain-Maurin and Marion [5.176] have observed n-type semiconduction (cf. Fig. 5.30, p. 153) [5.177]. Gibart [5.178] has found a Pauli paramagnetism (0.4×10^{-6}) and metallic behavior for monoclinic and orthorhombic ReO$_2$. Goodenough et al. [5.179] have provided an MO diagram of energy levels.

Goodenough [5.180] has suggested correcting the critical distance T—T' for the C4 structure, in relation to the number of electrons Z and the spin atomic component S:

$$R_c(\text{Å}) \sim 3 - 0.03(Z-22) - 0.04S(S+1) \tag{5.9}$$

and adding 0.88 or 1.36 Å for the second or third transition series. This gives, R being the experimental T—T' distance:

	TiO$_2$	VO$_2$	CrO$_2$	β-MnO$_2$
R_c	3.0	2.94	2.86	2.76
R	2.96	2.88	2.92	2.86

There should then be a metallic character for TiO$_2$ (when the d band is occupied, in other words for the reduced oxide) and VO$_2$, but a semiconductor character for CrO$_2$ and β-MnO$_2$. The problem concerning CrO$_2$ may be removed by complicating the MO diagram (Fig. 5.21). As already mentioned, it is an extreme case.

5.4. T$_2$O$_5$ AND TO$_3$ OXIDES

Group IVA metals do not exceed valency 4, and VA metals give oxides with valency 5 or 4. On the other hand, a mixture of valencies 6 and 5 produces a complex situation for VIA metals, as Table 5.1 shows [5.181]. Oxides (magnetic and perhaps semiconducting) probably even exist between CrO$_2$ and CrO$_3$ [5.182, 5.183]. The situation becomes straightforward again for VIIA metals, where only valencies 7 (Tc$_2$O$_7$, Re$_2$O$_7$) and 6 (ReO$_3$, DO$_9$ cubic structure) are observed. In all these crystals (except CrO$_3$), a distorted O$_6$ octahedron may be considered to be round the T atom. In Fig. 5.28, these octahedra are represented by their projections on a plane perpendicular to the quaternary axis, and are assumed to form chains along this axis. Their type of arrangement then makes it possible to visualize the structures of ReO$_3$, MoO$_3$, Mo$_n$O$_{3n-1}$, and V$_2$O$_5$. In this last case, the 6 V—O distances are 1.54, 1.77, 1.88, 1.88, 2.02, and 2.81 Å, and only 5 neighbors are generally considered, so that it is difficult to discuss the AOs used, although several bonds have a covalent character (the sum of the ionic radii is 1.99 Å).

5.4. T_2O_5 and TO_3 Oxides

TABLE 5.1

CrO_3	α-MoO_3 pale yellow (orthorhombic layer structure)	α-WO_3 pale yellow ($DO_{10} \sim DO_9$ structure)
	$\left.\begin{array}{l}\beta'\text{-}Mo_9O_{26}\\ \beta\text{-}Mo_8O_{23}\end{array}\right\}$ blue-violet	β-$W_{20}O_{58}$ blue ($\sim DO_9$)
($CrO_{2.6}$) ($CrO_{2.38-2.48}$) CrO_2 (C4)	γ-Mo_4O_{11} violet-red δ-MoO_2 (C4)	γ-$W_{18}O_{49}$(W_4O_{11}) violet-red δ-WO_2 (C4)

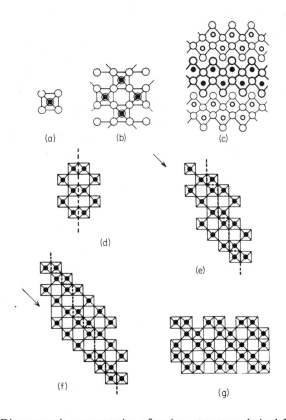

Fig. 5.28. Diagrammatic representation of various structures derived from DO_9. The O_6 octahedra chains with their opposite corners in common are represented by their projection on a perpendicular plane (a). They can have common corners, as in ReO_3 (DO_9) (b) or common edges as in MoO_3 (c). Structures derived from the latter, by slip, correspond to compositions T_nO_{3n} (d), T_nO_{3n-1} (e), and T_nO_{3n-2} (f). The very idealized structure of V_2O_5 is given in (g) (based on Wells [5.181]).

V_2O_5 (monoclinic), α-Nb_2O_5, and Ta_2O_5 (orthorhombic) are semiconducting [5.8]. Boros [5.184] has confirmed it for V_2O_5 on monocrystals, mentioning that conductivity is independent of atmosphere, and finding an infrared absorption threshold that corresponds to a forbidden band gap of 0.4 eV. The conductivity variation in relation to temperature is given by Simard et al. [5.185], Clark and Berets [5.186], Kachi et al. [5.31] (Fig. 5.19), Patrina and Ioffe [5.187] and, more precisely, for a crystal with a high degree of purity refined by a melted zone, by Haemers [5.188], who has found $E_G = 0.42$ eV (0.21 eV below 150°K) with a melting point at 690°C. Its effect on the oxidation catalysis of SO_2 into SO_3 has been studied by Butler [5.189] and Surnev and Bliznakov [5.190]. Butler mentions the effects of the atmosphere at 400°C. Properties in the liquid state have been investigated by Yurkov [5.191], Manakov et al. [5.192] and Zolyan and Regel [5.193]. Manakov et al. have shown that the semiconducting character remains, but changes to p-type (instead of n in the solid state). They give mobilities at 1000°C as 33 (electrons) and 189 (holes) $cm^2 V^{-1}sec^{-1}$. Zolyan and Regel mention sign changes in the Seebeck effect. Ioffe and Patrina [5.194], after measuring the EPR, have discussed the interpretation of the Boros spectrum, and suggested a transfer mechanism between V(IV) and V(V), with $\mu \sim 10^{-3}$ $cm^2 V^{-1}sec^{-1}$. Allersma et al. [5.195] have found an activation energy of 0.2 eV at room temperature, and mentioned the very high resistivity of vitreous V_2O_5. In conclusion, it may be said that in reduced samples an excitation mechanism across a forbidden band gap of 0.42 eV is superimposed on a transfer mechanism, with an activation energy of 0.2 eV. In view of the low symmetry of the vanadium environment and the resulting advanced division of the d level, these transfers probably take place in an elementary sublevel that is empty for V(V) (d^0) and full for V(IV) (d^1).

Nb_2O_5 is diamagnetic and melts at 1500°C. The negative temperature coefficient of its resistivity was applied very early on to produce thermistors. It presents several allotropic varieties [5.8]. Greener et al. [5.196], working with α-Nb_2O_5 monocrystals between 300 and 900°C, have found an activation energy of 1.63 eV, a carrier scattering by lattice vibrations and a conductivity variation in relation to the oxygen pressure in $P^{-1/4}$ (O vacancies, n-type). Manakov et al. [5.192] have mentioned a mobility of 22 $cm^2 V^{-1}sec^{-1}$ for electrons at 1200°C, and have shown that the semiconductor character survives in the liquid state. Valetta [5.197] has studied the valency induction of tungsten in $(Nb_{1-x}W_x)_2O_5$ ($0.0025 < x < 0.15$). Greener and Hirthe [5.198], working with monocrystals of reduced α-Nb_2O_5, have found energies of 0.2 and 0.9 eV and, with very reduced sintered samples, energies that are almost nil. They have suggested an excitation mechanism from an impurity level (O vacancies) to a d band. Kofstad [5.199] has suggested, instead, a law in $P^{-1/6}$. Janninck and Whitmore [5.200] have studied α-Nb_2O_{5-x} ($0.001 < x$

5.4. T_2O_5 and TO_3 Oxides

< 0.137) and confirmed the existence of a narrow d band. Greener *et al.* [5.201] have confirmed their results. Janninck and Whitmore [5.202] have measured the Seebeck effect and mentioned an effective mass of $4m$. Yahia [5.93] has measured the mobility, using a special method. The insulating oxide Ta_2O_5, isomorphous with α-Nb_2O_5, melts at 1800°C and becomes semiconducting by reduction (O vacancies) or valency induction. It is diamagnetic. Its dielectric, photoelectric and thermoelectric properties were studied before the war [5.8] (see also Nagasawa [5.203]). Later work, done entirely on films (see p. 154), involved the Ta–Ta_2O_5 rectifier effect and the dielectric properties of anodic oxide deposits. The absorption threshold is said to correspond to a 4.6 eV forbidden band gap [5.204].

Before dealing with TO_3 oxides, a few words should be said about ferroelectric phenomena, already alluded to in connection with β-MnO_2, and a more lengthy account of which is necessary in connection with WO_3. Ten classes of crystal symmetry possess at least one polar axis, and present in this direction a spontaneous polarization as function of temperature (pyroelectricity). In certain compounds, the pyroelectric structure is unstable: the spontaneous polarization can then be reversed by applying a sufficient opposing electrical field (*ferroelectric* phase) and a structure of higher symmetry replaces it when the temperature rises (*paraelectric* phase). The reversing of the polarization P by the effect of an alternating field E is accompanied by an hysteresis cycle, which disappears at the phase change (ferroelectric "Curie point") (cf. Fig. 5.24). The superficial analogy with ferromagnetism explains the terminology for the phenomenon observed in two classes of materials: "soft" materials, where the dipoles appearing in groups like SO_4^{2-} are aligned by the cooperative effect of ordered hydrogen bonds, and "hard" materials, where the dipoles appearing in the O_6 octahedra (displacement of the central T atom from its equilibrium position) are coupled by the T—O bonds. Only the second class is of interest to us here. The T atom (Ti, V, Nb, Ta, etc.) should theoretically have an empty d level, according to an empirical rule worked out by Smolenskii and confirmed for WO_3, but not for β-MnO_2. For further information readers may refer to the classic work by Jona and Shirane [5.205].

The straightforward typical structure of TO_3 oxides is the DO_9 cubic structure of ReO_3, represented by Fig. 5.29a. Chains of O_6 octahedra with their opposite corners in common, exist along the quaternary axes. If metallic M atoms occupy all the cubes thus represented, there is a change to the $E2_1$ structure of perovskite $CaTiO_3$, in which a large number of MTO_3 ternary compounds crystallize, including many ferroelectrics (Fig. 5.29b). The DO_9 structure and the DO_{10} structure of WO_3, which is very close to DO_9 and, in particular, involves the same type of O_6 octahedron arrangement, can therefore be looked on as perovskites with vacancies, distorted in the case of □WO_3. If the M sites are only partly occupied, one obtains intermediate structures between DO_9 and $E2_1$, those of "tungsten bronzes" (see p. 154). Finally, the layer structure of MoO_3 is more or less derived from DO_9, as well as those of the β- and γ-phases of the Mo–O, W–O, and (Mo, W)–O systems, in which one finds blocks of the original structure, rather as in the Ti_nO_{2n-1} series (Fig. 5.28).

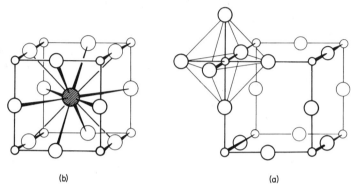

Fig. 5.29. DO_9 structure of ReO_3 (a) and $E2_1$ structure of perowskite $CaTiO_3$ (b) (based on Wells [5.181]).

MoO_3 has a susceptibility of 0.02×10^{-6}, and melts at 795°C. Its very low conductivity (6.10^{-13} mhos cm^{-1}) increases in the molten state [5.8]. Stähelin and Busch [5.206] have pointed out that it becomes semiconducting through reduction (0.5 eV) (cf. also [5.174]). Work with thin layers (see p. 154) has given activation energies of 0.56 and 1.83 eV (intrinsic) [5.207]. WO_3 has a susceptibility of 0.2×10^{-6}, and melts at 1470°C. It takes on a bluish tint by reduction, from $WO_{2.98}$ on. It has a high dielectric constant, and reaches a maximum at room temperature (10^5) [5.8]. It undergoes several phase changes, quasi orthorhombic or triclinic at room temperature, pseudomonoclinic, then orthorhombic up to 740°C and quadratic beyond this [5.208–5.210]. Magascuva [5.211], and then Okada et al. [5.212] and Matthias [5.213], have mentioned its ferroelectricity (later disputed [5.214]). Resistivity presents irregularities at phase changes [5.215, 5.216]. In a comprehensive study, Sawada [5.217] has indicated several (intrinsic) activation energies, notably 2.2 eV below 740°C, confirmed by the optical absorption threshold. The Seebeck effect is very slight. Later research by the same author [5.218] gives 10^{18} holes cm^{-3}, with a mobility of 36 cm^2 V^{-1}sec^{-1}, but casts doubt on the resistivity curves. Choain-Maurin and Marion [5.176] have observed that the resistance varies between 600 and 900°C in relation to the oxygen pressure and temperature. The α phase has (extrinsic) activation energies of 1 and 1.3 eV. β and γ are said to be metallic (Fig. 5.30). Crowder and Sienko [5.219] have measured the Hall and Seebeck effects of the γ phase and found $\mu \sim 15$ cm^2 V^{-1}sec^{-1} and

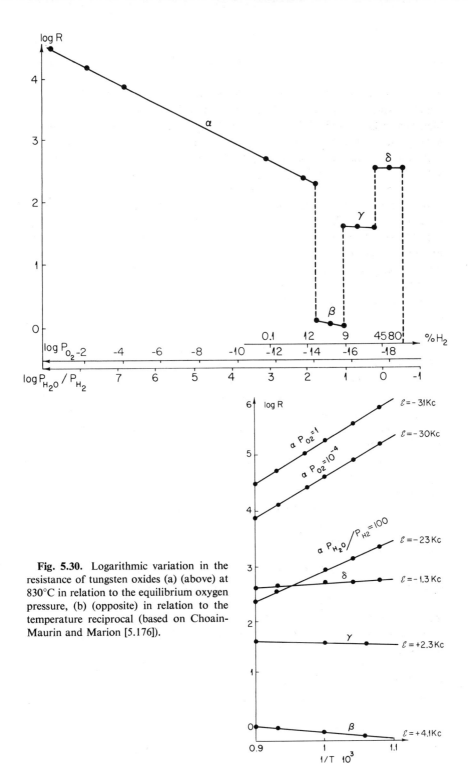

Fig. 5.30. Logarithmic variation in the resistance of tungsten oxides (a) (above) at 830°C in relation to the equilibrium oxygen pressure, (b) (opposite) in relation to the temperature reciprocal (based on Choain-Maurin and Marion [5.176]).

$m^* \sim 1.25m$ with a donor impurity level (interstitial W, 0.04 eV). ReO_3 is metallic, and Feretti et al. [5.220] have given an MO diagram of energy levels. Sleight and Gillson [5.221] have studied WO_3–ReO_3 solutions. In conclusion, MoO_3 and WO_3 (d^0) are, like TiO_2, semiconductors by excitation, but their reduction causes atoms of valency 5 (d^1) and a transfer mechanism in a d sublevel to appear. For ReO_3, there is a partly filled t_{2g}^1 subband.

Oxide bronzes may be defined as crystals in which the insertion of interstitial metals does not basically alter the structure and interatomic bonds. Chemists are familiar with M_xTiO_2 (M = Na, K), $M_xV_2O_5$ (M = Li, Na, Cu, Ag) and particularly M_xWO_3 (M = Li, Na, Cu, Ag, Tl), which were the first ones to be studied. They have in common their color, low resistivity, and homogeneity domain [5.2]. Ornatskaya [5.222] has mentioned the semiconductor character of vanadium bronze monocrystals. The (extrinsic) activation energies are 0.1 eV up to 350–400°C, and then 0.3 eV; $N \sim 10^{19}$ cm^{-3}, $\mu \sim 0.5$ cm^2 V^{-1}sec^{-1}, $m^* \sim 0.4m$, $\alpha \sim 100$ μV deg^{-1}. Hardy et al. [5.223] have found, for $Li_xV_2O_5$, energies of 0.14 ($0.22 < x < 0.62$, β phase) and 0.86 eV ($0.88 < x < 1$, γ phase). Hagenmuller et al. [5.224] have mentioned $2S = 0.9$ for V(IV) (M = Li, Ag). Ozerov's digest [5.225] summarizes early work on tungsten bronzes, which are all metallic. Conroy and Sienko [5.226], Sienko [5.227] and Sienko and Mazumder [5.228] have studied the insertion of Ag and Tl, suggesting that the electrons occupied the 5d band of tungsten. Ellerbeck et al. [5.229] and Shanks et al. [5.230] have studied the insertion of Na, and shown that conductivity increases rapidly as its content rises. Crowder and Sienko [5.219] have attributed to M the same role as interstitial W in γ-WO_3. Mackintosh [5.231], and then Fuchs [5.232], have given their interpretations. It is suggested that alkaline bronzes are supraconducting [5.233]. Sleight and Gillson [5.221] have mentioned rhenium bronzes (M = Na, K).

To end this chapter, one might mention the work done on thin layers: TiO_2 reduced on glass [5.234], anodic oxidation of Ti and Nb, photoelectric effect [5.235], anodic oxidation of Nb, reflectitivity and dielectric constant [5.236], ditto with counterelectrode, p–n photocell [5.237], ditto, contacts [5.238], Ta_2O_5 by anodic oxidation, optical absorption and photoconduction by Apker and Taft [5.204], ditto with Au counterelectrode, tunnel effect, field ionization by Mead [5.239], ditto, electron bombardment effect [5.240], ditto with MnO_2 counterelectrode, p–i–n junction [5.241], ditto [5.242], ditto, 0.6 eV by Smyth et al. [5.243], ditto, effect of humidity [5.244], ditto, conductivity [5.245], ditto a.c. dielectric properties [5.246], MoO_3, optical absorption, dielectric constant, resistivity, irradiation by Deb [5.207].

In this chapter we have met with diamagnetic semiconductors like TiO_2, V_2O_5, WO_3 (d^0). For the others, it has frequently been seen that the formation of T—T′ bond pairs eliminated any magnetic moment by balancing the spins of the two bachelor d electrons (Ti_2O_3, V_2O_3, and VO_2 low temperature). When these pairs do not form, there is generally metallic behavior (V_2O_3 and VO_2 high temperature, CrO_2, ReO_2, ReO_3) or antiferromagnetic semiconductor behavior if the $t_{2g}\alpha$ sublevel is full (Cr_2O_3, β-MnO_2). $Mn_xCr_{1-x}O_2$ solutions, however, prove that one can escape this dilemma by means of complex magnetic structures in mixed or ternary oxides.

REFERENCES

[5.1] F. J. Morin, "Semiconductors" (N. B. Hannay, ed.), p. 600. Academic Press, London and New York, 1963.
[5.2] A. D. Wadsley, "Non-stoichiometric Compounds" (L. Mandelcorn, ed.), p. 98. Academic Press, New York, 1964.
[5.3] A. D. Pearson, *J. Phys. Chem. Solids* **5**, 316 (1958).
[5.4] S. M. Ariya and N. I. Bogdanova, *Fiz. Tverd. Tela* **1**, 1022 (1959).
[5.5] A. A. Samokhvalov and A. G. Rustamov, *Fiz. Tverd. Tela* **5**, 1202 (1963).
[5.6] S. P. Denker, *J. Appl. Phys.* **37**, 142 (1966).
[5.7] S. N. L'vov, V. F. Nemchenko, G. V. Samsonov, and T. S. Verkhoglyadova, *Ukr. Fiz. Zh.* **8**, 1372 (1963).
[5.8] P. Pascal, "Nouveau traité de chimie minérale." Masson, Paris, 1956–1963.
[5.9] F. J. Morin, *Phys. Rev. Lett.* **3**, 34 (1959).
[5.10] S. Kawano, K. Kosuge, and S. Kachi, *J. Phys. Soc. Japan* **21**, 2744 (1966).
[5.11] S. P. Denker, *J. Phys. Chem. Solids* **25**, 1397 (1964).
[5.12] C. G. Shull, W. A. Strauser, and E. O. Wollan, *Phys. Rev.* **83**, 333 (1951).
[5.13] R. C. Miller and R. R. Heikes, *J. Chem. Phys.* **28**, 348 (1958).
[5.14] E. J. W. Verwey and J. H. De Boer, *Rec. Trav. Chim. Pays Bas* **55**, 541 (1936).
[5.15] V. G. Bhide and R. H. Dani, *Physica* **27**, 821 (1961).
[5.16] M. Rosenberg, P. Nicolau, and I. Bunget, *Phys. Status Solidi* **14**, K65 (1966).
[5.17] H. Krebs, "Grundzüge der anorganischen Kristallchemie." Enke, Stuttgart, 1968 (*Engl. transl.:* McGraw-Hill, Maidenhead, 1968).
[5.18] M. Foëx and J. Loriers, *C.R. Acad. Sci.* (*Paris*) **226**, 901 (1948).
[5.19] B. F. Naylor, *J. Amer. Chem. Soc.* **68**, 1077 (1948).
[5.20] S. F. Adler and P. W. Selwood, *J. Amer. Chem. Soc.* **76**, 346 (1954).
[5.21] M. Foëx and J. Wucher, *C.R. Acad. Sci.* (*Paris*) **241**, 184 (1955).
[5.22] J. Yahia and H. P. R. Frederikse, *Phys. Rev.* **123**, 1257 (1961).
[5.23] T. Kawakubo, T. Yanagi, and S. Nomura, *J. Phys. Soc. Japan* **15**, 2102 (1960).
[5.24] B. I. Al'shin and D. N. Astrov, *Zh. Eksperim. Teor. Fiz.* **44**, 1195 (1963).
[5.25] S. C. Abrahams, *Phys. Rev.* **130**, 2230 (1963).
[5.26] L. K. Keys and L. N. Mulay, *Appl. Phys. Lett.* **9**, 248 (1966).
[5.27] M. Foëx, *C.R. Acad. Sci.* (*Paris*) **223**, 1126 (1946).
[5.28] M. Foëx, S. Goldsztaub, R. Wey, J. Jaffray, R. Lyand, and J. Wucher, *J. Rech. Cent. Nat. Rech. Sci. Bellevue* [*Paris*] **4**, 237 (1952).
[5.29] S. Teranishi and K. Tarama, *J. Chem. Phys.* **27**, 1217 (1957).
[5.30] G. Goodman, *Phys. Lett.* **9**, 305 (1962).
[5.31] S. Kachi, T. Takada, and K. Kosuge, *J. Phys. Soc. Japan* **18**, 1839 (1963).
[5.32] G. A. Acket and J. Volger, *Physica*, **28**, 277 (1962).
[5.33] I. G. Austin, *Phil. Mag.* **7**, 961 (1962).
[5.34] S. Minomura and H. Nagasaki, *J. Phys. Soc. Japan* **19**, 131 (1964).
[5.35] T. Shinjo, K. Kosuge, S. Kachi, H. Takaki, M. Shiga, and Y. Nakamura, *J. Phys. Soc. Japan* **21**, 193 (1966).

[5.36] T. Shinjo and K. Kosuge, *J. Phys. Soc. Japan* **21**, 2622 (1966).
[5.37] D. J. M. Bevan, J. P. Shelton and J. S. Anderson, *J. Chem. Soc.* 1729 (1948).
[5.38] K. Hauffe and J. Block, *Z. Phys. Chem.* **198**, 232 (1951).
[5.39] B. N. Brockhouse, *J. Chem. Phys.* **21**, 961 (1953).
[5.40] J. Roche and J. Jaffray, *C.R. Acad. Sci. (Paris)* **240**, 2212 (1955).
[5.41] D. N. Astrov, *Zh. Eksperim. Teor. Fiz.* **38**, 984 (1960); **40**, 1035 (1961).
[5.42] S. Foner and M. Hanabusa, *J. Appl. Phys.* **34**, 1246 (1963).
[5.43] G. Schultz and M. Bober, "Zur Physik u. Chemie der Kristallphosphore" (H. Ortmann and H. Witzmann, eds.). Akad. Verlag, Berlin, 1960.
[5.44] W. A. Fischer and H. Dietrich, *Z. Phys. Chem.* **41**, 205 (1964).
[5.45] J. A. Crawford and R. W. Vest, *J. Appl. Phys.* **35**, 2413 (1964).
[5.46] J. F. G. De La Banda, J. A. Pajares Somoano, and J. Soria Ruiz, *An. Real. Soc. Espan. Fis. Quim.* **61 (A)**, 197, 311 (1965); **62 (A)**, 311 (1966).
[5.47] W. C. Hagel, *J. Appl. Phys.* **36**, 2586 (1965).
[5.48] J. B. Goodenough, *Phys. Rev.* **117**, 1442 (1960).
[5.49] D. Adler and J. Feinleib, *Phys. Rev. Lett.* **12**, 700 (1964).
[5.50] J. Feinleib and W. Paul, *Proc. Int. Conf. Solids High Press. 1st, Tucson, 1965* Academic Press, New York, 1965.
[5.51] G. J. Hyland and A. W. B. Taylor, *J. Phys. Soc. Japan* **21**, 819 (1966).
[5.52] F. A. Grant, *Rev. Mod. Phys.* **31**, 646 (1959).
[5.53] R. G. Breckenridge and W. R. Hosler, *Phys. Rev.* **91**, 793 (1953).
[5.54] D. C. Cronemeyer and M. A. Gilleo, *Phys. Rev.* **82**, 975 (1951).
[5.55] D. C. Cronemeyer, *Phys. Rev.* **87**, 876 (1952).
[5.56] R. A. Parker and J. H. Wasilik, *Phys. Rev.* **120**, 1631 (1960).
[5.57] B. H. Soffer, *J. Chem. Phys.* **35**, 940 (1961).
[5.58] A. Von Hippel, J. Kalnajs, and W. B. Westphal, *J. Phys. Chem. Solids* **23**, 779 (1962).
[5.59] K. G. Srivastava, *Phys. Rev.* **119**, 520 (1960).
[5.60] E. H. Greener and D. H. Whitmore, *J. Appl. Phys.* **32**, 1320 (1961).
[5.61] F. Cardon, *Verh. Kon. Vlaam. Akad. Wetensch. Lett. Schone Kunsten. Belg. Kl. Wetensch.* **69** (1962).
[5.62] J. A. Van Raalte, *J. Appl. Phys.* **36**, 3365 (1965).
[5.63] S. Anderson, B. Collen, U. Kuylenstierna, and A. Magneli, *Acta Chem. Scand.* **11**, 1641 (1957).
[5.64] S. Anderson, B. Collen, G. Krusse, U. Kuylenstierna, A. Magneli, H. Pestmalis, and S. Asbrink, *Acta Chem. Scand.* **11**, 1653 (1957).
[5.65] S. Anderson, *Acta Chem. Scand.* **14**, 1161 (1960).
[5.66] T. Hurlen, *Acta Chem. Scand.* **13**, 365 (1959).
[5.67] H. P. R. Frederikse, *J. Appl. Phys. suppl.* **32**, 2211 (1961).
[5.68] D. S. Tannhauser, *Solid State Commun.* **1**, 223 (1963).
[5.69] M. D. Earle, *Phys. Rev.* **61**, 56 (1942).
[5.70] B. I. Boltaks, F. I. Vasenin, and A. E. Salunina, *Zh. Tekh. Fiz.* **21**, 532 (1951).
[5.71] S. I. Gorelik, *Zh. Eksperim. Teor. Fiz.* **21**, 826 (1951).
[5.72] K. Hauffe, "Reaktionen in und an festen Stoffen," p. 136. Springer, Berlin, 1955.
[5.73] H. K. Henisch, Electronic Eng. (GB) 320 (Oct. 1946).
[5.74] J. P. Suchet, B.F.1.053.235 (1952); U.S. Patent 2.714.096 (1955).
[5.75] S. Kataoka and T. Suzuki, *Bull. Electrotech. Lab. Tokyo* **18**, 732 (1954).
[5.76] W. Kleber, H. Peibst and W. Schröder, *Z. Phys. Chem.* **215**, 63 (1960).
[5.77] L. E. Hollander, T. J. Diesel, and G. L. Vick, *Phys. Rev.* **117**, 1469 (1960).
[5.78] L. E. Hollander and P. L. Castro, *Phys. Rev.* **119**, 1882 (1960).
[5.79] H. P. R. Frederikse, W. R. Hosler and J. H. Becker, *Proc. Int. Conf. Semicond. Phys. Prague 1960* p. 868. Czech J. Phys. (1961).

[5.80] K. Sakata, *J. Phys. Soc. Japan* **16**, 1026 (1961).
[5.81] R. R. Hasiguti, K. Minami, and H. Yonemitsu, *J. Phys. Soc. Japan* **16**, 2223 (1961); *Trans. Nat. Res. Inst. Metals (Japan)* **4**, 155 (1962).
[5.82] R. R. Hasiguti, N. Kawamiya, and E. Yagi, *J. Phys. Soc. Japan* **19**, 573 (1964).
[5.83] L. E. Hollander and P. L. Castro, *J. Appl. Phys.* **33**, 3421 (1962).
[5.84] G. A. Acket and J. Volger, *Physica* **29**, 225 (1963).
[5.85] J. Yahia, *Phys. Rev.* **130**, 1711 (1963).
[5.86] R. Keezer, *J. Appl. Phys.* **35**, 1866 (1964).
[5.87] R. Keezer, J. Mudar, and D. E. Brown, *J. Appl. Phys.* **35**, 1868 (1964).
[5.88] E. H. Greener, F. J. Barone, and W. M. Hirthe, *J. Amer. Ceramic Soc.* **48**, 623 (1965).
[5.89] R. N. Blumenthal, J. Coburn, J. Baukus, and W. M. Hirthe, *J. Phys. Chem. Solids* **27**, 643 (1966).
[5.90] V. N. Bogomolov and P. M. Shavkunov, *Fiz. Tverd. Tela* **5**, 2027 (1963).
[5.91] V. N. Bogomolov and V. P. Zhuze, *Fiz. Tverd. Tela* **5**, 3285 (1963).
[5.92] G. A. Acket and J. Volger, Phys. Letters 8, 244 (1963).
[5.93] J. Yahia, *J. Phys. Chem. Solids* **25**, 881 (1964).
[5.94] J. H. Becker and W. R. Hosler, *Phys. Rev.* **137A**, 1872 (1965).
[5.95] G. A. Acket, Thesis, Rijksuniversiteit, Utrecht (1965).
[5.96] G. Perny and R. Lorang, *J. Chim. Phys. (Fr.)* **63**, 833 (1966).
[5.97] G. A. Acket and J. Volger, *Physica* **32**, 1680 (1966).
[5.98] J. Yahia, *Phys. Lett.* **23**, 425 (1966).
[5.99] V. N. Bogomolov and V. P. Zhuze, *Fiz. Tverd. Tela* **8**, 2390 (1966).
[5.100] Ya. M. Ksendzov, *Zh. Tekh. Fiz.* **20**, 117 (1950).
[5.101] M. G. Harwood, *Brit. J. Appl. Phys.* **16**, 1493 (1965).
[5.102] H. Grunewald, *Ann. Phys. (Leipzig)* **14**, 121 (1954).
[5.103] R. J. J. Zijlstra, F. J. Leeuwerik, and T. G. M. Kleinpenning, *Phys. Lett.* **23**, 185 (1966).
[5.104] R. Cauville, *C.R. Acad. Sci. (Paris)* **229**, 1228 (1949).
[5.105] R. Cauville, State thesis, Paris (1952).
[5.106] Ya. M. Ksendzov, *Izv. Akad. Nauk SSSR, Ser. Fiz.* **22**, 237 (1958).
[5.107] O. Ishikura and T. Sakata, *Jap. J. Appl. Phys.* **3**, 498 (1964).
[5.108] A. K. Ivukina and Ya. I. Panova, *Fiz. Tverd. Tela* **6**, 2857 (1964).
[5.109] G. P. Bogoroditskii, V. Kristya, and Ya. I. Panova, *Fiz. Tverd. Tela* **9**, 253 (1967).
[5.110] O. W. Johnson, *Phys. Rev.* **136**, A284 (1964).
[5.111] R. W. G. Wyckoff, "Crystal Structures," 2nd ed., Vol. 1. Wiley (Interscience), New York, 1963.
[5.112] N. Perakis and J. Wucher, *C.R. Acad. Sci. (Paris)* **235**, 354 (1952).
[5.113] J. Jaffray and D. Dumas, *J. Rech. Cent. Nat. Rech. Sci. Bellevue [Paris]* **27**, 360 (1954).
[5.114] K. Kosuge, T. Takada, and S. Kachi, *J. Phys. Soc. Japan* **18**, 318 (1963).
[5.115] C. H. Neuman, A. W. Lawson, and R. F. Brown, *J. Chem. Phys.* **41**, 1591 (1964).
[5.116] K. Kosuge, S. Kachi, H. Nagasaki, and S. Minomura, *J. Phys. Soc. Japan* **20**, 178 (1965).
[5.117] T. Kawakubo, *J. Phys. Soc. Japan* **20**, 516 (1965).
[5.118] P. F. Bongers, *Solid State Commun.* **3**, 275 (1965).
[5.119] G. J. Hyland and A. W. B. Taylor, *J. Phys. Soc. Japan* **21**, 819 (1966).
[5.120] J. Umeda, S. Ashida, H. Kusumoto, and K. Narita, *J. Phys. Soc. Japan* **21**, 1461 (1966).
[5.121] T. Ohashi and A. Watanabe, *J. Amer. Ceramic Soc.* **49**, 519 (1966).
[5.122] I. Kitahiro, T. Ohashi, and A. Watanabe, *J. Phys. Soc. Japan* **21**, 2422 (1966).

[5.123] I. Kitahiro and A. Watanabe, *J. Phys. Soc. Japan* **21**, 2423 (1966).
[5.124] A. S. Barker Jr., H. W. Verleur, and H. J. Guggenheim, *Phys. Rev. Lett.* **17**, 1286 (1966).
[5.125] K. Kosuge, *J. Phys. Soc. Japan* **22**, 551 (1967).
[5.126] C. Guillaud, A. Michel, J. Benard, and M. Fallot, *C.R. Acad. Sci. (Paris)* **219**, 58 (1944).
[5.127] F. J. Darnell and W. H. Cloud, Coll. Int. Comp. Ox. Elem. Trans. 126 (Bordeaux 1964), C.N.R.S., Paris (1965).
[5.128] F. Hulliger, *Helv. Phys. Acta* **32**, 615 (1959).
[5.129] B. Kubota and E. Hirota, *J. Phys. Soc. Japan* **16**, 345 (1961).
[5.130] K. Siratori and S. Iida, *J. Phys. Soc. Japan* **15**, 2362 (1960).
[5.131] T. J. Swoboda, P. Arthur Jr., N. L. Cox, J. N. Ingraham, A. L. Oppegard, and M. S. Sadler *J. Appl. Phys. suppl.* **32**, 374 (1961).
[5.132] D. S. Chapin, J. A. Kafalas, and J. M. Honig, *J. Phys. Chem.* **69**, 402 (1965).
[5.133] R. Druilhe and J. Bonnerot, *C.R. Acad. Sci. (Paris)* **263B**, 55 (1966).
[5.134] R. M. Chrenko and D. S. Rodbell, *Phys. Lett.* **A24**, 211 (1967).
[5.135] D. S. Rodbell, J. M. Lommel, and R. C. De Vries, *J. Phys. Soc. Japan* **21**, 2430 (1966).
[5.136] R. Druilhe and J. P. Suchet, *Czech. J. Phys.* **B17**, 337 (1967).
[5.137] J. Brenet, *Bull. Soc. Fr. Minéral. Cristallogr.* **73**, 409 (1950); **77**, 797 (1954).
[5.138] H. Bizette and B. Tsaï, Coll. Int. Polarisat. Matière 164, C.N.R.S., Paris (1949).
[5.139] R. A. Erickson, *Phys. Rev.* **85**, 745 (1952).
[5.140] R. A. Erickson, quoted by Yoshimori [5.141].
[5.141] A. Yoshimori, *J. Phys. Soc. Japan* **14**, 807 (1959).
[5.142] J. Brenet and A. Heraud, *C.R. Acad. Sci. (Paris)* **226**, 413 (1948).
[5.143] J. Amiel, J. Brenet, and G. Rodier, *C.R. Acad. Sci. (Paris)* **227**, 60 (1948).
[5.144] H. Bizette, *J. Phys. Radium* **12**, 161 (1951).
[5.145] J. N. Das, *Z. Phys.* **151**, 345 (1958).
[5.146] J. P. Chevillot and J. Brenet, *C.R. Acad. Sci. (Paris)* **248**, 776 (1959).
[5.147] V. G. Bhide, R. V. Damle, and R. H. Dani, *Physica* **25**, 579 (1959).
[5.148] V. G. Bhide and R. V. Damle, *Physica* **26**, 33, 513 (1960).
[5.149] V. G. Bhide and M. S. Multani, *J. Sci. Ind. Res.* **19B**, 312 (1960).
[5.150] Y. L. Yousef and B. S. Farag, *Physica* **31**, 706 (1965).
[5.151] J. S. Wiley and H. T. Knight, *J. Electrochem. Soc.* **111**, 656 (1964).
[5.152] H. Zeilmaker and C. Drotschmann, *Rec. Trav. Chim. Pays-Bas* **86**, 545 (1966).
[5.153] R. Druilhe, *C.R. Acad. Sci. (Paris)* **263**, 653 (1966).
[5.154] J. P. Chevillot and J. Brenet, *C.R. Acad. Sci. (Paris)* **249**, 1869 (1959).
[5.155] P. Gibart, State thesis, Strasbourg (1966).
[5.156] K. Siratori and S. Iida, *J. Phys. Soc. Japan* **15**, 210 (1960).
[5.157] G. Villers and R. Druilhe, *C.R. Acad. Sci. (Paris)* **264**, 843 (1967).
[5.158] R. Druilhe, Mémoire Conserv. Nat. Arts Métiers, Paris (1967).
[5.159] J. M. Dixon, L. D. La Grange, U. Merten, C. F. Miller, and J. T. Porter II, *J. Electrochem. Soc.* **110**, 276 (1963).
[5.160] E. D. Whitney, *J. Electrochem. Soc.* **112**, 91 (1965).
[5.161] A.-M. Anthony, *C.R. Acad. Sci. (Paris)* **256**, 5130 (1963).
[5.162] A.-M. Anthony and J. Renon, *C.R. Acad. Sci. (Paris)* **256**, 1718 (1963).
[5.163] A.-M. Anthony, F. Cabannes, and J. Renon, *Ann. Phys. (Paris)* **9**, 1 (1964).
[5.164] R. W. Vest, N. M. Tallan, and W. C. Tripp, *J. Amer. Ceramic Soc.* **47**, 635 (1964).
[5.165] R. W. Vest and N. M. Tallan, *J. Amer. Ceramic Soc.* **48**, 472 (1965).
[5.166] N. M. Tallan, R. W. Vest, and H. C. Graham, *Mater. Sci. Res.* **2**, 33 (1965).
[5.167] F. A. Kröger, *J. Amer. Ceramic Soc.* **49**, 215 (1966).

References

[5.168] A.-M. Anthony, A. Guillot, and P. Nicolau, *C.R. Acad. Sci. (Paris)* **262B**, 896 (1966).
[5.169] C. E. Curtiss, L. M. Doney, and J. R. Johnson, *J. Amer. Ceramic Soc.* **37**, 458 (1954).
[5.170] E. H. Greener, Ph.D. thesis, Northwestern Univ., Evanston, Illinois (1960).
[5.171] P. O. Kolchin and N. V. Sumarokova, *At. Energ.* **10**, 168 (1961).
[5.172] R. F. Janninck and D. H. Whitmore, *J. Phys. Chem. Solids* **27**, 1183 (1966).
[5.173] R. H. Griffith, P. R. Chapman, and P. R. Lindars, *Discuss. Faraday Soc.* 258 (1950).
[5.174] C. Pettus, *Proc. IEEE* **53**, 98 (1965).
[5.175] R. C. Vickery and J. C. Hipp, *J. Appl. Phys.* **37**, 2926 (1966).
[5.176] C. Choain-Maurin and F. Marion, *C.R. Acad. Sci. (Paris)* **259**, 4700 (1964).
[5.177] C. M. Nelson, G. E. Boyd, and W. T. Smith, *J. Amer. Chem. Soc.* **76**, 348 (1954).
[5.178] P. Gibart, *C.R. Acad. Sci. (Paris)* **259**, 4237 (1964); **261**, 1525 (1965).
[5.179] J. B. Goodenough, P. Gibart, and J. Brenet, *C.R. Acad. Sci. (Paris)* **261**, 2331 (1965).
[5.180] J. B. Goodenough, Coll. Int. Comp. Ox. Elém. Trans. 162 (Bordeaux 1964), C.N.R.S., Paris (1965).
[5.181] A. F. Wells, "Structural Inorganic Chemistry," 3rd ed. Oxford Univ. Press (Clarendon), London and New York, 1962.
[5.182] S. M. Ariya, S. A. Chukarev, and V. B. Glushkova, *Zh. Obshch. Khim. SSSR* **23**, 1241 (1953).
[5.183] O. Glemser, U. Hauschild, and F. Trüpel, *Z. Anorg. Chem.* **277**, 113 (1954).
[5.184] J. Boros, *Z. Phys.* **126**, 721 (1949).
[5.185] G. Simard, J. Steger, R. Arnott, and L. Siegel, *Ind. Eng. Chem.* **47**, 1424 (1955).
[5.186] H. Clark and J. Berets, *J. Chem. Phys.* **34**, 204 (1960).
[5.187] I. B. Patrina and V. A. Ioffe, *Fiz. Tverd. Tela* **6**, 3227 (1964).
[5.188] J. Haemers, *C.R. Acad. Sci. (Paris)* **259**, 3740 (1964).
[5.189] J. D. Butler, *Trans. Faraday Soc.* **56**, 1842 (1960).
[5.190] L. Surney and G. B. Bliznakov, *C.R. Acad. Bulg. Sci.* **17**, 1107 (1964).
[5.191] V. A. Yurkov, *Zh. Eksp. teor. Fiz.* **22**, 223 (1952).
[5.192] A. I. Manakov, O. A. Esin, and B. M. Lepinskikh, *Dokl. Akad. Nauk. SSSR* **142**, 1124 (1962).
[5.193] T. S. Zolyan and A. R. Regel, *Fiz. Tverd. Tela* **6**, 1520 (1964).
[5.194] V. A. Ioffe and I. B. Patrina, *Fiz. Tverd. Tela* **6**, 3045 (1965).
[5.195] T. Allersma, R. Hakim, T. N. Kennedy, and J. D. Mackenzie, *J. Chem. Phys.* **146**, 154 (1967).
[5.196] E. H. Greener, D. H. Whitmore, and M. E. Fine, *J. Chem. Phys.* **34**, 1017 (1961).
[5.197] R. Valetta, *J. Chem. Phys.* **37**, 67 (1962).
[5.198] E. H. Greener and W. M. Hirthe, *J. Electrochem. Soc.* **109**, 600 (1962).
[5.199] P. Kofstad, *J. Phys. Chem. Solids* **23**, 1571 (1962).
[5.200] R. F. Janninck and D. H. Whitmore, *J. Chem. Phys.* **37**, 2750 (1962).
[5.201] E. H. Greener, G. A. Fehr, and W. M. Hirthe, *J. Chem. Phys.* **38**, 133 (1963).
[5.202] R. F. Janninck and D. H. Whitmore, *J. Chem. Phys.* **39**, 179 (1963).
[5.203] S. Nagasawa, *Denki Kagaku (J. Electrochem. Soc. Japan)* **18**, 158 (1950).
[5.204] L. Apker and E. A. Taft, *Phys. Rev.* **88**, 58 (1952).
[5.205] F. Jona and G. Shirane, "Ferroelectric Crystals." Pergamon Press, Oxford, 1962.
[5.206] P. Stähelin and G. Busch, *Helv. Phys. Acta* **23**, 530 (1950).
[5.207] S. K. Deb and J. A. Chopoorian, *J. Appl. Phys.* **37**, 4818 (1966).
[5.208] M. Foëx, *C.R. Acad. Sci. Paris* **220**, 917 (1945); **228**, 1335 (1949).
[5.209] R. Ueda and T. Ichinokawa, *Phys. Rev.* **80**, 1106 (1950).
[5.210] J. Wyart and M. Foëx, *C.R. Acad. Sci. Paris* **232**, 2459 (1951).
[5.211] S. Magascuva, *Denki Kagaku (J. Electrochem. Soc. Japan)* **16**, 13 (1948).

[5.212] T. Okada, K. Hirakawa, and F. Irie, *Busseiron Kenkyu* (*Res. Chem. Phys.* [*Kyoto*]) **15**, 49 (1949); *Proc. Phys. Soc. Japan* **4**, 143 (1949).
[5.213] B. T. Matthias, *Phys. Rev.* **76**, 430 (1949).
[5.214] J. Hirsch, Rep. Brit. Electr. Res. Assoc., Ref. L/T 351 (1956).
[5.215] S. Sawada, R. Ando, and S. Nomura, *Phys. Rev.* **84**, 1054 (1951).
[5.216] S. Sawada, *Phys. Rev.* **91**, 1010 (1953).
[5.217] S. Savada, *J. Phys. Soc. Japan* **11**, 1237 (1956).
[5.218] S. Sawada and G. C. Danielson, *Phys. Rev.* **113**, 803 (1959).
[5.219] B. L. Crowder and M. J. Sienko, *J. Chem. Phys.* **38**, 1576 (1963).
[5.220] A. Feretti, D. B. Rogers, and J. B. Goodenough, *J. Phys. Chem. Solids* **26**, 2007 (1965).
[5.221] A. W. Sleight and J. L. Gillson, *Solid State Commun.* **4**, 601 (1966).
[5.222] Z. I. Ornatskaya, *Fiz. Tverd. Tela* **6**, 1254 (1964).
[5.223] A. Hardy, J. Galy, A. Casalot, and M. Pouchard, *Coll. Int. Comp. Ox. Elem. Trans.* (*Bordeaux 1964*) p. 18, C.N.R.S., Paris 1965.
[5.224] P. Hagenmuller, J. Galy, M. Pouchard, and A. Casalot, *Mater. Res. Bull.* **1**, 45 (1966).
[5.225] R. P. Ozerov, *Usp. Khimii* **24**, 951 (1955).
[5.226] L. E. Conroy and M. J. Sienko, *J. Amer. Chem. Soc.* **79**, 4048 (1957).
[5.227] M. J. Sienko, *J. Amer. Chem. Soc.* **81**, 5556 (1959).
[5.228] M. J. Sienko and B. R. Mazumder, *J. Amer. Chem. Soc.* **82**, 3508 (1960).
[5.229] L. D. Ellerbeck, H. R. Shanks, P. H. Sidles, and G. C. Danielson, *J. Chem. Phys.* **35**, 298 (1961).
[5.230] H. R. Shanks, P. H. Sidles, and G. C. Danielson, *Advan. Chem.* **30**, 237 (1963).
[5.231] A. R. Mackintosh, *J. Chem. Phys.* **38**, 1991 (1963).
[5.232] R. Fuchs, *J. Chem. Phys.* **42**, 3781 (1965).
[5.233] A. R. Sweedler, C. J. Raub, and B. T. Matthias, *Phys. Lett.* **15**, 108 (1965).
[5.234] S. Yamaguchi, *Nuovo Cimento* **11**, 876 (1959).
[5.235] J. Rupprecht, *Naturwissenschaften* **47**, 127 (1960).
[5.236] L. Young, *Can. J. Chem.* **38**, 1141 (1960).
[5.237] E. L. Chopra and L. C. Bobb, *Proc. Inst. Elect. Electron. Eng.* (*US*) **51**, 1784 (1963).
[5.238] B. Yu. Lototskii and L. K. Chirkin, *Fiz. Tverd. Tela* **8**, 1967 (1966).
[5.239] C. A. Mead, *J. Appl. Phys.* **32**, 646 (1961); *Phys. Rev.* **128**, 2088 (1962).
[5.240] N. L. Yasnopol'skii, A. P. Balashova, and A. E. Shabel'nikova, *Radiotekh. Elektron.* **7**, 1665 (1962).
[5.241] Y. Ishikawa, Y. Sasaki, Y. Seki, and S. Inowaki, *J. Appl. Phys.* **34**, 867 (1963).
[5.242] C. L. Standley and L. I. Maissel, *J. Appl. Phys.* **35**, 1530 (1964).
[5.243] D. M. Smyth, G. A. Shirn, and T. B. Tripp, *J. Electrochem. Soc.* **110**, 1264, 1271 (1963); **111**, 1331 (1964).
[5.244] N. Schwartz and M. Gresh, *J. Electrochem. Soc.* **112**, 295 (1965).
[5.245] G. P. Klein, *J. Electrochem. Soc.* **113**, 348 (1966).
[5.246] C. Cherki, R. Coelho, and J. L. Mariani, *Solid State Commun.* **4**, 411 (1966).

Chapter 6 | **Oxides of the Metals Fe, Co, Ni, Cu, and Homologues**

6.1. Mössbauer Effect

In view of the importance of the Mössbauer effect in studying all the compounds into which iron can be introduced, it is necessary to devote an entire section to it at the beginning of this chapter. The Mössbauer effect provides valuable information on the electron clouds surrounding the nuclei, i.e., on the nature of the interatomic links and the rate of filling of the d levels. What is said is based mainly on the recent short book by Flinn [6.1], but readers wishing for more details may also refer to the works by Gol'danskii [6.2] and Wertheim [6.3].

The phenomena of natural and artificial radioactivity has shown that atomic nuclei are not constituted in an absolutely immutable way. The chemistry of nuclear reactions involves the absorption or emission of the particles α (proton), β (electron) or γ (photon) by the nuclei (see Harvey [6.4], for example). During these reactions, the nuclei pass through unstable energy states. The disintegration reaction of the artificial isotope of Co with mass number 57 is of very special interest for the investigation of the Mössbauer effect. This is an unstable nucleus, and it absorbs an electron inside its own atom (K capture) to change into ^{57}Fe.

$$e_K^- + {}^{57}_{27}\text{Co} \longrightarrow {}^{57}_{26}\text{Fe}^* \tag{6.1}$$

$$^{57}_{26}\text{Fe}^* \longrightarrow {}^{57}_{26}\text{Fe} + \gamma \tag{6.2}$$

Fe* indicates an excited state of the nucleus, endowed with an energy of 136.4 keV compared with the fundamental state. In addition, for 91% of the nuclei, one can distinguish a second excited state with an energy of 14.4 keV. Isotope 57 of Co has a half-life of 270 days. It emits γ rays at three energies: 136.4, 122, and 14.4 keV. Let us assume that on their trajectory these photons meet other ^{57}Fe nuclei (2% in natural iron). Some of these nuclei may absorb a photon in changing to a higher energy state, and then reemit it immediately in any direction. This is gamma fluorescence. Mössbauer [6.5] discovered that below a certain temperature the recoil energy of a fraction of the nuclei in a solid was too low to excite lattice vibrations. Absorption and reemission therefore take place without recoil, i.e., the nucleus affected remains part of the whole solid. This is the Mössbauer effect, or gamma fluorescence without recoil. In the case of ^{57}Fe and photons of 14.4 keV (1 Å), 77% of the nuclei behave in this way at room temperature. The total energy difference tolerated by resonance is extraordinarily low ($14.4 \pm 4 \times 10^{-12}$ keV or $3.5 \times 10^{12} \pm 1$ MHz).

Such a precise matching of the energies in the processes of emission by the source and absorption by the sample implies that the solids are strictly identical and at rest in relation to each other. Identity is essential because the emission and absorption frequencies are affected by any difference in the electronic environment of the atomic nucleus involved (or in the environment of this atom). Relative rest is also necessary because the Doppler–Fizeau effect brings about a change in the apparent frequency of an acoustic or electromagnetic wave as a function of its propagation speed in relation to an observer. But it so happens that the frequency differences caused by the two phenomena described above are comparable. For instance, the difference resulting from the electron structure of the ions Fe^{2+} and Fe^{3+} corresponds to a relative speed of approximately 1 mm sec^{-1}. The first use of the Mössbauer effect in solids chemistry will therefore consist of reestablishing the resonance by balancing the effect of a different environment by a relative shift in the source and the absorbing sample. Figure 6.1 shows chemically the system used. A meter sensitive to gamma photons and an electronic apparatus capable of selecting the impulses corresponding to the 14.4 keV ones are placed in the extension of the source-sample direction, and the position of the source is varied periodically. The number of impulses is recorded in different channels, each corresponding to a relative speed (or absorbed photon frequency). The results form the Mössbauer frequency spectrum of the sample. For a given sample and temperature, resonance absorption is in proportion to the density of the atoms studied in the sample. Approximately 10 mg of natural iron per square centimeter of thin-layer preparation is needed to obtain a spectrum. Figure 6.2 compares the spectrum obtained with the one that would be obtained from a sample identical with the source. The difference in the resonance peaks is the "isomeric shift," from the word "isomer," which in nuclear chemistry refers to nuclei formed from the same elementary particles but with different energies.

6.1. Mössbauer Effect

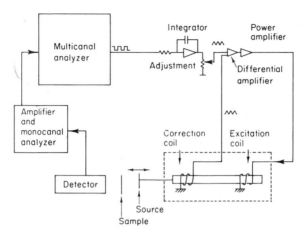

Fig. 6.1. Diagrammatic representation of an apparatus to study the Mössbauer effect by periodical speed variation. [Based on Flinn, in "Experimental Methods of Material Research." (H. Herman, ed.), Fig. 2, p. 169. Wiley (Interscience), New York, 1967. Used with permission of the copyright owner.]

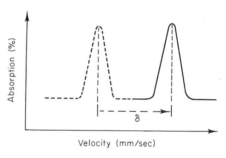

Fig. 6.2. Mössbauer spectrum of a sample identical to the source (in dotted lines) and of any sample; δ indicates the shift in the resonance frequency.

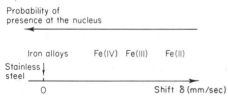

Fig. 6.3. Diagrammatic relationship between the scale of probabilities of presence in the nucleus and that of isomeric shifts.

The isomeric shift allows the effect of the environment of the Fe nucleus in different solids to be measured on the same scale. This effect is in fact the probability of presence in the nucleus of the electrons surrounding it. The remarks in Chapter 1 show that it arises from two types of electrons: s localized electrons (the AOs of which have an above-zero probability of presence in the nucleus) and delocalized electrons, i.e., free electrons. In Fe and its metal alloys, the predominant influence of the latter leads to high probabilities in the nucleus.

For reasons of convenience, because the source generally used in the first years of research into the Mössbauer effect was a thin sheet of stainless steel containing a little ^{57}Co, the corresponding probability has often been taken as zero on the scale. The lower probabilities observed in semiconductor compounds are therefore reflected in a shift in the opposite direction (Fig. 6.3). The isomeric shift was first observed by Kistner and Sunyar [6.6]. The values for the nuclei of Fe atoms in iron oxides and alloys are given below [6.7]:

$$\text{oxides Fe(II) } (3d^6) \quad \delta = 1.3 \text{ to } 1.5 \text{ mm sec}^{-1}$$
$$\text{Fe(III) } (3d^5) \quad \delta = 0.4 \text{ to } 0.6 \text{ mm sec}^{-1}$$
$$\text{Fe(IV) } (3d^4) \quad \delta = 0.1 \text{ mm sec}^{-1} \text{ (ferrates)}$$

$$\text{alloys } (3d^7 4s) \quad \delta = -0.1 \text{ to } +0.2 \text{ mm sec}^{-1}$$

The screen effect of the 3d electrons, which reduce the effect of the s localized electrons on the nucleus, may be more easily understood when one notes that valency II (3d^6) in reality corresponds to the lowest probability.

Atomic nuclei in which the symmetry of the electrical charges is not spherical have an electric moment (called "quadrupolar") which can interact with an electrical field gradient. The nucleus of ^{57}Fe, in its ground state, possesses a spin quantum number of $\frac{1}{2}$ and a spherical charge distribution. In the excited state, on the other hand, the spin number is $\frac{3}{2}$ and distribution is no longer spherical. In the absence of a magnetic field, the energy of the system formed by the nucleus, its electrons and neighboring atoms therefore depends on the relative orientation of the nucleus and the electrical field gradient created by its environment. It is shown that two separate excited states appear: this is a nuclear effect which could be compared to the Stark effect in a nonhomogeneous field. Figure 6.4a shows schematically the two $\Delta E \pm \varepsilon$ transitions which result; ε measures the quadrupolar interaction. The electrical field gradient in the ^{57}Fe nucleus results both from the asymmetry of the d layer of this atom (except for the symmetrical configurations $t_{2g}^3 \alpha \, e_g^2 \alpha$ and $t_{2g}^3 \alpha \, t_{2g}^3 \beta$) and from the possible distortion of the ligand octahedron. The result is a spectrum with two peaks.

Some nuclei, like that of ^{57}Fe, also have a magnetic moment that interact with an intense magnetic field created by their environment. In this case, the six possible energy states are all separate, and Fig. 6.4b shows, schematically, the six allowed transitions. This is the nuclear Zeeman effect. The internal field of the nucleus, or hyperfine field, is deduced by comparing the distance apart of the lines with those of pure iron at room temperature ($H_h = 330$ kOe). For a polycrystalline magnetic sample, the relative intensities of the absorption lines are theoretically 3–2–1–1–2–3. This is the case shown in Fig. 6.5. The direction of the hyperfine field is opposite to that of the magnetic moment observed in the atom, and it arises principally from the spin polarization of the

6.1. Mössbauer Effect

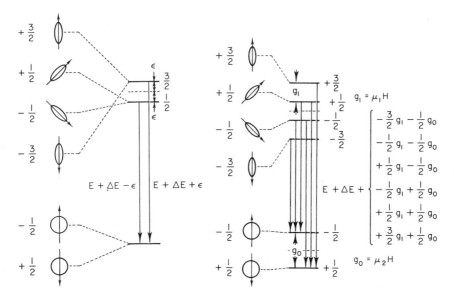

Fig. 6.4. Diagrammatic representation of the possible transitions in the nucleus (a) (left) in the presence of an electrical field gradient, (b) (right) in the presence of an intense magnetic field. [Based on Flinn in "Experimental Methods of Material Research" (H. Herman, ed.), Fig. 6, p. 174 and Fig. 8, p. 176. Wiley (Interscience), New York, 1967. Used with permission of the copyright owner.]

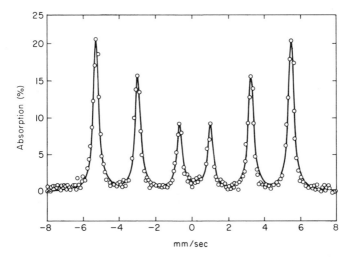

Fig. 6.5. Mössbauer spectrum of pure iron at room temperature. [Based on Flinn in "Experimental Methods of Material Research" (H. Herman, ed.), Fig. 9, p. 177. Wiley (Interscience), New York, 1967. Used with permission of the copyright owner.]

s electrons by the atom's magnetic moment. Its absolute value varies in the same direction as the number of bachelor d electrons, but may be modified by other contributions, such as that of free electrons.

Reference has already been made to the Mössbauer effect in Chapter 5 in connection with recent attempts to interpret the metal–semiconductor transitions of V_2O_3 and VO_2. In the first case (Section 5.1), the work done by Shinjo *et al.* involved a solution of 98.5% V_2O_3 and 1.5% of Fe_2O_3 enriched with 85% 57 isotope. The authors have observed a hyperfine field of 315 kOe at 131°K, and at room temperature an isomeric shift of 0.8 mm sec^{-1} and no quadrupole interaction. In the second case (Section 5.3), the work done by Kosuge involved solutions of 0.5, 1, and 2% Fe_2O_3, 85% enriched. The spectrum reveals the presence of trivalent iron ($\delta \sim 0.5$ mm sec^{-1}). Figure 6.6

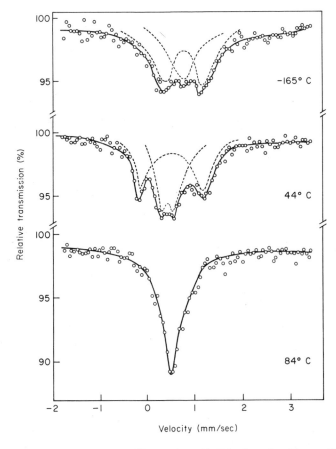

Fig. 6.6. Mössbauer spectrum of ^{57}Fe in the oxide VO_2 (based on Kosuge [5.125]).

6.2. TO Oxides

also shows that in the low temperature stable monoclinic phase, there are two separate sites, one of which presents an appreciable quadrupole effect, but the hyperfine field is nonexistent. Many other examples of Mössbauer spectra will be met in connection with iron oxides and chalcogenides.

6.2. TO OXIDES

The oxides MnO, FeO, CoO, and NiO crystallize in the structure of rock salt and are antiferromagnetic. The magnetic order introduces a quadratic distortion for CoO and a rhombohedral distortion for the others. The phase change at T_N is marked by an expansion irregularity (Fig. 6.7). Testing with neutrons has been done by Shull et al. [6.9] and then by Roth [6.10]. It provides an experimental verification of the superexchange interactions envisaged by Kramers [6.11]. In MnO ($t_{2g}^0 \beta$) and NiO ($t_{2g}^3 \beta$), the moments are parallel to the (111) plane. In FeO ($t_{2g}^1 \beta$), they are perpendicular to them. In CoO ($t_{2g}^2 \beta$), they are parallel to direction [117]. In the first two, the dominant magnetic interaction is that of the spins, and $2S = 5$ and 2. In FeO and CoO, the spin–orbit interaction is important.

The position of wüstite FeO in the Fe–O phase diagram is given in Fig. 6.8. Its formula is in fact $\square_x Fe_{1-x}O$, and the phase is not stable for $x < 0.05$. It can reach $x = 0.25$ at $1000°C$. A number of trivalent iron atoms are therefore present (cf. Section 3.3), but the presence of iron atoms in tetrahedral sites has been pointed out, as well as the conglomeration of defects. Distortion occurs below $T_N = 198°K$. Conductivity at $1000°C$ depends on the oxygen pressure

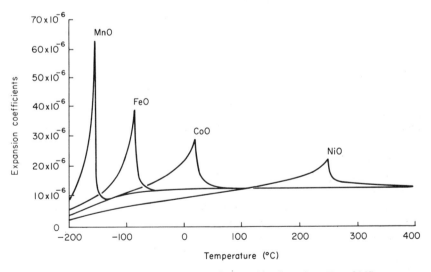

Fig. 6.7. Expansion irregularities of TO oxides (based on Foëx [6.8]).

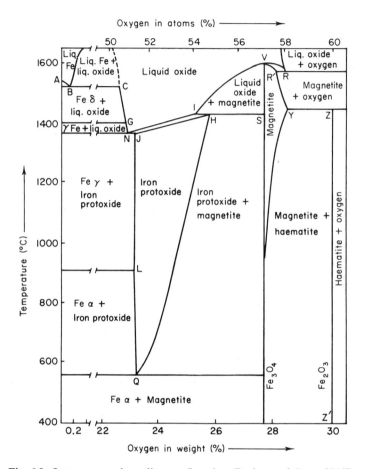

Fig. 6.8. Iron–oxygen phase diagram (based on Darken and Gurry [6.12]).

[6.13]. The paramagnetic susceptibility is said to give $2S = 3.1$ [6.14], while the result obtained by neutron diffraction depends on hypotheses concerning the defects [6.15]. The first research into conductivity [6.16, 6.17] was taken up again by Marion [6.18] and Aubry and Marion [6.19]. It is 90 to 200 mhos cm^{-1}, and varies extensively in terms of x. In Section 3.3, Tannhauser's [6.20] work was mentioned, and Fig. 3.10 showed the satisfactory correspondence obtained with the hypothesis of semiconductivity by transfers among di- and trivalent iron atoms, proportional to the product of $[Fe(II)_{oct}][Fe(III)_{oct}]$ concentrations. Figure 6.9 indicates an activation energy of 0.067 eV. These results are confirmed by Ariya and Brach [6.21]. Research using the Mössbauer effect, carried out by Shirane et al. [6.7], shows a quadrupolar effect of the Fe(II) with $\varepsilon = 0.3$ mm sec^{-1}, and the presence of Fe(III) (Fig. 6.10). These

6.2. TO Oxides

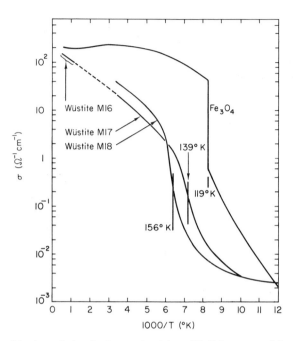

Fig. 6.9. Logarithmic variation in the conductivity of FeO in terms of the temperature reciprocal (based on Tannhauser [6.20]).

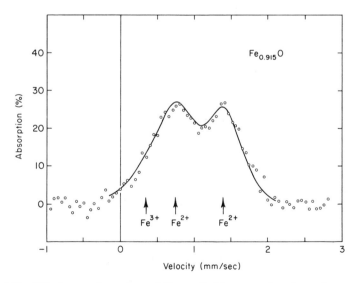

Fig. 6.10. Mössbauer absorption of $Fe_{0.915}O$. The curve has been drawn on the assumption that Fe(II)/Fe(III) = 8 (based on Shirane *et al.* [6.7]). The ratio would be 5 if the tetrahedral sites were empty.

authors attribute it to a distortion of the ligand octahedron, but we feel that it may also come from the asymmetry of the d layer. Geiger et al. [6.22] have shown that the effect of the oxygen pressure at 1000°C conforms to the law of mass action (except at the phase limits). To sum up, the transfer mechanism is very probable, and in accordance with the p-type observed for low values of x (Fig. 6.11).

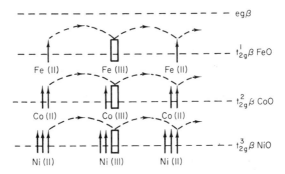

Fig. 6.11. Diagram of the transfer mechanism in the $t_{2g}\beta$ localized level for FeO, CoO, and NiO.

The oxide CoO is more stable, and melts at 1800°C. The structure is distorted between $-190°$ and $-70°C$ (dilatometric irregularity at $T_N = 19°C$). Its conductivity depends, like that of FeO, on the oxygen pressure, and it has catalytic properties [6.13]. The existence of a spin–orbit interaction is illustrated by the values $2S = 4$ [6.23] and 3.8 [6.10] (instead of 3) which would be obtained, by ignoring it, from the paramagnetic susceptibility and neutron diffraction. The semiconductor character (p-type) in the presence of an excess of oxygen and the effect of the disorder on the Seebeck effect have been mentioned by Croatto [6.24], while Young et al. [6.25] have studied the effect of pressure and attribute it to a lowering of the hopping frequency in a transfer mechanism. Roilos and Nagels [6.26] have measured the resistivity of pure monocrystals. The activation energies are 0.7 or 1.3 eV. Rao and Smakula [6.27] have found a dielectric constant ε' of 13 at room temperature, independent of the frequency. In a very comprehensive study of polycrystals, Fischer and Tannhauser [6.28] have shown that the conductivity above 900°C varies in relation to the oxygen pressure to the power of $\frac{1}{4}$, then $\frac{1}{6}$, and have concluded that once- or twice-ionized cobalt vacancies exist. An indirect assessment of the mobility of holes above 1000°C shows that it increases exponentially, and reaches 0.4 cm^2 V^{-1}sec^{-1} at 1350°C, with an activation energy of 0.3 eV. Zhuze and Shelykh [6.29] have measured the Hall effect in monocrystals, and recorded an exponential increase in mobility with temperature, indicating a semiconductibility mechanism by transfers between Co(II) and Co(III), as in

6.2. TO Oxides

wüstite. Shelykh et al. [6.30] feel however that mobility no longer plays an essential role at high temperatures. Austin et al. [6.31] have confirmed the results obtained by Zhuze et al.: the mobility has activation energies of 0.073 ($T < T_N$) and 0.024 eV ($T > T_N$) (Fig. 6.12). Bruck and Tannhauser [6.32] have discussed the assessment of the mobility by a combination of conductimetric and gravimetric measurements (used by Yahia for TiO_2 and Fischer et al. for CoO). Austin et al. [6.33] have found a μ_E/μ ratio ~ 6. The two mobilities decrease above 150°C. They have discussed all the results and referred to the polaron theory. To sum up, the transfer mechanism is very likely below 150°C (Fig. 6.11).

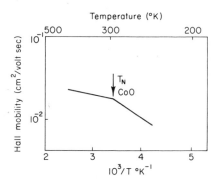

Fig. 6.12. Logarithmic variation in the Hall mobility of a monocrystal of CoO in terms of the temperature reciprocal (based on Austin et al. [6.31]).

The oxide NiO melts at 2000°C but, being less stable than CoO, decomposes from 1000°C upwards. Its semiconductor character has long been known, and varies in relation to the oxygen pressure P. Conductivity changes from 10^{-8} for NiO to 10^{-3} mhos cm^{-1} for $Ni_{0.995}O$ [6.13]. $T_N = 250°K$ for the low-temperature form and 375°C for the cubic form [6.34]. Wright and Andrews [6.35] have obtained polycrystals by the oxidation of a sheet of nickel, and measured the conductivity and Seebeck and Hall effects in relation to temperature (Fig. 6.13). The activation energies are 0.3 and 0.6 eV at low temperatures, and 2 eV at high temperatures The unit volume contains 7×10^{22} Ni cm^{-3} (~ 100 times fewer holes, in accordance with the stoichiometric state). Hogarth [6.36] has studied the effect of P on the activation energy. Johnston and Cronemeyer [6.37] have observed absorption peaks in monocrystals at 0.24 and beyond 1 eV. The interpretation of these results has been questioned by Morin [6.38] and Young [6.25], in relation to a transfer mechanism at the d level. For the link with magnetic properties, cf. [6.39, 6.40]. Newman and Chrenko [6.41] have found an absorption threshold corresponding to a 4 eV forbidden band. Sewell [6.42] and Klinger [6.43] have mentioned hops activated by phonons or "small polarons." Further research into the effect of oxygen has been done by Mitoff [6.44] [concentration of nickel vacancies $0.11 P^{1/6} \exp(-17,800 RT)$ per NiO,] Koide and Takei [6.45] (resistivity for 0.02 to 0.12 vacancies per NiO atom), Saltsburg et al. [6.46, 6.47] (transitory adsorption of O) and Hauffe [6.48] (exchanges with the atmosphere), while the Hall effect has been studied by Fujime et al. [6.49] (2.7×10^{14} holes cm^{-3} of $\mu = 3.7 \times 10^{-4}$ on a monocrystal at room temperature), Zhuze and Shelykh [6.50] (μ decreases from 300 to 500°C, resulting in a d band) and Bosman et al. [6.51] (change of sign at T_N). Rao and Smakula [6.27] have found a dielectric constant ε' of 12 at room temperature where $f > 10^5$ Hz. Snowden

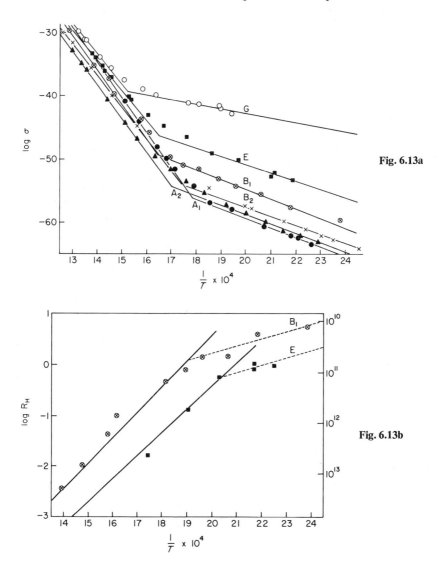

Fig. 6.13a

Fig. 6.13b

6.2. TO Oxides

et al. [6.52, 6.53] have studied resistivity between 10^3 and 10^{10} Hz and reached the conclusion of a hopping mechanism with a frequency of 3×10^9 Hz. Nachman et al. [6.54] have reached the same conclusion from measurements of the Hall effect, but they point out irregularities ($T < T_N$). Kzendzov and Drabkin [6.55] have found $E_G = 3.7$ eV at high temperatures (conductivity/ temperature and photoconductivity). Vernon and Lowell [6.56] have suggested that there is a slight distortion. Makarov et al. [6.57] have measured μ_E on monocrystals, finding 0.14 (electrons) and 0.3 cm^2 V^{-1}sec^{-1} (holes) $\pm 25\%$, of the same order as μ. To sum up, the transfer mechanism seems again the most likely (Fig. 6.11).

The oxide CuO does not crystallize in the B1 structure, but in a very different monoclinic structure, where the coordination seems to be 4 [6.58]. Its semiconductor character and the high Seebeck effect are known [6.59–6.61], and depend on the conditions of sintering [6.62]. The magnetic susceptibility shows a flattened maximum at around 540°K [6.63, 6.64], while the specific heat [6.65] and neutron diffraction [6.66] indicate that $T_N = 230$°K. Caglioti et al. [6.67] have confirmed the absence of distortion below T_N. Young et al. [6.25] have calculated the hole mobility from the conductivity and Seebeck effect, and they explain its decrease at high pressures by a lower diffusion rate of self-trapped holes. O'Keefe and Stone [6.68] have confirmed the susceptibility maximum (3.4×10^{-6}) and antiferromagnetism. Kholoday [6.69] has given the infrared

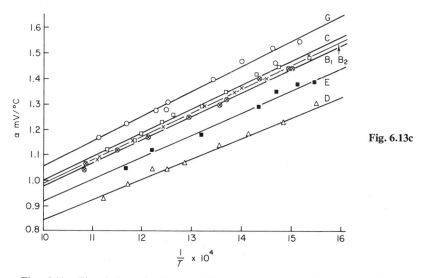

Fig. 6.13c

Fig. 6.13. Electrical properties of NiO: logarithmic variation in conductivity (a) (opposite: top) and Hall coefficient (b) (opposite: bottom) and linear variation in the Seebeck coefficient (c) (above), in terms of the temperature reciprocal for various samples (based on Wright and Andrews [6.35]).

absorption spectrum. These data, however, are insufficient to show the semi-conduction mechanism.

TO oxides provide many solid solutions with other oxides. Hauffe and Block [6.70], followed by Shimomura and Tsubokawa [6.71] and Schwab and Schmid [6.72] have studied the effect of various additions to NiO. See also [6.52] ($NiO+Cr_2O_3$), [6.22] ($FeO+Cr_2O_3$ or TiO_2), [6.73, 6.74] ($NiO+Ga_2O_3$), [6.75] (CaO–CuO). But the greatest amount of work has been done on the addition of Li_2O. Verwey et al. [6.76] described in 1950 the principle of valency induction on CoO–Li_2O (Fig. 6.14) and NiO–Li_2O (Fig. 6.15) systems (cf. Sections 2.4 and 3.3). The activation energy of samples sintered in air at 1200°C decreases, but without disappearing entirely, as the lithium content increases. Parravano [6.77] has confirmed the appearance of Ni(III) in a quantity equivalent to Li. Morin [6.38] has used the conductivity and Seebeck effect to calculate a hole mobility of less than 0.01 $cm^2\ V^{-1}sec^{-1}$, increasing exponentially with temperature. Heikes and Johnston [6.78] have studied $Li_xT_{1-x}O$ systems (T = Mn, Co, Ni, Cu) and have interpreted the properties by means of the hopping model (cf. Section 3.3) where $v_0 = 10^{13}$ (Mn), 5×10^{12} (Co), 3×10^{11} (Ni) and 5×10^{11} (Cu). Miller and Heikes [6.79] have completed research on Mn. Goodenough et al. [6.80] have extended work on Ni up to $x = 0.5$. Where $0 < x < 0.3$, there is antiferromagnetism, and doubling of the cell parameter below T_N. Where $0.3 < x < 0.5$, the structure is rhombohedral, and ferrimagnetism appears as a result of a partial ordering of the Li and Ni atoms in successive (111) planes. The resulting moment approaches 0.3 μ_B where $x = 0.43$, with $T_C \sim 200°K$. The conductivity does not appear to have been studied. Van Houten [6.81] has studied the internal friction of $Li_xNi_{1-x}O$ and concludes that it corresponds with the hop model. According to all these authors, then, the addition of Li_2O does not alter the nature of the transport properties of TO oxides.

In 1963 Ksendzov et al. [6.82] stated a different opinion, after measuring, for the first time, the Hall coefficient at various temperatures for $Li_xNi_{1-x}O$ monocrystals. The carrier density increases exponentially with temperature, suggesting an excitation mechanism, and the mobility increases with the Li content (impurity band). Zhuze and Shelykh [6.50] have confirmed these results. Koide [6.83] have returned to the measurement of resistivity/temperature in monocrystals, and Hauffe [6.48] has studied exchanges with the atmosphere. Roilos and Nagels [6.26], working with $Li_xT_{1-x}O$ (T = Co, Ni), have found the relatively high mobility of 0.25 $cm^2\ V^{-1}sec^{-1}$ for Ni, with the maximum occurring around 300°K. Akiyama [6.84] has dealt with magnetic interactions. Springthorpe et al. [6.85], on the basis of the resistivity of $Li_xNi_{1-x}O$ (0.2 to 3.2% Li) at low temperatures, mention hopping conduction in both an impurity level and a d level. Bosman and Crevecoeur [6.86], using alternating current, have measured the conductivity and Seebeck effect from

6.2. TO Oxides

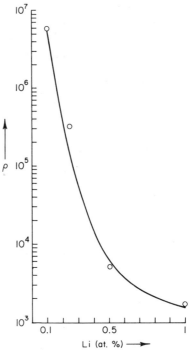

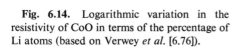

Fig. 6.14. Logarithmic variation in the resistivity of CoO in terms of the percentage of Li atoms (based on Verwey et al. [6.76]).

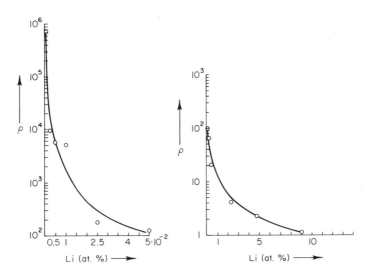

Fig. 6.15. Logarithmic variation in the resistivity of NiO in terms of the percentage of Li atoms (based on Verwey et al. [6.76]).

100 to 1300°K (Fig. 6.16). There is impurity conduction for $T < 170°K$, but variation in the two quantities depends on the number of carriers for $T > 170°K$. Austin et al. [6.87] have continued the research done by Springthorpe et al. Fischer and Wagner [6.88] and Van Daal and Bosman [6.89] have discussed the influence of structural defects on the Hall effect. Finally, Austin et al. [6.90] have done very comprehensive work on 0–0.6% Li solutions, and compared their results with earlier records. They conclude that the hypothesis put forward by Heikes et al. (d electron hops) is not borne out experimentally. The hops are caused by impurities ($T < 250°K$), and there is a narrow 3d band of small polarons at 0.2–1.2 eV (depending on the Li content) of the valence band, so that different mechanisms may be observed depending on the temperature. Recent work has also shown that crystals behaved differently depending on whether the Li is distributed uniformly or not.

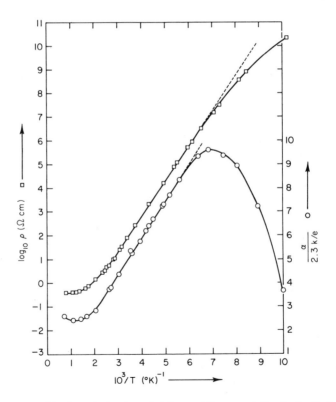

Fig. 6.16. Logarithmic variation in resistivity, and linear variation in the Seebeck effect (in units 2.3 k/e: 198 μV deg^{-1}) in terms of the temperature reciprocal in $Li_xNi_{1-x}O$ where $x = 0.88 \times 10^{-3}$ (based on Bosman and Crevecoeur [6.86]).

6.3. T_3O_4 Oxides

The mineral magnetite, Fe_3O_4, has been known since ancient times, and its ferromagnetic properties were at the origin of the study of magnetism. The stability range of the phase, shown in Fig. 6.8, is fairly narrow at room temperature. The $H1_1$ cubic structure is that of spinel $MgAl_2O_4$ (Fig. 6.17), where the compact arrangement of oxygen atoms provides one tetrahedral site (A) and two octahedral sites (B) per O_4 group. While the divalent atoms occupy the A sites in the "normal" structure, they can occupy half the B sites in an "inverted" structure [6.91]. Verwey and de Boer [6.92] have shown that magnetite is inverted and could be written as

$$[Fe(III)]_A[Fe(II)Fe(III)]_BO_4 \qquad (6.3)$$

The origin of the inverted structure probably lies in the partly covalent character of the bonds [6.93]. The Curie point T_C is 575°C, and the melting point is 1600°C. At room temperature the resistivity is very low, around 10^{-2} Ω cm, the Seebeck coefficient is 60 μV deg^{-1}, and the magnetization intensity differs depending on the axes of the crystal (Fig. 6.18). Variations in

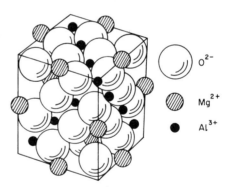

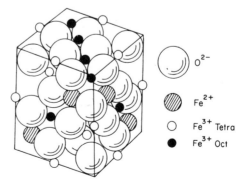

Fig. 6.17. Structures of spinel $MgAl_2O_4$ (a) (top), and magnetite Fe_3O_4 (b) (bottom).

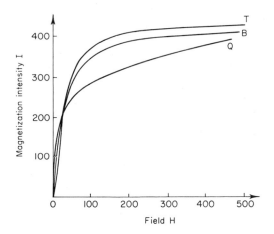

Fig. 6.18. Variation in the magnetization intensity in relation to the fields along the (B) binary [110], (T) ternary [111], and (Q) quaternary [100] axes of a cubic crystal of magnetite. [Based on Pascal, "Nouveau Traité de Chimie Minérale." Masson, Paris, 1956–1963.]

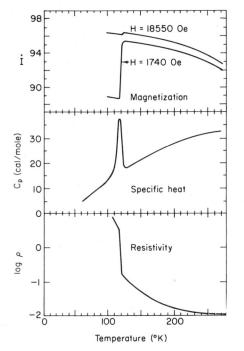

Fig. 6.19. Linear variations in the magnetization intensity and specific heat and logarithmic variation in resistivity in relation to temperature around 120°K. [Based on Pascal, "Nouveau Traité de Chimie Minérale." Masson, Paris, 1956–1963.]

6.3. T_3O_4 Oxides

the magnetization, specific heat, and resistivity with the temperature show a change towards 120°K (Fig. 6.19). Extrapolation to 0°K of the magnetization curves recorded above 120°K gives a saturation moment of 4.07 μ_B. The variation in resistivity, on either side of 120°K, is that of an electronic semiconductor [6.13].

Verwey et al. [6.94–6.96] have interpreted the change which Fe_3O_4 undergoes as follows: In the inverted spinel structure, the iron atoms of valencies 2 and 3 both occupy octahedral sites. If they are distributed in a disordered way, these sites are all strictly equivalent, and the transfer of the sixth Fe(II) d electron [which becomes Fe(III)] to an Fe(III) neighbor [which becomes Fe(II)] happens easily and quickly. The high conductivity observed above 120°K corresponds to this situation, with a transfer frequency of 10^{12} to 10^{13} Hz. In a non-stoichiometric or impure composition, the number of Fe(II) and Fe(III) atoms is no longer the same, reducing the number of possible transfers and consequently the conductivity. If, on the other hand, the iron atoms are distributed in the octahedral sites in an ordered way, the sites are no longer equivalent, and the density of transfers is considerably reduced. This is what happens below 120°K, where the conductivity drops suddenly by a factor of 100 in pure and stoichiometric samples [6.97]. A lower symmetry (orthorhombic) and anisotropy of the conductivity have also been observed [6.98, 6.99]. Neel [6.100] interprets the magnetic properties of Fe_3O_4 as follows. The iron atoms occupying the A sites form a ferromagnetic sublattice, and those occupying the B sites form a second, opposed to the first. If one accepts, with Verwey and de Boer [6.92], that the divalent atoms occupy the B sites (inverted structure), they then alone contribute to the resulting moment [cf. Eq. (6.3)]. The number of apparent bachelor electrons $2S$ per Fe_3O_4 formula is therefore 4, which corresponds excellently to experimental measurements (4.07 worked out from the saturation magnetization, and 3.92 from the paramagnetism). This work, which has now become classic, is closely bound up with the ferromagnetism theory and the interpretation of the properties of ferrites TFe_2O_4. On the subject of Fe–O–Fe magnetic interactions, see also Gilleo [6.101].

A few general surveys of electrical and magnetic properties [6.102–6.104] mention in particular an activation energy of around 0.06 eV below room temperature. Neutron reflection has enabled Shull [6.105] to confirm the opposite directions of the Fe moments in the two sublattices. Tombs and Rooksby [6.106] have defined 119°K as the precise point of crystallographic transformation (see also [6.98, 6.99, 6.107–6.109]). Permeabilities and complex dielectric constants have been measured at high frequencies [6.110–6.112]. McReynolds and Riste [6.113] and Hamilton [6.114] have observed neutron diffraction, between the room temperature and T_C, and in the neighborhood of 119°K, respectively. The ordered structure at low temperatures, suggested by Verwey et al., has been confirmed. The Hall effect has been measured by

Samokhvalov and Fakidov [6.115] who have found $N \sim 10^{20}$ electrons cm^{-3} at room temperature (Fig. 6.20). The magnetoresistance, which is always negative, follows a parabolic law (Fig. 6.21). The same effective field governs the two galvanomagnetic effects. Shortly afterwards, Lavine [6.116] studied a very pure synthetic crystal, and found $R_0 = -1.8 \times 10^{11}$ V cm A^{-1}Oe^{-1} (-1.8×10^{-3} cm^3 C^{-1}), $N = 3.5 \times 10^{21}$ electrons cm^{-3} and $\mu = 0.45$ cm^2 V^{-1}sec^{-1}. The result concerning mobility has been confirmed [6.117]. Samokhvalov and Fakidov [6.118] next measured the variation between 0 and 100°C, in fields

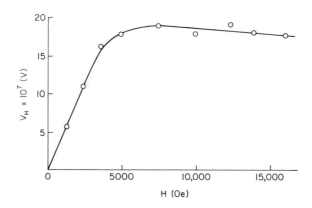

Fig. 6.20. Variation in the Hall voltage $V_H = R_0(H_i + 4\pi\alpha M)Ib$ in relation to the external field H in a sample of $R_0 = 0.02$ cm^3 C^{-1}; R_0 and αR_0 are the ordinary and extraordinary coefficients ($\alpha < 0$); H_i is the field inside the sample, M the magnetization intensity, $I = 30$ mA the density of the primary current, and b the thickness of the sample. The extraordinary term is saturated at around 7 KOe (based on Samokhvalov and Fakidov [6.115]).

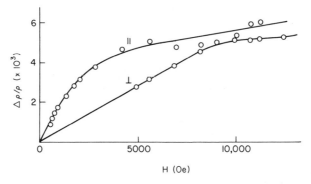

Fig. 6.21. Variation in the longitudinal and transverse magnetoresistances $\Delta\rho/\rho = A + B(H_i + 4\pi\alpha M)^2$ in relation to the external field H in the same sample. A graph in terms of $H_{\mathrm{eff}}^2 = (H_i + 4\pi\alpha M)^2$ would give straight lines with $\alpha_\perp = 7$ and $\alpha_\| = 23$ (based on Samokhvalov and Fakidov [6.115]).

6.3. T_3O_4 Oxides

reaching 20 kOe, of the ordinary and extraordinary Hall constants. They claim that the first, rather imprecise measurement corresponds to a forbidden band (?) of 0.03–0.05 eV, and the second is proportional to a power of the resistivity. Lavine [6.119] has studied more particularly the extraordinary Hall effect (Fig. 6.22), and shown that theories worked out for iron–nickel alloys do not apply. We shall return in Part Three to the frequently complicated apparatuses and techniques used for such measurements, as well as theories for the interpretation of the results. Miles et al. [6.120], and then Smith [6.121] (Fig. 6.22) have measured the resistivity up to T_C. Samokhvalov and Fakidov [6.122] have shown that the change at 119°K is spread out over a wide range of temperatures (maximum Seebeck effect at 95°K). Zotov [6.123–6.125] has found that the resistivity below 112°K and the magnetoresistance are affected by cooling in a magnetic field. Westwood et al. [6.126] have studied the reflection spectrum.

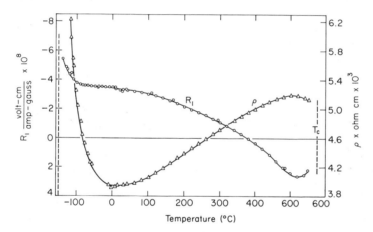

Fig. 6.22. Variations in the resistivity and extraordinary Hall coefficient in Fe_3O_4 in relation to temperature (based on Calhoun [6.99], Lavine [6.119], and Smith [6.121]).

Haubenreisser [6.127] suggests interpreting these results by a transfer mechanism (electron hops) taking the electron–phonon interactions into account. He has calculated a resistivity minimum at 600°K (instead of 300°K, Fig. 6.22). Schröder [6.128, 6.129] has studied the effect of stoichiometry on the shape of a resistivity/temperature curve. He has found that the minimum is shifted towards high temperatures, and the maximum less accentuated when the Fe(II) content is reduced. Tannhauser [6.20] has compared the behavior of Fe_3O_4 with that of other iron oxides, and concluded that there is a transfer mechanism, as for FeO. He later turned to the ordered structure (Fig. 6.23) proposed by Verwey et al. at low temperatures, showing that a complex defect

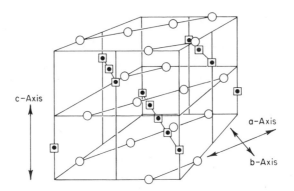

Fig. 6.23. Order of the electronic structure of Fe_3O_4 below 119°K. Only the iron atoms in the octahedral sites are shown. Blank circles represent Fe(III) atoms, squares containing a black dot (sixth d electron) represent Fe(II) atoms (based on Tannhauser [6.130]).

intervened in the transfer mechanism [6.130]. Yamaguchi [6.131] has interpreted the electron diffraction spectra. The directions of the iron atomic moments are parallel to the (111) planes.

Here again, the Mössbauer effect seems to offer an interesting experimental approach. Using a stainless steel source, Bauminger et al. [6.132] have obtained spectra which they interpret as the superimposition of the Zeeman effects in two groups of iron nuclei with different parameters. At 300°K, the lines of one of the groups are characteristic of Fe(III) in ionic compounds ($\delta = 0.45$ mm sec^{-1}, $H_h = 500$ KOe) and also coincide with those of γFe_2O_3 (cf. Section 6.4). The lines of the second ($\delta = 0.7$ mm sec^{-1}, $H_h = 450$ KOe) correspond to parameters that lie in between those of Fe(II) and Fe(III) in the same compounds. Comparison of their intensity shows that approximately half the Fe(III) nuclei give the lines of the first group, and the rest of the iron nuclei the lines of the second. There is therefore clearly a fast exchange of electrons between the Fe(II) and Fe(III) atoms in the B sites, in accordance with the hypothesis put forward by Verwey et al., and this exchange takes place at a much higher frequency than that of the nuclear phenomenon (approximately 10^8 Hz), since a single average value for the resonance parameters is observed. At 85°K, the lines in one of the groups involve Fe(II) ($\delta = 1.15$ mm sec^{-1}, $H_h = 450$ KOe) and the authors assume that the lines of the second correspond to all the Fe(III) atoms, which would have the same parameters in A and B sites ($\delta = 0.65$ mm sec^{-1}, $H_h = 510$ KOe), as in fact happens in γFe_2O_3 and $NiFe_2O_4$. Ôno et al. [6.133, 6.134] have done a separate study of the same points, using a chromium source (which is more precise) and 57 isotope-enriched samples (Fig. 6.24). They have found slightly different parameters [300°K: A sites: $H = 495$ KOe, $\delta = 0.46$ mm sec^{-1}; B sites: $H = 470$ KOe, $\delta = 0.87$ mm sec^{-1}; 30 and 85°K: Fe(III): $H = 510$ KOe, $\delta = 0.6$ mm sec^{-1}; Fe(II): $H = 470$ KOe, $\delta = 0.88$ mm sec^{-1}] and in particular an above-zero quadrupole interaction, of around 0.1 mm sec^{-1} at 300°K and 0.2 (Fe(III)) or

6.3. T_3O_4 Oxides

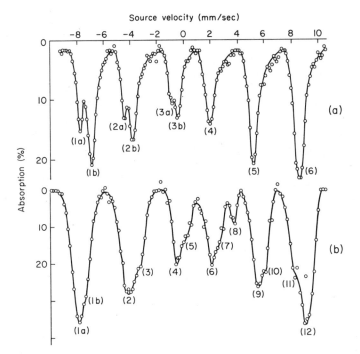

Fig. 6.24. Mössbauer spectra of Fe_3O_4 at 300 (thin absorbent layer) (a) and 85°K (thick absorbent layer) (b). Percentage absorption is in relation to the relative speed source-sample (based on Ito *et al.* [6.134]).

1.9 mm sec^{-1} (Fe(II)) at 85°K. Their interpretation of phenomena below the transformation temperature, tested on the spectrum of a monocrystal cooled in a magnetic field, is more complex, but does not call into doubt the separate resonance of Fe(II) and Fe(III).

Haussmannite Mn_3O_4, mentioned in Section 5.1, crystallizes in a quadratic structure derived from that of spinel by distortion. Verwey and de Boer [6.92] have pointed out that valencies 2 and 4 of Mn are much more stable than valency 3, and they attribute the high resistivity of these crystals (10^9 times that of Fe_3O_4) to an ordered formula $Mn(IV)Mn_2(II)O_4$ rather than an $Mn(II)Mn_2(III)O_4$ formula. Bhide and Dani [6.135] note that the Mn(II)–O–Mn(II) interactions have far less energy than Fe(III)–O–Fe(III) (cf. Gilleo [6.101]) and they put this down to the high resistivity of an $[Mn(II)]_A$ $[Mn(II)Mn(IV)]_B O_4$ formula. The magnetic structure is triangular [6.136]. Goodenough [6.137], finally, has shown that the cooperative Jahn–Teller effect resulting from $[Mn_2(III)]_B$ (d^4) causes a change in the Mn—O bonds. $Mn_xFe_{3-x}O_4$ solid solutions exist in every proportion. Their Seebeck effect has been studied by Simsa [6.138], and their resistivity by Miyata [6.139]. Lotgering [6.140], Rosenberg *et al.* [6.141] and Simsa and Zaveta [6.142]. The activation energy rises suddenly to 0.3 eV for $x = 1$ and reaches 0.65 eV for Mn_3O_4. Resistivity varies in a similar way, with a thermal hysteresis caused by the cubic–quadratic phase change. The transfer mechanism takes place between Fe or Mn atoms depending on the value of x. The oxide Co_3O_4 crystallizes, like Fe_3O_4, in the spinel structure, but is as poor a conductor as Mn_3O_4. It is paramagnetic and $\chi = 35 \times 10^{-6}$ at room temperature [6.13]. Verwey and de Boer [6.92], using electrical and crystallochemical arguments, have again suggested an ordered Co(IV) Co(II)$_2$O$_4$ formula here. The Co(IV) atom, although unknown in chemistry, should have a certain stability (d^5). Goodenough [6.137] has shown that [Co(III)$_2$]$_B$ has a $t_{2g}^6 e_g^0$ formula. $Co_xFe_{3-x}O_4$ solid solutions also exist. The oxide Ni_3O_4 has never been recorded.

Numerous substitutions are possible in magnetite, such as T(II)Fe(III)$_2$O$_4$ (normal or inverted ferrites), M(I)$_{0.5}$Fe(III)$_{2.5}$O$_4$ (M = alkaline metal), Fe(II)T(III)$_2$O$_4$ (notably chromites), T(IV)Fe(II)$_2$O$_4$ (T = Ti, Mo, etc) [6.13]. Consideration of them lies beyond the scope of this book.

6.4. T_2O_3 Oxides

Hematite α-Fe$_2$O$_3$ has the D5$_1$ rhombohedral structure of corundum α-Al$_2$O$_3$, already described in Section 5.1, in connection with the oxides Ti$_2$O$_3$, V$_2$O$_3$, and Cr$_2$O$_3$. It melts at 1350°C, and up to $T_C = 675$°C shows slight ferromagnetism with thermoremanence (Fig. 6.25) [6.13]. This phenomenon has been used to measure the intensity and direction of the Earth's magnetic field during the formation of rocks [6.143–6.145]. Morin [6.146] has discovered the magnetic transformation shown on Fig. 6.26, at 250°K (see also [6.147]). The magnetic properties have been reviewed by Chevallier [6.148], the orientation of the atomic moments, investigated using neutrons, by Shull et al. [6.149], and the structure in the neighborhood of T_C by Willis and Rooksby [6.150]. The semiconductor properties, which had been mentioned by Bevan et al. [6.151] (Fig. 6.27), have been reexamined by Morin [6.152] in relation to stoichiometry (iron excess n-type or defect p-type). The resistivity of a pure, stoichiometric sample at room temperature is estimated at 10^{14} Ω cm with $E_A = 1.17$ eV. An extraordinary Hall effect has been observed. The carrier density, worked out from the Seebeck effect, leads to the mobility being calculated at less than 10^{-2} cm^2 V^{-1}sec^{-1}, increasing exponentially with temperature. See also [6.107]. Optical absorption indicates $E_G = 1.9$ eV, and Morin [6.153] considers that there is a transfer mechanism, on which an excitation mechanism may be superimposed. Neel and Pauthenet [6.154] and Neel [6.155] have explained the magnetic properties by the superimposing of a classic antiferromagnetism ($T_N = 675$°C), and a parasitic ferromagnetism resulting from imperfect balancing of the atomic moments. Below 250°K, these are orientated along the ternary axis, and have a low isotropic resultant. Above it, they are free in the perpendicular plane, with a higher anisotropic resultant. This interpretation has been questioned by Haigh [6.156]. Corliss et al. [6.157] have studied Morin's transformation by neutron diffraction.

6.4. T_2O_3 Oxides

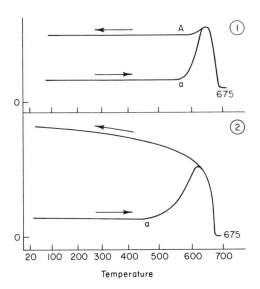

Fig. 6.25. Phenomenon of hot magnetization (thermoremanence) of hematite: (1) cold, (2) hot. [Based on Pascal, "Nouveau Traité de Chimie Minérale." Masson, Paris, 1956–1963.]

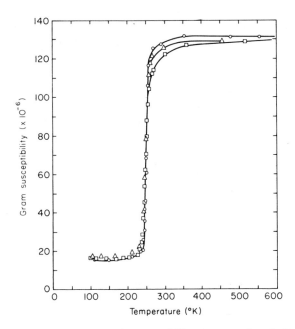

Fig. 6.26. Variation in the magnetic susceptibility per gram in relation to absolute temperature of α-Fe_2O_3 and α-Fe_2O_3–TiO_2 solutions: (O) Fe_2O_3, ($\triangle$) Fe_2O_3+0.05% Ti, ($\square$) Fe_2O_3+0.2% Ti (based on Morin [6.146]).

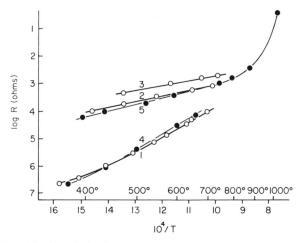

Fig. 6.27. Logarithmic variation in conductance in relation to the reciprocal of absolute temperature: 1 and 4, sintering in air; 2, 3, and 5, sintering in vacuum and partial oxidation in magnetite at 1000°C (based on Bevan et al. [6.151]).

Wucher [6.158] has interpreted the constant susceptibility of 20×10^{-6} above T_C. Finally, Dzyaloshinskii [6.159, 6.160] produces a general thermodynamic theory of the parasitic ferromagnetism of Fe_2O_3 and similar crystals. See also Gilleo [6.101] on Fe–O–Fe interactions.

Sewell [6.161] interprets the conduction in α-Fe_2O_3 as a hopping mechanism resulting from a high electron–phonon coupling. Nakau [6.162] reveals the anisotropy of the conduction and its variation at 250°K. Tannhauser [6.20] does not believe that the transfers between Fe atoms of different valencies play any important part. Martius et al. [6.126] have studied the infrared spectrum. Ôno et al. [6.133], using the Mössbauer resonance, have obtained the hyperfine field and the quadrupole interaction on each side of 250°K. The latter undergoes an abrupt variation there (Fig. 6.28), due to the change in angle between the electrical field gradient and the hyperfine field. Lin [6.163] has suggested nonparallelism of the magnetic domains, and Tasaki et al. [6.164] have shown that nonstoichiometry and impurities affect only the isotropic part of the parasitic ferromagnetism. This appears only when the moments are in the (111) plane, corresponding to Dzyaloshinskii's theory. Aharoni et al. [6.165] have found, by differential thermal analysis, $T_C = 725°C$, while Freier et al. [6.166], using the Mössbauer resonance, have observed the disappearance of the Zeeman effect at 690°C, and Robbrecht and Doclo [6.167], using a thermal expansion technique, have seen singular points at -40, -7, $+684$, and $+726°C$. Abrahams [6.168] has compared the magnetic structure proposed by Shull et al. [6.149] with that of Cr_2O_3 (Fig. 6.29). A very detailed study of the resistivity and Seebeck/temperature curves for a large number of samples has led Gardner, Tanner et al. [6.169, 6.170] to conclude that conduction is often dominated below 800°C by the effect of impurities or an excess of iron

6.4. T_2O_3 Oxides

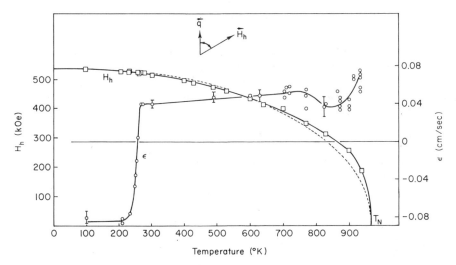

Fig. 6.28. Variation in the hyperfine field H_h and quadrupole interaction ε in relation to temperature. The dotted line represents the theoretical curve of H_h according to Weiss' theory (based on Ôno et al. [6.133]).

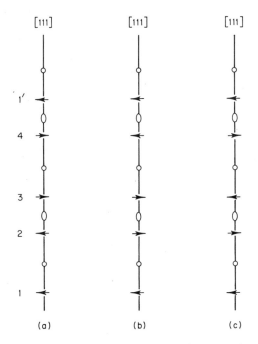

Fig. 6.29. Possible arrangements of the moments of 4 successive metallic atoms in the [111] direction of the structure of corundum. Circles represent inversion centers. Ellipses represent binary axes perpendicular to or in the plane of the figure: (a) α-Fe_2O_3, (b) not known, (c) Cr_2O_3 (based on Abrahams [6.168]).

($E_A = 0.1$ eV), while an intrinsic excitation mechanism intervenes above 1000°C (see also [6.171]). Imbert and Gerard [6.172] have continued the research done by Ôno et al. [6.133], and have shown that the orientation of the spins changes by 90° at 250°K. See also another theoretical study [6.173]. Finally, Smrcek et al. [6.174] have suggested that the magnetism which appears when hematite is heated might come from a partial transformation into the γ phase (see below).

α-Fe_2O_3 provides solid solutions or substitute crystals with many other oxides, notably with TiO_2 up to the isomorphous ilmenite $FeTiO_3$. In some cases valency induction occurs. The introduction of a tetravalent Ti atom, for instance, causes a valency change in an iron atom in accordance with the formula $Ti(IV)_x Fe(III)_{2-2x} Fe(II)_x O_3$ and the appearance of an n-type, as is confirmed by Morin [6.146, 6.152]. This material has been used to make thermistors (resistors with a negative temperature coefficient). Conduction takes place by transfers between Fe(II) and Fe(III), with very low mobility. Resistivity falls to a few ohm centimeters for 1% Ti atoms. Figure 6.30 shows the Seebeck effect irregularity observed in the neighborhood of the magnetic transformation. According to Grunewald [6.175], conduction is nevertheless intrinsic at high temperatures. Suchet [6.176] has studied the effect of the temperature of sintering of Fe_2O_3–TiO_2 ceramics in air, and has found a marked increase in resistivity at high temperatures, resulting from the suppression, at first on the surface and then deeper, of the valency induction by oxidation of the iron to valency III (Fig. 6.31). Haigh [6.177] has pointed out that Morin's transformation can still be observed for a substitution of 10% Ti atoms, and Kaye [6.178] has shown that its temperature drops in inverse proportion to the

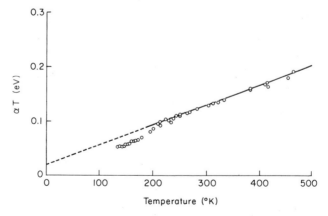

Fig. 6.30. Variation in the product of the Seebeck coefficient times absolute temperature in relation to absolute temperature. $Fe_2O_3 + 1\%$ Ti, sintering in oxygen (based on Morin [6.152]).

6.4. T_2O_3 Oxides

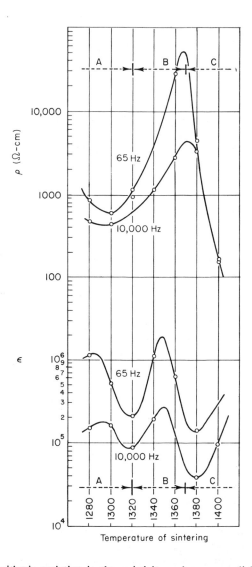

Fig. 6.31. Logarithmic variation in the resistivity and apparent dielectric constant in relation to the sintering temperature of ceramics (0.5% Ti atoms), (A) chemisorption of oxygen at the grain boundaries, (B) diffusion of oxygen from the surface of the grains (with rise in apparent dielectric constant), (C) transformation into magnetite (based on Suchet [6.176]).

Ti content. The addition of SnO_2 (*n*-type) has been investigated by Acket and Volger [6.179], and that of MgO (*p*-type) by Gardner et al. [6.180]. The addition of Al_2O_3, with which hematite forms solid solutions in all proportions, has been studied from the magnetic viewpoint [6.181, 6.182]. Other solid solutions exist with Cr_2O_3 (100%), Mn_2O_3 (50%), and V_2O_3 (30%). Tasaki et al. [6.164, 6.183], have studied the effect of various metal atoms on the magnetic properties of α-Fe_2O_3.

There is another iron sesquioxide, the γ phase, sometimes called maghemite, which crystallizes in a pseudocubic quadratic structure and presents a high magnetization intensity, with $T_C = 675°C$, like the α phase. Its magnetism is used in magnetic recording on flexible tapes. This structure derives from the structure of spinel, and corresponds to a magnetite with vacancies on the B sites (an oxide γ-Mn_2O_3 derives in the same way from Mn_3O_4). It is obtained by oxidation of magnetite at low temperature, is unstable, and changes toward hematite by heating to 300–350°C. The addition of alkaline oxides stabilizes it [6.13]. The substitution of aluminum or trivalent chromium also gives very stable substances [6.184]. Solutions exist in all proportions with γ-Al_2O_3 [6.185]. Martinet [6.107] has found a resistivity/temperature law corresponding to $E_A \sim 0.07$ eV, as for Fe_3O_4. Neel [6.100] and Henry and Boehm [6.186], by measuring the average moment, and Ferguson and Hass [6.187], by neutron diffraction, confirm the existence of vacancies at B sites in the structure. Ferguson et al. give the formula $Fe(III)_{21.33}\square_{2.67}O_{32}$. Kushiro [6.188] has studied the effect of pressure on the $\gamma \rightarrow \alpha$ transformation temperature. Ueda and Hasegawa [6.189] have continued research into the position of vacancies, using X rays and neutrons, and have shown that some degree of order exists (see also [6.190]).

The existence of the oxide Co_2O_3 is uncertain, and Ni_2O_3 has never been recorded.

To sum up, whenever Fe(II) and Fe(III) atoms are neighbors in equivalent crystallographic sites, in $Fe_{1-x}O$, Fe_3O_4 above 120°K or $Ti_xFe_{2-x}O_3$, a semiconduction mechanism by the electron transfer is observed between these atoms. On the other hand, when Fe(III) exists alone, in pure, stoichiometric α-Fe_2O_3, it is the excitation mechanism that appears as soon as the temperature is high enough for the effect of the unavoidable lattice defects to be negligible.

For the first time in this chapter we have dealt with typical magnetic semiconductors, Fe_3O_4 and Fe_2O_3, that have been the object of detailed research. This is why they have been described in more detail, with particular insistence on their magnetic properties, which are inseparable from their electrical conduction. Magnetite in particular, with its substitute derivatives (ferrites), provides an extremely interesting field for study. It is clear, however, that everything has not been understood yet in the properties of these crystals. This will be covered in Part Three, and particularly in Chapter 10, when we deal with the extraordinary Hall effect.

REFERENCES

[6.1] P. A. Flinn, "Experimental Methods of Materials Research" (H. Herman, ed.). Wiley (Interscience), New York, 1967.
[6.2] V. I. Gol'danskii, "Effekt Messbauera i ego primeneniya v khimii." Inst. Khim. Fiz. Akad. Nauk SSSR, Moscow, 1963; English transl. Consultants Bureau, New York, 1964.
[6.3] G. K. Wertheim, "Mössbauer Effect, Principles and Applications." Academic Press, New York, 1964.
[6.4] B. G. Harvey, "Introduction to Nuclear Physics and Chemistry." Prentice Hall, Englewood Cliffs, New Jersey, 1965.
[6.5] R. L. Mössbauer, *Z. Phys.* **151**, 124 (1958).
[6.6] O. C. Kistner and A. W. Sunyar, *Phys. Rev. Lett.* **4**, 342 (1960).
[6.7] G. Shirane, D. E. Cox, and S. L. Ruby, *Phys. Rev.* **125**, 1158 (1962).
[6.8] M. Foëx, *Bull. Soc. Chim.* **19**, 373 (1952).
[6.9] C. G. Shull, W. A. Strauser, and E. O. Wollan, *Phys. Rev.* **83**, 333 (1951).
[6.10] W. L. Roth, *Phys. Rev.* **110**, 1333 (1958).
[6.11] H. Kramers, *Physica* **18**, 101 (1952).
[6.12] L. S. Darken and R. W. Gurry, *J. Amer. Chem. Soc.* **68**, 798 (1946).
[6.13] P. Pascal, "Nouveau Traité de Chimie Minérale." Masson, Paris, 1956–1963.
[6.14] H. Bizette and B. Tsaï, *C.R. Acad. Sci. Paris* **217**, 390 (1943).
[6.15] W. L. Roth, *Acta Crystallogr.* **13**, 140 (1960).
[6.16] W. A. Fischer and H. Von Ende, *Arch. Eisenhüttenw.* **21**, 297 (1950).
[6.17] J. W. Tomlinson and H. Inouye, *J. Chem. Phys.* **20**, 193 (1952).
[6.18] F. Marion, State thesis, Nancy (1955).
[6.19] J. Aubry and F. Marion, *C.R. Acad. Sci. Paris* **241**, 1778 (1955).
[6.20] D. S. Tannhauser, *J. Phys. Chem. Solids* **23**. 25 (1962).
[6.21] S. M. Ariya and B. Ya. Brach, *Fiz. Tverd. Tela* **5**, 3496 (1963).
[6.22] H. Geiger, R. L. Levin, and J. B. Wagner Jr., *J. Phys. Chem. Solids* **27**, 947 (1966).
[6.23] F. Trombe, *J. Phys. Radium* **12**, 170 (1951).
[6.24] U. Croatto, *Ric. Sci.* **19**, 1324 (1949).
[6.25] A. P. Young, W. B. Wilson, and C. M. Schwartz, *Phys. Rev.* **121**, 77 (1961).
[6.26] M. Roilos and P. Nagels, *Solid State Commun.* **2**, 285 (1964).
[6.27] K. V. Rao and A. Smakula, *J. Appl. Phys.* **36**, 2031 (1965).
[6.28] B. Fisher and D. S. Tannhauser, *J. Chem. Phys.* **44**, 1663 (1966).
[6.29] V. P. Zhuze and A. I. Shelykh, *Fiz. Tverd. Tela* **8**, 629 (1966).
[6.30] A. I. Shelykh. K. S. Artemov, and V. E. Shvaiko-Shvaikovskii, *Fiz. Tverd. Tela* **8**, 883 (1966).
[6.31] I. G. Austin, A. J. Springthorpe, and B. A. Smith, *Phys. Lett.* **21**, 20 (1966).
[6.32] A. Bruck and D. S. Tannhauser, *J. Appl. Phys.* **37**, 3647 (1966).
[6.33] I. G. Austin, A. J. Springthorpe, B. A. Smith, and C. E. Turner, *Proc. Phys. Soc.* **90 (1)**, 157 (1967).
[6.34] C. Henry la Blanchetais, *J. Phys.* **12**, 765 (1951).
[6.35] R. W. Wright and J. O. Andrews, *Proc. Phys. Soc.* **62A**, 446 (1949).
[6.36] C. A. Hogarth, *Proc. Phys. Soc.* **64B**, 69 (1951).
[6.37] R. W. Johnston and D. C. Cronemeyer, *Phys. Rev.* **93**, 634 (1954).
[6.38] F. J. Morin, *Phys. Rev.* **93**, 1199 (1954).
[6.39] R. R. Heikes, *Phys. Rev.* **99**, 1232 (1955).
[6.40] E. Yamaka and K. Sawamoto, *Phys. Rev.* **112**, 1861 (1958).
[6.41] R. Newman and R. M. Chrenko, *Phys. Rev.* **114**, 1507 (1959).
[6.42] G. L. Sewell, *Proc. Phys. Soc.* **76**, 985 (1960).
[6.43] M. I. Klinger, *Izv. Akad. Nauk SSSR, Ser. Fiz.* **25**, 1342 (1961).

[6.44] S. P. Mitoff, *J. Chem. Phys.* **35**, 882 (1961).
[6.45] S. Koide and H. Takei, *J. Phys. Soc. Japan* **18**, 319 (1963).
[6.46] H. Saltsburg and D. P. Snowden, *Phys. Chem. Sol. Surfaces Conf.*, Providence, 1964, *Surface Sci.* **2**, 288 (1964).
[6.47] H. Saltsburg, D. P. Snowden, and M. C. Garrison, *J. Phys. Chem.* **68**, 3765 (1964).
[6.48] K. Hauffe, "Transition Metal Compounds," p. 37. Gordon & Breach, New York, 1964.
[6.49] S. Fujime, M. Murakami, and E. Hirahara, *J. Phys. Soc. Japan* **16**, 183 (1961).
[6.50] V. P. Zhuze and A. I. Shelykh, *Fiz. Tverd. Tela* **5**, 1756 (1963).
[6.51] A. J. Bosman, H. J. van Daal, and G. F. Knuvers, *Phys. Lett.* **19**, 372 (1965).
[6.52] D. P. Snowden, H. Saltsburg, and J. H. Pereue Jr. *J. Phys. Chem. Solids* **25**, 1099 (1964).
[6.53] D. P. Snowden and H. Saltsburg, *Phys. Rev. Lett.* **14**, 497 (1965).
[6.54] M. Nachman, F. G. Popescu, and J. Rutter, *Phys. Status Solidi* **10**, 519 (1965).
[6.55] Ya. M. Ksendzov and I. A. Drabkin, *Fiz. Tverd. Tela* **7**, 1884 (1965).
[6.56] M. W. Vernon and M. C. Lowell, *J. Phys. Chem. Solids* **27**, 1125 (1966).
[6.57] V. V. Makarov, Ya. M. Ksendzov, and V. I. Kruglov, *Fiz. Tverd. Tela* **9**, 663 (1967).
[6.58] R. W. G. Wyckoff, "Crystal Structures," 2nd ed., Vol. 1. Wiley (Interscience), New York, 1965.
[6.59] L. Meyer-Schützmeister, *Z. Phys.* **129**, 148 (1951).
[6.60] M. Perrot, G. Peri, J. Robert, J. Tortosa, and A. Sauze, *C.R. Acad. Sci. Paris* **242**, 2519 (1956).
[6.61] G. Peri, M. Perrot, and J. Robert, *J. Phys. Radium* **18**, 282 (1957).
[6.62] F. Oprea and P. Balta, *Bull. Inst. Politech. Bucarest* **21**, 73 (1959).
[6.63] H. Bizette and B. Tsaï, *C.R. Acad. Sci. Paris* **241**, 182 (1955).
[6.64] N. Perakis, A. Serres, and T. Karantassis, *J. Phys. Radium* **17**, 134 (1956).
[6.65] Jih-Heng Hu and H. L. Johnston, *J. Amer. Chem. Soc.* **75**, 2471 (1953).
[6.66] B. N. Brockhouse, *Phys. Rev.* **94**, 781 (1954).
[6.67] G. Caglioti, F. P. Ricci, A. Santoro, and V. Scatturin, *J. Phys. Soc. Japan suppl.* **17**, B-II, 348 (1962).
[6.68] M. O'keefe and F. S. Stone, *J. Phys. Chem. Solids* **23**, 261 (1962).
[6.69] G. A. Kholoday, *Ukr. Fiz. Zh.* **10**, 1036 (1965).
[6.70] K. Hauffe and J. Block, *Z. Phys. Chem.* **196**, 438 (1951).
[6.71] Y. Shimomura and I. Tsubokawa, *J. Phys. Soc. Japan* **9**, 19 (1954).
[6.72] G. M. Schwab and H. Schmid, *J. Appl. Phys. suppl.* **33**, 426 (1962).
[6.73] E. G. Schlosser, *Z. Elektrochem.* **65**, 453 (1961).
[6.74] H. P. Rooksby and M. W. Vernon, *Brit. J. Appl. Phys.* **17**, 227 (1966).
[6.75] S. Gocan, *An. Stiint. Univ. "Al. I. Cuza" Iasi (Ser. noua) I (Roum.)* **9**, 503 (1963).
[6.76] E. J. W. Verwey, P. W. Haaijman, F. C. Romeijn, and G. W. van Oosterhout, *Philips Res. Rep.* **5**, 173 (1950).
[6.77] G. Parravano, *J. Chem. Phys.* **23**, 5 (1955).
[6.78] R. R. Heikes and W. D. Johnston, *J. Chem. Phys.* **26**, 582 (1957).
[6.79] R. C. Miller and R. R. Heikes, *J. Chem. Phys.* **28**, 348 (1958).
[6.80] J. B. Goodenough, D. G. Wickham, and W. J. Croft, *J. Appl. Phys.* **29**, 382 (1958); *J. Phys. Chem. Solids* **5**, 107 (1958).
[6.81] S. van Houten, *Semicond. Conf. (Exeter 1962)*, p. 197. Inst. Phys. & Phys. Soc., London 1962.
[6.82] Ya. M. Ksendzov, L. N. Ansel'm, L. L. Vasil'eva, and V. M. Latysheva, *Fiz. Tverd. Tela* **5**, 1537 (1963).
[6.83] S. Koide, *J. Phys. Soc. Japan* **18**, 1699 (1963).
[6.84] M. Akiyama, *J. Phys. Soc. Japan* **20**, 182 (1965).

[6.85] A. J. Springthorpe, I. G. Austin, and B. A. Austin, *Solid State Commun.* **3**, 143 (1965).
[6.86] A. J. Bosman and C. Crevecoeur, *Phys. Rev.* **144**, 763 (1966).
[6.87] I. G. Austin, A. J. Springthorpe, and B. A. Smith, *Phys. Lett.* **21**, 20 (1966).
[6.88] B. Fisher and J. B. Wagner Jr., *Phys. Lett.* **21**, 606 (1966).
[6.89] H. J. van Daal and A. J. Bosman, *Phys. Lett.* **23**, 525 (1966).
[6.90] I. G. Austin, A. J. Springthorpe, B. A. Smith, and C. E. Turner, *Proc. Phys. Soc.* **90 (1)**, 157 (1967).
[6.91] T. F. W. Barth and E. Posnjack, *Z. Kristallogr.* **82**, 325 (1932).
[6.92] E. J. W. Verwey and J. H. de Boer, *Rec. Trav. Chim. Pays-Bas* **55**, 531 (1936).
[6.93] E. J. W. Verwey and E. L. Heilmann, *J. Chem. Phys.* **15**, 174 (1947).
[6.94] J. H. de Boer and E. J. W. Verwey, *Proc. Phys. Soc.* **59A**, 59 (1937).
[6.95] E. J. W. Verwey and P. W. Haayman, *Physica* **8**, 979 (1941).
[6.96] E. J. W. Verwey, P. W. Haayman, and F. C. Romeijn, *J. Chem. Phys.* **15**, 181 (1947).
[6.97] E. J. W. Verwey, *Nature* **144**, 327 (1939).
[6.98] S. C. Abrahams and B. A. Calhoun, *Acta Crystallogr.* **6**, 105 (1953).
[6.99] B. P. Calhoun, *Phys. Rev.* **94**, 1577 (1954).
[6.100] L. Neel, *Ann. Phys. (Paris)* **12**, 137 (1948).
[6.101] M. A. Gilleo, *Phys. Rev.* **109**, 777 (1958).
[6.102] T. Okamura and Y. Torizuka, *Sci. Rep. Res. Inst. Tohôku Univ., Ser. A* **2**, 352 (1950).
[6.103] C. A. Domenicali, *Phys. Rev.* **78**, 458 (1950).
[6.104] M. A. Grabovskii, *Izv. Akad. Nauk SSSR, Ser. Geofiz.* **61** (1951).
[6.105] C. G. Shull, *Phys. Rev.* **81**, 626 (1951).
[6.106] N. C. Tombs and H. P. Rooksby, *Acta Crystallogr.* **4**, 474 (1951).
[6.107] J. Martinet, *C.R. Acad. Sci. Paris* **234**, 2167 (1952).
[6.108] M. Bernard and J. Jaffray, *J. Phys. Radium* **13**, 705 (1952).
[6.109] H. P. Rooksby and B. T. M. Willis, *Acta Crystallogr.* **6**, 565 (1953).
[6.110] J. B. Birks, *Proc. Phys. Soc.* **B63**, 65 (1950).
[6.111] S. G. Salikhov, *Izv. Akad. Nauk. SSSR, Ser. Fiz.* **18**, 456 (1954).
[6.112] I. Nagy, D. Pallagi and L. Pal, *Acta Phys. Sci. Hungar.* **6**. 341 (1956).
[6.113] A. W. McReynolds and T. Riste, *Phys. Rev.* **95**, 1161 (1954).
[6.114] W. C. Hamilton, *Phys. Rev.* **110**, 1050 (1958).
[6.115] A. A. Samokhvalov and I. G. Fakidov, *Fiz. Metal. Metalloved.* **4**, 249 (1957).
[6.116] J. M. Lavine, *Phys. Rev.* **114**, 482 (1959).
[6.117] W. Mann, *Ann. Phys. (Leipzig)* (series 7) **3**, 122 (1959).
[6.118] A. A. Samokhvalov and I. G. Fakidov, *Fiz. Tverd. Tela* **2**, 414 (1960).
[6.119] J. M. Lavine, *Phys. Rev.* **123** 1273 (1961).
[6.120] P. A. Miles, W. B. Westphal, and A. von Hippel, *Rev. Mod. Phys.* **29**, 293 (1957).
[6.121] D. O. Smith, not published, quoted by Lavine [6.119].
[6.122] A. A. Samokhvalov and I. G. Fakidov, *Fiz. Metal. Metalloved.* **7**, 465 (1959).
[6.123] T. D. Zotov, *Fiz. Tverd. Tela* (Sb. II) **8**, 10 (1959).
[6.124] T. D. Zotov, *Fiz. Metal. Metalloved.* **9**, 48 (1960).
[6.125] T. D. Zotov, *Kristallografiya* **9**, 929 (1964).
[6.126] W. D. Westwood, A. G. Sadler, and D. C. Lewis, *J. Can. Ceram. Soc.* **33**, 138 (1964).
[6.127] W. Haubenreisser, *Phys. Status Solidi* **1**, 619 (1961).
[6.128] H. Schröder, *Phys. Status. Solidi* **1**, K152 (1961).
[6.129] H. Schröder, *Monatsber. Deut. Akad. Wiss. Berlin* **3**, 615 (1961).
[6.130] D. S. Tannhauser, *Phys. Kondens. Mater.* **3**, 146 (1964).

[6.131] S. Yamaguchi, *Naturwissenschaften* **49**, 252 (1962).
[6.132] R. Bauminger, S. G. Cohen, A. Marinov, S. Ofer, and E. Segal, *Phys. Rev.* **122**, 1447 (1961).
[6.133] K. Ôno, Y. Ishikawa, A. Ito, and E. Hirahara, *J. Phys. Soc. Japan suppl.* **17** (B-I), 125 (1962).
[6.134] A. Ito, K. Ôno, and Y. Ishikawa, *J. Phys. Soc. Japan* **18**, 1465 (1963).
[6.135] V. G. Bhide and R. H. Dani, *Physica* **27**, 821 (1961).
[6.136] T. Nagamiya, *Ann. Rep. Sci. Works Fac. Sci. Osaka Univ.* **8**, 1 (1960).
[6.137] J. B. Goodenough, "Magnetism and the Chemical Bond," Wiley (Interscience), New York, 1963.
[6.138] Z. Simsa, *Czech. J. Phys.* **13**, 471 (1963).
[6.139] N. Miyata, *J. Phys. Soc. Japan* **16**, 206 (1961).
[6.140] F. K. Lotgering, *J. Phys. Chem. Solids* **25**, 195 (1964).
[6.141] M. Rosenberg, P. Nicolau, and I. Bunget, *Phys. Status Solidi* **14**, K65 (1966).
[6.142] Z. Simsa and K. Zaveta, *Czech. J. Phys.* **B13**, 471 (1966).
[6.143] L. Neel, *Ann. Phys. Paris* **4**, 249 (1949).
[6.144] L. Neel, *Ann. Geophys. Paris* **5**, 99 (1949).
[6.145] J. Roquet, *Ann. Geophys. Paris* **10**, 225 (1954).
[6.146] F. J. Morin, *Phys. Rev.* **78**, 819 (1950).
[6.147] C. Guillaud, *J. Phys. Radium* **12**, 489 (1951).
[6.148] R. Chevallier, *J. Phys. Radium* **12**, 172 (1951).
[6.149] C. G. Shull, W. A. Strauser, and E. O. Wollan, *Phys. Rev.* **83**, 333 (1951).
[6.150] B. T. M. Willis and H. P. Rooksby, *Proc. Phys. Soc.* **B65**, 950 (1952).
[6.151] D. J. M. Bevan, J. P. Shelton, and J. S. Anderson, *J. Chem. Soc.* 1729 (1948).
[6.152] F. J. Morin, *Phys. Rev.* **83**, 1005 (1951).
[6.153] F. J. Morin, *Phys. Rev.* **93**, 1195 (1954).
[6.154] L. Neel and R. Pauthenet, *C.R. Acad. Sci. Paris* **234**, 2172 (1952).
[6.155] L. Neel, *Rev. Mod. Phys.* **25**, 58 (1953).
[6.156] G. Haigh, *Phil. Mag.* (Ser. 8) **2**, 877 (1957).
[6.157] L. M. Corliss, J. M. Hastings, and J. E. Goldman, *Phys. Rev.* **93**, 893 (1954).
[6.158] J. Wucher, *C.R. Acad. Sci. Paris* **241**, 288 (1955).
[6.159] I. E. Dzyaloshinskii, *Zh. Eksp. Teor. Fiz.* **32**, 1547 (1957).
[6.160] I. E. Dzyaloshinskii, *J. Phys. Chem. Solids* **4**, 241 (1958).
[6.161] G. L. Sewell, *Proc. Phys. Soc.* **76** (Part 6), 985 (1960).
[6.162] T. Nakau, *J. Phys. Soc. Japan* **15**, 727 (1960).
[6.163] S. T. Lin, *J. Phys. Soc. Japan suppl.* **17** (B-I). 226 (1962).
[6.164] A. Tasaki, K. Siratori, and S. Iida, *J. Phys. Soc. Japan suppl.* **17** (B-I), 235 (1962).
[6.165] A. Amaroni, E. H. Frei, and M. Schieber, *Phys. Rev.* **127**, 439 (1962).
[6.166] S. Freier, M. Greenshpan, P. Hillman, and H. Shechter, *Phys. Lett.* **2**, 191 (1962).
[6.167] G. G. Robbrecht and R. J. Doclo, *Phys. Lett.* **3**, 85 (1962).
[6.168] S. C. Abrahams, *Phys. Rev.* **130**, 2230 (1963).
[6.169] R. F. G. Gardner, F. Sweett, and D. W. Tanner, *J. Phys. Chem. Solids* **24**, 1175, 1183 (1963).
[6.170] D. W. Tanner, F. Sweett, and R. F. G. Gardner, *Brit. J. Appl. Phys.* **15**, 1041 (1964).
[6.171] R. S. Roth and J. L. Waring, *Ann. Mineralogist* **49**, 242 (1964).
[6.172] P. Imbert and A. Gerard, *Bull. Soc. Belge Phys.* (Ser 4) **1**, 3 (1964).
[6.173] P. J. Flanders and S. Shtrikman, *Solid State Commun.* **3**, 285 (1965).
[6.174] K. Smrcek, O. Cejchan, and J. Chvatik, *Acta Tech. Hungar.* **54**, 61 (1966).
[6.175] H. Grunewald, *Ann. Phys. Leipzig* **14**, 129 (1954).
[6.176] J. Suchet, *J. Phys. Radium* **18**, 10A (1957).
[6.177] G. Haigh, *Phil. Mag.* (Ser 8) **2**, 505 (1957).

[6.178] G. Kaye, *Proc. Phys. Soc.* (Part I) **80**, 238 (1962).
[6.179] G. A. Acket and J. Volger, *Physica* **32**, 1543 (1966).
[6.180] R. F. G. Gardner, R. L. Moss, and D. W. Tanner, *Brit. J. Appl. Phys.* **17**, 55 (1966).
[6.181] P. W. Selwood, L. Lyon, and M. Ellis, *J. Amer. Chem. Soc.* **73**, 2310 (1951).
[6.182] Ya. V. Vasil'ev and G. A. Shcherbakova, *Fiz. Tverd. Tela* **5**, 1090 (1963).
[6.183] A. Tasaki and S. Iida, *J. Phys. Soc. Japan* **16**, 1697 (1961).
[6.184] A. Michel, "Coll. Int. Réactions Etat Sol." p. 99. C.N.R.S., Paris, 1948.
[6.185] V. Cirilli, *Gazz. Chim. Ital.* **80**, 347 (1950).
[6.186] W. E. Henry and M. J. Boehm, *Phys. Rev.* **101**, 1253 (1956).
[6.187] G. A. Ferguson and M. Hass, *Phys. Rev.* **112**, 1130 (1958).
[6.188] I. Kushiro, *J. Geomagn. Geoelect.* **11**, 148 (1960).
[6.189] R. Ueda and K. Hasegawa, *J. Phys. Soc. Japan, suppl.* **17** (B-II), 391 (1962).
[6.190] M. Yuzuri, Y. H. Kang, and Y. Goto, *J. Phys. Soc. Japan suppl.* **17** (B-I), 253 (1962).

Chapter 7 | **Transition Metal Chalcogenides**

7.1. Main Structures

The main crystallographic structures found in this class of compounds have symbols $B8_1$ (nickel–arsenide), C6 (cadmium hydrate), C7 (molybdenum disulfide), C2 (pyrite) and C18 (marcasite). We shall deal mainly with the standard structures $B8_1$ and C2, already described in Chapter 4, but which have special aspects for chalcogenides. Readers wishing for more detailed information about them, as well as rarer structures (B10, B13, B17, B34, C19, C27, C37, $D7_2$), may refer to the general works by Wells [7.1] and Wyckoff [7.2].

The lattice of the structure of nickel–arsenide NiAs was described in Section 4.1. In the case of chalcogenides, this structure frequently contains metal vacancies, in accordance with the general formula $\square_x T_{1-x} X$. The vacancies, generally ordered, affect only one T atom plane in two, perpendicular to axis c [7.3–7.5]. The $B8_1$ structure very often does not exist when vacancies are absent: this is the case particularly for VS, CoS, CoSe, CrTe, MnTe, FeTe, CoTe. It is usually more or less pure in the neighborhood of the TX composition, and then an overstructure appears, followed by a monoclinal distortion when the level of 54% X atoms is reached (approximately T_7X_8). It is particularly marked for CrS, FeS, TiSe, CrSe, FeSe, CoSe, NiSe, CrTe The distortion occasionally stops at 60% X atoms, for example in the case of Cr_2S_3, Cr_2Se_3, Cr_2Te_3, and Fe_2Te_3. In certain cases, x can reach 0.5. There is then a $\square TX_2$ formula, in which one T atom plane in two, perpendicular to axis c, is empty (Fig. 7.1). This is the C6 structure of cadmium hydrate $Cd(OH)_2$. One of the varieties of iodide CdI_2 also crystallizes in this structure.

7.1. Main Structures

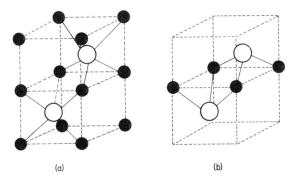

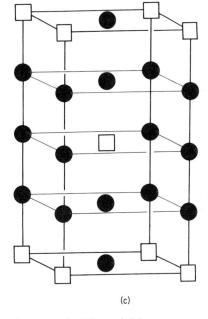

Fig. 7.1. B8$_1$ structure of NiAs (a) and C6 structure of Cd(OH)$_2$ (b); alternation of partly filled and full (001) planes for an intermediate T$_3$X$_4$ compound (c). The black dots represent the metal atoms and the blank squares the vacancies. The X atoms have been left out of the bidimensional cell (c).

In some cases there is continuous transition between the B8$_1$ and C6 structures, from $x = 0$ to $x = 0.5$: TiSe to TiSe$_2$, TiTe to TiTe$_2$, NiTe to NiTe$_2$ [7.2, 7.6]. Digests concerning the phases of the B8$_1$ or neighboring structures have been published by Haraldsen [7.7] and Kjekshus and Pearson [7.8].

It was seen in Section 2.1 that Krebs' [7.9] research revealed the role of the p orbitals in M—X or T—X bonds of compounds crystallizing in the B1 structure. For B8$_1$ structure compounds, where the ratio of axis c to axis a is around 1.6 to 1.7, Pearson [7.10] proposes a similar bond scheme. In particular, he points out that the resonating p^3 bonds are compatible with coordination of the T atoms, and that the occupation of the d states worked out from this scheme corresponds to the magnetic properties ($2S = 5$ for MnTe, 4 for CrSe and FeS), while this would not be so in the case of a scheme using d–s–p hybrid orbitals. When c/a falls to around 1.5, a metallic bond is superimposed on this chart, corresponding to d electron transfers in a narrow band. However surprising this may seem, Pearson's hypothesis has been confirmed experimentally. It was known that the position of the $K\beta_1$ emission line of X spectra was subject to the influence of T—X bonds of the compound being studied [7.11]. To define this influence, Men'shikov and Nemnonov [7.12, 7.13] have studied the spectra of a whole series of chromium compounds. Next, in a major analysis of their results [7.14], they showed that the d electrons played a very small role in the T—X bonds, due mainly to resonating p^3 bonds (Fig.

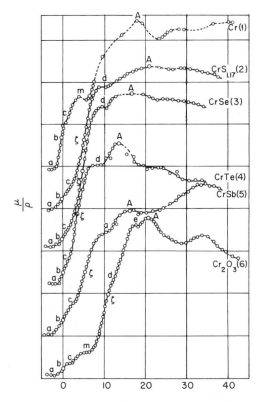

Fig. 7.2. Variation in the absorption coefficient (in arbitrary units) in relation to the energy of the radiation quantum X for the K line of chromium in different compounds. The initial absorption field (abc), resulting from the 1s–3d transitions, is much weaker than in the metal chromium. On the other hand, the existence of a shoulder (d), resulting from the 1s–3p transitions, shifts the absorption front ζ of chromium in the oxide towards high wavelengths (based on Men'shikov and Nemnonov [7.14]).

7.1. Main Structures

7.2). More recently, detailed work by Barstad et al. [7.15] on the structure of nickel tellurides has shown that the Te—Ni—Te angles do not vary by more than ten minutes or so when one moves from divalent nickel (NiTe, d^8) to tetravalent nickel (NiTe$_2$, d^6).

The transfer mechanism in compounds with multiple valencies was described in Section 2.4, and in Section 3.3 the example of wüstite $Fe_{1-x}O$ or, to be more accurate, $\square_x Fe^{2}_{1-3x} Fe^{3+}_{2} O$, was examined. It has just been seen that structure $B8_1$ may be considered, as an initial approximation, as a distorted B1 structure, and that there is no essential difference between them, neither as regards the nature of the T—X bonds, nor even as regards the X—T—X angles, since these do not differ much from the angles of the B1 structure. It is therefore possible, when considering the $B8_1$ structure, to carry out the same calculation of the fraction of T atoms that has changed valencies. The author [7.16] has shown that arguments based on the number of d electrons in each atom lead to the same result as arguments based on the ionic charge, which is

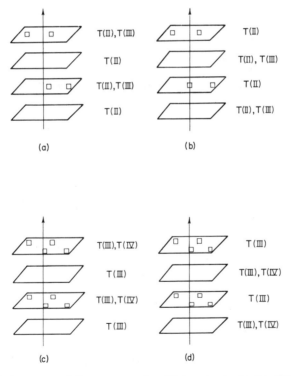

Fig. 7.3. Various cases of the presence of multiple valencies in the planes containing vacancies (a, c) or full planes (b, d). $0 < x < \frac{1}{3}$ (a, b) and $\frac{1}{3} < x < \frac{1}{2}$ (c, d) (based on Suchet [7.16]).

not valid for strongly covalent crystals. We can write that a fraction x of the metalloid atoms have a nonbonding electronic pair, while the rest of these atoms each contribute 3 electrons (shown in parentheses) to 3 bonding pairs (Lewis pairs):

$$(1-x)\,T^{d^k s^2} + X^{s^2 p^4} \longrightarrow \square_x X_x^{s^2 p^6} + T_{1-x}^{d^{k'} s^0 (p^3)} X_{1-x}^{s^2 (p^3)} \qquad (7.1)$$

The invariance of the total number of d, s, and p electrons then gives the average number k' of d electrons per T atom:

$$k' = k - 2x/(1-x) \qquad (7.2)$$

which will be an integer if $x = 0$ ($k' = k$), $\frac{1}{3}(k' = k-1)$ or $\frac{1}{2}(k' = k-2)$. For any other value, two different valencies appear. Figures 7.3 and 7.4 give an idea of the forecasts that can be made of the transfers of d electrons within the (001) planes when the vacancies are ordered (when the planes with vacancies and full planes are not equivalent). The estimated minimum values can be observed qualitatively on polycrystals, when the homogeneity domain is wide (tellurides), since the transfers which take place within the (001) planes are certainly not negligible compared with those taking place parallel to axis c (cf. orientation of the t_{2g} orbitals, Section 4.1).

Figure 7.1b shows that the C6 structure is made up of layers parallel to the (001) planes of the corresponding B8$_1$ structure. If the T atoms of the layer

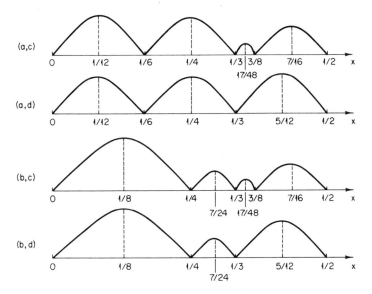

Fig. 7.4. Possible transfer density within the (001) planes (in arbitrary units) in relation to x, in the different cases of Fig. 7.3 (based on Suchet [7.16]).

7.1. Main Structures

are placed in the plane of the figure, one obtains the representation given in Fig. 7.5a. Several variants of this arrangement exist, depending on the exact position of the X atoms (C19, C27). A neighboring layer arrangement, shown in Fig. 7.5b, is that of the C7 structure of MoS_2. Each T atom in it is still surrounded by 6X atoms, but the X—T—X angles diverge more from 90° [7.1, 7.2]. Each double layer of X atoms is articulated with the following one, to form the sequences AABBAABB..., AABBCCAABBCC..., or AABBAA-CCAABBAACC..., corresponding to patterns of 2, 3, or 4 layers, respectively, where the T atoms may occupy different sites. Certain heavy metal compounds have structures of this type, close to C7 [7.17]. Table 7.1 summarizes all the TX_2 compounds crystallizing in these structures. It has been assumed that the vacancies take part in the covalent bonds as valency 0 atoms, and the corresponding electronic formula of the T atom has been given, with the electrons involved in the T—X Lewis pairs in parentheses; $2S$ indicates the number of bachelor nonbonding d electrons that could be calculated on this basis. Only the formulae in the first and last columns are reliable, however, because of the absence of magnetic moment in most of the intervening compounds. Those with structure C7 have been discussed by Pauling [7.18]. The compounds on this table are generally semiconductors or pseudometals.

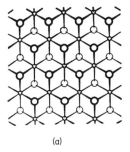
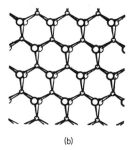

(a)   (b)

Fig. 7.5. Layer structures C6 (a) and C7 (b). The small circles represent T atoms in the plane of the figure, and the large circles the X atoms above and below (based on Wells [7.1]).

It was seen in Section 4.1 that the e_g–s–p hybridization of the metal AOs is general in the C2 and C18 structures. Table 7.2 brings together the known compounds and classifies them in relation to the number of d electrons, according to Hulliger and Mooser [7.19, 7.20]. The electron formula of the T atom has been added, and the expected $2S$ number in each case. In fact, while this number has been confirmed experimentally for the first two columns, the same is not true for the last two, in which no magnetic moment is apparent (except CoS_2, ferromagnetic, cf. Section 7.2), and the conductivity is of metallic type. The d electrons in the highest level are therefore probably

TABLE 7.1

Known compounds	Structure	2S	T Electronic formula	
			Level before higher level	Higher level
TiS$_2$ TiSe$_2$ TiTe$_2$	C6	0	$3d^0$ ...	$4s^0(p^3)$
ZrS$_2$ ZrSe$_2$ ZrTe$_2$	C6	0	$4d^0$ ...	$5s^0(p^3)$
HfS$_2$ HfSe$_2$	C6	0	$5d^0$ ...	$6s^0(p^3)$
VSe$_2$	C6	1	$3d^1$ ...	$4s^0(p^3)$
NbS$_2$	C19	1	$4d^1$ ...	$5s^0(p^3)$
TaS$_2$ TaTe$_2$	Distorted C6	1	$5d^1$ ...	$6s^0(p^3)$
MoS$_2$ MoSe$_2$ MoTe$_2$	C7	(2)	?	?
WS$_2$ WSe$_2$ WTe$_2$?	C7	(2)	?	?
ReS$_2$ ReSe$_2$?	C7	(3)	?	?
FeTe$_2$?	C6	(2)	$3t_{2g}^4 e_g^0$ ...	$4s^0(p^3)$
CoSe$_2$? CoTe$_2$?	C6	1	$3t_{2g}^5 e_g^0$ ...	$4s^0(p^3)$
β-RhTe$_2$	C6	1	$4t_{2g}^5 e_g^0$ ...	$5s^0(p^3)$
IrTe$_2$?	C6	1	$5t_{2g}^5 e_g^0$ ...	$6s^0(p^3)$
NiSe$_2$? NiTe$_2$	C6	0	$3t_{2g}^6 e_g^0$ ...	$4s^0(p^3)$
PdSe$_2$ PdTe$_2$	C6	0	$4t_{2g}^6 e_g^0$ ...	$5s^0(p^3)$
PtS$_2$ PtSe$_2$ PtTe$_2$	C6	0	$5t_{2g}^6 e_g^0$ ...	$6s^0(p^3)$

7.1. Main Structures

TABLE 7.2

Known compounds	Structure	$2S$	T Electronic formula Level before higher level	Higher level
MnS_2 $MnSe_2$ $MnTe_2$	C2	5	$3d^5$ ⋯	$(4sp^3d^2)$
FeS_2 $FeSe_2$ $FeTe_2$	C2/C18 C18 C18	0	$3d^6(d^2$ ⋯	$4sp^3)$
RuS_2 $RuSe_2$ $RuTe_2$	C2	0	$4d^6(d^2$ ⋯	$5sp^3)$
OsS_2 $OsSe_2$ $OsTe_2$	C2	0	$5d^6(d^2$ ⋯	$6sp^3)$
CoS_2 $CoSe_2$ $CoTe_2$	C2 C2/C18 C18	1	$3d^6(d^2$ ⋯	$4sp^3)d^1$
RhS_2 $RhSe_2$ $\alpha\text{-}RhTe_2$	C18 C2	1	$4d^6(d^2$ ⋯	$5sp^3)d^1$
$IrS_{2.8}$ $IrSe_{2.8}$ $IrTe_{2.8}$	Pseudo-pyrite	?	?	?
NiS_2 $NiSe_2$	C2	2	$3d^6(d^2$ ⋯	$4sp^3)d^2$
PdS_2 $PdSe_2$	Distorted C2	2	$4d^6(d^2$ ⋯	$5sp^3)d^2$
(CuS_2)	C2	(3)	$3d^6(d^2$ ⋯	$4sp^3)d^3$

entirely delocalized for Co and Ni compounds and their homologues. The type of bond peculiar to the first column results from the stability of the d^5 electronic configuration [7.21].

Dudkin's and Goodenough's criteria for forecasting electrical properties have already been used in Sections 4.1 and 4.4. Dudkin's predicts the semiconducting character of MnS and MnSe in the B1 structure and of CrS, FeS, CrSe, FeSe, VTe, CrTe, MnTe, and FeTe in the $B8_1$ structure. Goodenough's, for compounds of the first transition series with the $B8_1$ structure, predicts semiconduction in TiS, TiSe, TiTe, VTe, CrSe, CrTe, and MnTe. A doubt remains for VS, VSe, FeS, and FeSe. However, it may be noted that the ratio c/a is abnormally high for TiS (1.93) and high for TiSe (1.76) and VS (1.75), marking a transition between the interstitial compounds and the real $B8_1$ structure.

7.2. Sulfides

Apart from MnS, one variety of which has a B1 structure, all monosulfides in the first transition series have $B8_1$ structures. In the other series, this same structure is found (for NbS) along with others (B1 for ZrS, B34 for PdS, B17 for PtS). A classification of disulfides was proposed in Section 7.1. The intermediate sulfides, when they exist, generally have a fairly distorted $B8_1$ structure with vacancies (except Fe_3S_4, Co_3S_4, and Ni_3S_4, which have the $H1_1$ spinel structure or the neighboring $D7_2$ structure). A few higher sulfides have been identified: TiS_3, ZrS_3, HfS_3, and VS_4, which are monoclinic, MoS_3 and WS_3, which have an unknown structure, and Tc_2S_7 and Re_2S_7, which are quadratic [7.6, 7.22].

Titanium sulfides form a single $B8_1$ phase for $0.1 < x < 0.33$ [7.23]. McTaggart [7.24] has found resistivities of less than $10^{-3}\,\Omega\,\mathrm{cm}$ for TiS and TiS_2. Grimmeiss et al. [7.25] have obtained them in the monocrystal state, finding metallic properties for TiS and considering Ti_3S_4 and TiS_2 as impure semiconductors. TiS_3 is semiconducting with $\alpha = -500\,\mu V\,\mathrm{deg}^{-1}$, $E_G = 0.9$ eV (optical absorption) and $e_d = 0.14$ eV (resistivity/temperature curve). Greenaway et al. [7.26] have studied the optical properties of TiS_2. Zirconium sulfides crystallize in a distorted B1 structure for $0 < x < 0.33$. ZrS_2 is semiconducting and its photoelectric properties have been studied by Zhuze and Ryvkin [7.27], E_G (indirect) $= 1.68$ eV [7.26]. ZrS_3 is also semiconducting with $\alpha = -500\,\mu V\,\mathrm{deg}^{-1}$, $E_G = 2.17$ eV (optical absorption) and $e_d = 0.82$ eV (resistivity/temperature curve). A photoconductivity peak is observed at 2.3 eV [7.25]. Hafnium sulfides have a $B8_1$ phase, but only HfS_2 and HfS_3 have been studied; E_G (indirect) $= 1.96$ eV for the former [7.26] and $E_G = 2.8$ eV (photoconductivity 2.85 eV) and e_d or $e_a = 0.57$ eV for the latter [7.25]. Vanadium sulfides have two $B8_1$ phases for $0.05 < x < 0.18$ and $0.25 < x < 0.33$. Brunie, Chevreton et al. have studied the order of vacancies in V_5S_8 [7.28] and V_3S_4 [7.29]. A $B8_1$ phase also exists in niobium sulfides.

Chromium sulfides seem to have three fairly distorted $B8_1$ phases for $0 < x < 0.15$ (disordered vacancies), $0.18 < x < 0.25$ and $x = 0.33$ (ordered vacancies). Cr_2S_3 is a catalyst. Slight ferromagnetism exists for $0.13 < x < 0.25$, with $T_C = -120°C$ (Fig. 7.6) [7.6, 7.22]. Fakidov and Grazhdankina [7.30, 7.31] have found a resistivity of $4 \times 10^{-4}\,\Omega$ cm for $x = 0.15$, varying abruptly, like the specific heat, at $T_N = 28°C$. They claim that there are semiconductor properties at $0 < x < 0.11$. Susceptibility measurements give an average $2S$ of 1.6 at room temperature and 2.3 at $600°K$ (instead of 4, as for a

7.2. Sulfides

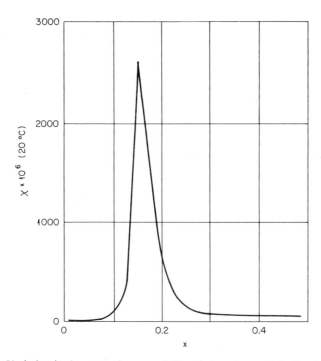

Fig. 7.6. Variation in the magnetic susceptibility of chromium sulfides in relation to the sulfur content. [Based on Pascal, "Nouveau Traité de Chimie Minérale." Masson, Paris, 1956–1963.]

ferromagnetic material). Watanabe and Tsuya [7.32] have studied the magnetic and electrical properties for $x = 0.15$ and 0.25, and have concluded that the behavior is metallic. Yosida [7.33] believes that CrS is antiferromagnetic, and that magnetization appears with nonstoichiometry. Smirnov [7.34] has found a magnetization maximum for $x = 0.15$. Yuzuri et al. [7.35], after a comprehensive magnetic study, conclude that ferrimagnetism exists. Kamigaichi et al. [7.36] have found a monoclinic phase (CrS) for $0 < x < 0.11$, coexisting with the $B8_1$ phase and responsible for p-type semiconduction, while the $B8_1$ phase itself is metallic (Fig. 7.7). They claim that this latter phase exists alone above 350°C or where $x > 0.11$; $T_N = 180$ (CrS) and -140°C (Cr_7S_8) [7.37]. For Dwight et al. [7.38], the metallic phase Cr_5S_6 ($x = 0.17$) is characterized by ferrimagnetism, with a transformation at 158°K and $T_C = 305$°K. The effect of the pressure shows that the magnetism is bound up with the order of the vacancies [7.39, 7.40]. For the investigation with neutrons, see [7.41, 7.42]. According to Bertaut et al. [7.43], the di- and trivalent atoms are ordered in accordance with case b on Fig. 7.3, the spins are parallel to direction [10$\bar{1}$], and $T_N = 260$°K (Fig. 7.8).

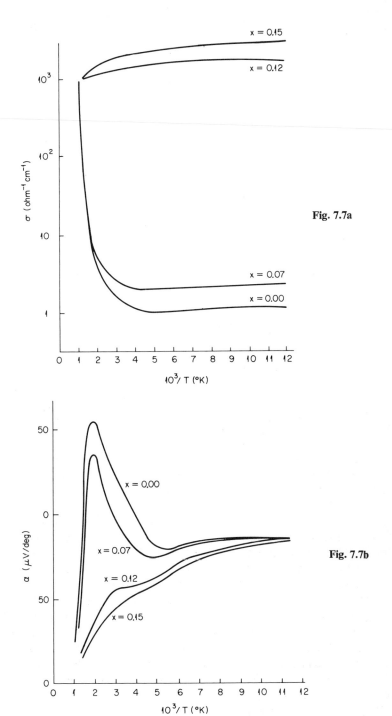

Fig. 7.7a

Fig. 7.7b

7.2. Sulfides

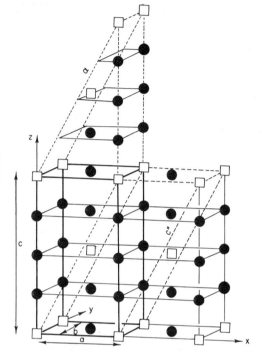

Fig. 7.8. Order of vacancies in Cr_3X_4 compounds (bidimensional cell): (□) vacancies, (●) Cr (based on Bertaut et al. [7.43]).

Molybdenum disulfide (molybdenite) has been the object of much research. It is a semiconductor at low temperatures, but becomes metallic above 200°C [7.6], and presents a diamagnetic [7.44] and electric [7.45] anisotropy and rectifier [7.46] and photoconductivity [7.47] effects; cf. also [7.48] (effect of pressure) and [7.49] (activation energies of n- and p-types). Mansfield and Salam [7.50], working with natural crystals, have found a mobility $\mu_H \sim 100$ cm^2 V^{-1}sec^{-1}, varying like $T^{-3/2}$, $\alpha \sim 500$ to 700 μV deg^{-1}, and $e_a = 0.09$ eV (cf. conductivity and Hall coefficient in Fig. 7.9); $E_G = 1.1$ eV [7.51]. The photoconductivity [7.52], field effect [7.53], dielectric constant (5.6 at 8 MHz) [7.54] and optical absorption [7.55] have been measured. Tungsten disulfide is stable between $WS_{1.86}$ and $WS_{2.3}$, and presents a rectifier effect [7.46] with $E_G = 1.1$ eV [7.51]. Manganese monosulfide includes three antiferromagnetic varieties $\alpha(B1)$, $\beta(B3)$, and $\gamma(B4)$, with $2S = 5$ [7.56–7.59]. The magnetic structure of the α phase has been studied by Corliss et al. [7.60], and its optical transmission by Batsanov and Kopytina [7.61]. The disulfide is antiferromagnetic with

---

Fig. 7.7. Logarithmic variation in the conductivity (a) and linear variation in the Seebeck coefficient (b) of $Cr_{1-x}S$ chromium sulfides (based on Kamigaichi et al. [7.36]).

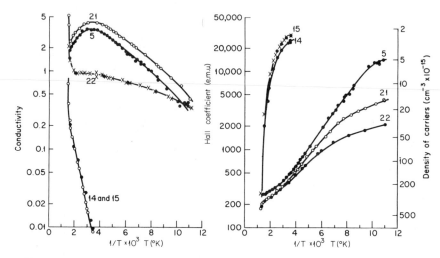

Fig. 7.9. Logarithmic variation in the conductivity (a) and Hall coefficient (b) of natural monocrystals of molybdenite (based on Mansfield and Salam [7.50]).

$2S = 5$ [7.62, 7.63], and presents an order of the third type [7.64, 7.65]. El'iott [7.21] has shown that a special bond involves hybridization of the 4d sublevel (cf. Table 7.2). Rhenium disulfide presents a rectifier effect with $E_G = 1.1$ eV [7.51].

Iron sulfides form a homogeneous phase between $Fe_{1.14}S$ and $Fe_{0.875}S$ (Fe_7S_8), including the stoichiometric composition FeS (troilite). We are interested here in compositions where $0 < x < 0.125$ (pyrrhotites), which have a $B8_1$ structure with vacancies, possibly distorted by the order of the vacancies. Two transformation temperatures have been observed, marked by sudden variations in the parameter c: $T_\alpha = 138°C$ ($x = 0$), which decreases if x increases, is accompanied by the possible disappearance of the overstructure, and $T_\beta = T_N = 325°K$ is accompanied by the disappearance of the magnetic order. For $0.05 < x < 0.1$, no overstructure exists at room temperature, but an embryonic organization of the vacancies appears at $150 < T_\gamma < 250°C$, and disappears at $T_\varepsilon \leqslant T_\beta$, accompanied by a magnetic susceptibility peak. For $0.1 < x < 0.125$, there is a monoclinal distortion and a resulting magnetic moment (Fig. 7.10) [7.6, 7.22]. Their semiconducting properties have been stated [7.67], and later denied [7.32]. Hirone and Tsuya [7.68] and Yosida [7.33, 7.69] explain the ferrimagnetism by the presence of two sublattices, one of which contains trivalent iron atoms (Fig. 7.11). Bertaut [7.3, 7.70, 7.71] has shown that the ordered state of the vacancies is more stable than the disordered state, and he has studied their distribution on the iron planes of even order. Benoit [7.72] has observed the disappearance of order in Fe_7S_8 at 560°C. Pauthenet [7.73] has pointed out a slow change in the direction of spins

7.2. Sulfides

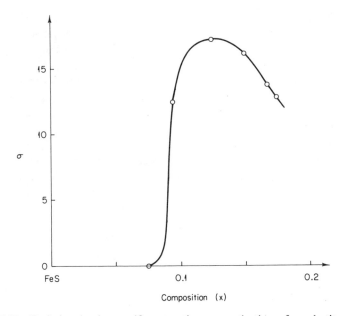

Fig. 7.10. Variation in the specific saturation magnetization of pyrrhotites $Fe_{1-x}S$ in terms of the parameter x, $T = 290°K$ and $H = 2 \times 10^4 G$ (based on Benoit [7.66]).

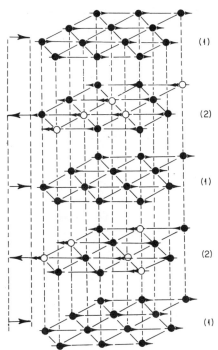

Fig. 7.11. Possible structure of two sublattices with iron atoms in pyrrhotites. The black circles represent Fe(II) atoms and the white circles Fe(III) atoms. The small arrows indicate the direction of the atomic moments, and the large ones that of the resultants in each sublattice (based on Hirone and Tsuya [7.68]).

towards 50°K. Neel [7.74, 7.75] has criticized Hirone's and Yosida's model, and he attributes the ferrimagnetism to the order of the vacancies, which are too far apart to interact when $x < 0.08$. Benoit [7.66, 7.76] has summarized earlier work, discussed the effect of the distribution of the Fe(II) and Fe(III) atoms on the average resulting moment of Fe_7S_8, and shown that the Fe(III) atoms have a tendency to take up position in the planes with vacancies (case a on Fig. 7.3). Hirone et al. [7.77, 7.78] have found T_β (315°C) and the variation of T_α in relation to x, by specific heat measurements.

Electrical research began in 1955. Benoit [7.66] finally confirmed the semiconductor character (Fig. 7.12) and suggested a mixed transfer and excitation mechanism. Kamigaichi [7.79] and Kamigaichi et al. [7.80] have found an irregularity at T_α (Fig. 7.13). The lowest conductivity along axis c may be linked to the direction of the t_{2g} orbitals (cf. Section 4.1). Another irregularity has been found at T_α [7.81]. Simultaneous investigation of the magnetic susceptibility and conductivity by Hirahara and Murakami [7.82] has shown that spins are parallel to c below T_α, and perpendicular above. The conductivity irregularity is said to result from the difference in the magnetic scattering in the two cases. Hihara [7.83] has continued the work: T_α falls from 138 to 115 ($x = 0.04$) and 110°C ($x = 0.05$), and the p-type semiconduction is unaffected by the ferrimagnetism. Finally, Fujime et al. [7.84] have measured the Hall effect, and shown that the number of carriers remains constant at T_α, where the conductivity irregularity results solely from a variation in mobility along axis c. For the effect of thermal expansion, see [7.85]. A three-electron Fe(II)— Fe(II) bond is said to occur in the base plane [7.86]. Neutron diffraction investigation by Sparks et al. [7.42, 7.87, 7.88] on the composition $x = 0.04$ confirms that the moments are contained in the base plane at room temperature and turn between 185°C and T_α (c increases by 1% and a decreases by 0.5% below this). Spontaneous magnetization [7.89] and the magnetic anisotropy [7.90] have been studied. Bin and Pauthenet [7.91] have worked out the fraction of Fe(III) atoms of each sublattice from this. The effect of pressure confirms that the magnetism is bound up with the order of the vacancies [7.39]. Ôno et al. [7.92], followed by Horita and Hirahara [7.93] have again confirmed, using the Mössbauer effect, the direction of the spins on each side of T_α, with $H_h = 320$ kOe. Hafner et al. [7.94] have found a higher quadrupole interaction. Theodossiou [7.95] has continued the work of Fujime et al., and found a change from n- to p-type, suggesting a higher hole than electron mobility. He also claims that the ordinary Hall coefficient R_0 varies significantly between 250°K and room temperature.

The pyrite FeS_2 is semiconducting, and Marinace [7.96] has found, for the n-type, $\rho = 0.07$ Ω cm, $R = -7$ cm^3 C^{-1}, $\mu = 100$ cm^2 V^{-1} sec^{-1} and, for the p-type, $\rho = 2$ Ω cm, $R = 3$ cm^3 C^{-1}, $\mu = 2$ cm^2 V^{-1} sec.$^{-1}$. At low temperatures, $e_a = 0.1$ eV and e_d is very low. At high temperatures, $E_G = 1.2$ eV

7.2. Sulfides

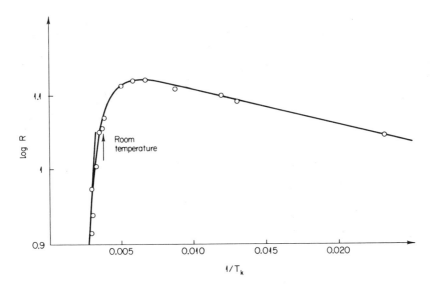

Fig. 7.12. Logarithmic variation of resistance in relation to the temperature reciprocal (based on Benoit [7.66]).

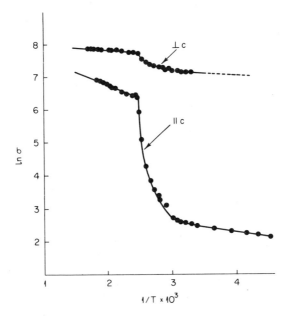

Fig. 7.13. Logarithmic variation in the conductivity of a $Fe_{1-x}S$ monocrystal (x very low) in relation to the temperature reciprocal in the two main directions (based on Kamigaichi et al. [7.80]).

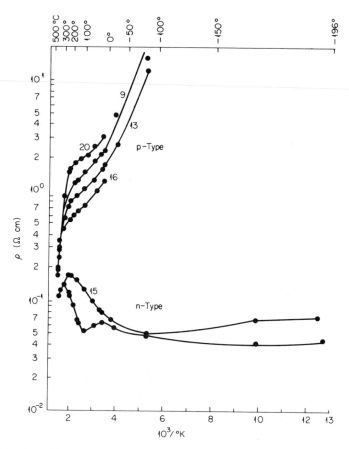

Fig. 7.14. Logarithmic variation in the resistivity of pyrite in relation to the temperature reciprocal (based on Marinace [7.96]).

(Fig. 7.14). For the neutron investigation, cf. [7.97]. Marcasite is 200 times more conducting [7.6]. Cobalt sulfides form an unstable $B8_1$ phase below 460°C for $0.02 < x < 0.08$, not including CoS [7.98, 7.99]. It is antiferromagnetic with $T_N = 85$°C [7.66] and $2S = 1$; the Co is reported to have valency 4 with Co—Co bonds [7.100]. The electrical investigation has been carried out by Kuznetsov [7.101]. Semiconductor properties are reported to exist in the liquid state [7.102]. $Co_xFe_{1-x}S$ solid solutions are ferrimagnetic up to $x = 0.5$ [7.103]. The disulfide CoS_2 is ferromagnetic, with $T_C = 110$°K [7.62] and $2S \sim 1$ [7.63, 7.66]. Nickel sulfides form a $B8_1$ phase at high temperatures and a stable B13 rhombohedral phase at room temperature (millerite). This millerite is slightly paramagnetic ($2S = 0$) [7.66] and the nickel is reported

7.3. Selenides 213

to have valency 4 [7.100]. Its semiconducting properties have been investigated by Hauffe and Flindt [7.104]. The B13–B8$_1$ phase change modifies the conductivity [7.105]. See also Kuznetsov [7.101]. The neutron investigation by Sparks and Komoto [7.42, 7.106] shows that the B8$_1$ phase, quenched from 700°K, is antiferromagnetic, with $T_N = 263°$K. There is a contraction of 1% on c and 0.3% on a, for $T > T_N$. The spins are parallel to axis c, with ferromagnetic coupling in the base plane; $2S \sim 2$ [7.106, 7.107]. Semiconductor properties are reported to exist in the liquid state [7.102]. The Li$_x$Ni$_{1-x}$S solid solutions exist in the B8$_1$ structure [7.108]. The disulfide NiS$_2$ is paramagnetic, with $2S = 2$ [7.62, 7.63, 7.66]. The palladium sulfides Pd$_4$S and Pd$_{2.2}$S seem to be metallic [7.109]. Copper disulfide has been prepared at very high pressure [7.110].

The semiconduction mechanism in sulfides with B8$_1$ structure is uncertain. We believe that it usually involves transfers. The variation in mobility along axis c during the T_α transformation of pyrrhotites proves this. The transfer density should therefore confirm the forecasts of Fig. 7.4, if the sites of all the planes containing vacancies are equivalent. For a more complex order, however, a minimum may occur for $x = \frac{1}{8}$. In any case, transfers parallel to axis c cannot be ignored, and neither can the underlying excitation mechanism, at high temperatures. In di- and trisulfides, only the excitation mechanism seems to be involved.

7.3. SELENIDES

Apart from MnSe, one variety of which has the B1 structure, all monoselenides in the first transition series have the B8$_1$ structure. However, it does not seem to exist in those of the other series, which are less familiar. A classification of diselenides was proposed in Section 7.1. Nonstoichiometric selenides, when they exist, generally have a more or less distorted B8$_1$ structure with vacancies. A few higher selenides have been identified: monoclinic ZrSe$_3$ and HfSe$_3$, possibly MoSe$_3$, which is of unknown structure, and Re$_2$Se$_7$, which is probably quadratic.

For $0 < x < 0.5$, titanium selenides form a single B8$_1$–C6 phase, the susceptibility of which varies between 2 and 1×10^{-4} [7.6]. Grimmeiss et al. [7.25] consider monocrystals TiSe and TiSe$_2$ as impure semiconductors. Chevreton et al. have studied the structures with vacancies of Ti$_3$Se$_4$ [7.111] and Ti$_5$Se$_8$ [7.112]. Greenaway et al. [7.26] have studied the optical properties of TiSe$_2$. Zirconium selenides crystallize in a distorted B1 structure. ZrSe$_2$ is semiconducting, and its photoelectric properties have been studied by Zhuze and Ryvkin [7.27, 7.113] and Putseiko [7.114]. ZrSe$_3$ is also semiconducting with $\alpha = -1$ mV deg^{-1}, $E_G = 1.25$ eV (optical absorption) and $e_d = 0.55$ eV (resistivity/temperature) [7.25]. Hafnium diselenide is also semiconducting, with E_G (indirect) = 1.13 eV [7.26]. Vanadium selenides form three phases with the structures $B8_1$ (VSe), $B8_1$ with vacancies (V_2Se_3) and C6 (VSe_2) with a magnetic susceptibility maximum for V_2Se_3 [7.6]. The structures with vacancies of V_3Se_4 [7.111] and V_5Se_8 [7.112] have been studied. VSe_2 is semiconducting, with photoelectric properties [7.114]. Niobium and tantalum diselenides have layer structures [7.17]. Tantalum diselenide is antiferromagnetic, with $T_N = 130°K$ [7.115]. The electrical properties of $TaSe_2-WSe_2$ solid solutions have been studied [7.116].

Chromium selenides seem to have three $B8_1$ phases for $0 < x < 0.13$, $0.16 < x < 0.25$ and $0.31 < x < 0.33$, the second showing a monoclinal distortion. CrSe is antiferromagnetic [7.6, 7.22]. Tsubokawa [7.117] and Lotgering and Gorter [7.118] have observed an irregularity in the susceptibility curve (T_N?) between 250 and 300°K (Fig. 7.15). Corliss et al. [7.119], using the neutron diffraction system, have found that the spins have an "umbrella-shaped" structure (Fig. 7.16) and that $2S = 2$. Men'shikov et al. [7.14] have shown that the d electrons do not contribute in any significant way to the Cr—Se bonds. Masumoto et al. [7.120] have found irregularities in the susceptibility and conductibility curves at 180 and 305°K. Chevreton and Bertaut [7.121] have investigated the order of vacancies in Cr_5Se_8 and Cr_3Se_4. Bertaut et al. [7.41, 7.43, 7.122] have used neutron diffraction to investigate the role of vacancies in Cr_3Se_4, which they compare with $MnBr_2$. The di- and trivalent atoms are ordered in accordance with the case b in Fig. 7.3, with the spins parallel to the [10$\bar{1}$] direction and $T_N = 80°K$. A change occurs at very low temperatures. Chevreton et al. [7.123] deny the existence of stoichiometric CrSe, and, in research on compressed powders, have found resistivity/temperature curves that are characteristic of semiconductors for Cr_3Se_4 (0.09 eV) and Cr_2Se_3 (0.16 eV) and ambiguous, with thermal hysteresis, for Cr_7Se_8. In a more comprehensive study, Masumoto [7.124] has shown that for $0 < x < 0.09$, the conductivity increases suddenly by a factor of 10 above 305°K, with thermal hysteresis (Fig. 7.17). He attributes this irregularity to the disappearance of distortion of the $B8_1$ structure, probably connected with a Jahn–Teller effect. $CrSe_xTe_{1-x}$ solid solutions have been studied by Tsubokawa [7.117] and Lotgering et al. [7.118], who find ferromagnetism appearing for $x < 0.8$.

The semiconducting character of molybdenum diselenide has been mentioned by Brixner [7.125] and Brixner and Teufer [7.126], who have found an n-type and a high Seebeck effect. Hicks [7.127] has mentioned $\alpha = -900$ μV deg^{-1}, and $\rho = 20$ Ω cm at 100°C, and $\mu = 15$ cm^2 V^{-1}sec^{-1} at room temperature. Tungsten diselenide is also semiconducting, with a p-type [7.125], an optical absorption threshold at 7800 Å ($E_G = 1.6$ eV) [7.128], $\alpha = +560$ μV deg^{-1} and $\rho = 0.57$ Ω cm at 100°C [7.127]. Brixner [7.116] has studied $W_{1-x}Ta_xSe_2$ solid solutions, and has shown that the compositions with

7.3. Selenides

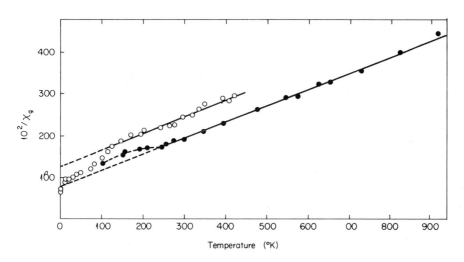

Fig. 7.15. Variation in the magnetic susceptibility reciprocal of CrSe in relation to absolute temperature (based on Tsubokawa (O) [7.117] and Lotgering and Gorter (●) [7.118]).

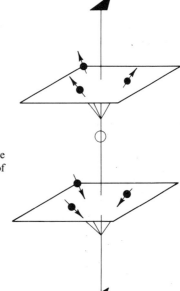

Fig. 7.16. "Umbrella-shaped" structure of the atomic moments in two successive (001) planes of CrSe (based on Corliss *et al.* [7.119]).

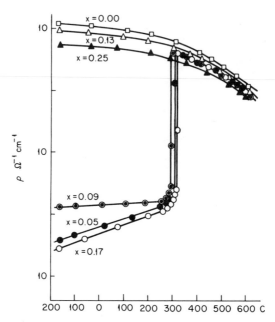

Fig. 7.17. Logarithmic variation in the conductivity of chromium selenides in relation to temperature (based on Masumoto [7.124]).

$0.01 < x < 0.05$ are excellent thermoelements with factors of merit f/T of 5 to 9×10^{-4} (cf. Section 4.3). Hicks [7.127] confirms this result, quoting a factor of 5×10^{-4} between 500 and 1000°C for $x = 0.01$, as well as $\mu = 100$ cm^2 V^{-1}sec^{-1} at room temperature. He has also studied $Mo_{1-x}Ta_xSe_2$ solutions, and the effect of various additions. It seems likely that the sharp drop in resistivity caused by the addition of tantalum is due to a valency induction. The resistivity and Seebeck effect in $MoSe_2$–WSe_2 solid solutions have been investigated by Revolinsky and Beerntsen [7.129]. The activation energy remains around 0.1 eV. All compositions (including $MoSe_2$) have been found to be of p-type. Marked anisotropy has been observed, as is to be expected in the C7 structure. Champion [7.130] has confirmed these results, and dealt with the quaternary system $MoSe_2$–WSe_2–$MoTe_2$–WTe_2.

Manganese monoselenide has only one stable variety, which has the B1 structure and is antiferromagnetic. $T_N = 140°K$ with thermal hysteresis [7.6]. For the neutron study, cf. [7.131]. The semiconductor character, pointed out by Enji et al. [7.132], has been studied by Palmer [7.133] who has found $\rho = 3 \times 10^3 \ \Omega$ cm, with thermal hysteresis, and a nonmeasurable Hall effect. McGuire and Heikes [7.134] have found that $Li_xMn_{1-x}Se$ solid solutions are ferromagnetic between 70 and 110°K for $x > 0.07$ [the Mn(III) moments are

7.3. Selenides

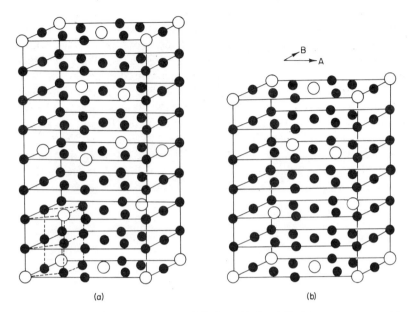

Fig. 7.18. Cells of the overstructures of Fe_7Se_8: (a) 4c triclinic overstructure, stable at low temperatures, (b) 3c hexagonal overstructure, stable at high temperatures. The white circles represent vacancies. The selenium atoms are not shown. The elementary cell is shown in dotted lines (based on Okazaki [7.140]).

balanced] and have metallic properties. For the theoretical explanation cf. [7.135]. The diselenide is antiferromagnetic with a complex order [7.64, 7.65], and a special bond [7.21].

Iron selenides form a $B8_1$ phase for $0 < x < 0.25$ (Fe_3Se_4), but are stable only at high temperatures for $0 < x < 0.125$ (Fe_7Se_8) and give way by annealing below 335°C to a B10 phase. A monoclinic distortion due to the order of the vacancies exists for $0.14 < x < 0.25$. The $B8_1$ phase is antiferromagnetic ($x = 0$), but the presence of iron vacancies makes the balance imperfect, and causes ferrimagnetism for $x > 0$ [7.6, 7.22]. The magnetic properties have been studied by Hirone et al. [7.136, 7.137]; $T_N \sim 150°K$ where x is low, while for Fe_7Se_8 $T_C = 174°K$, and there is magnetization of 0.2 μ_B. The order of the vacancies has been investigated very comprehensively by Okazaki and Hirakawa [7.138], who have described overstructures for Fe_7Se_8 and Fe_3Se_4 similar to those of Fe_7S_8 and Cr_3S_4. But Abrikosov [7.139] has found a monoclinic distortion in Fe_7Se_8 at room temperature, confirmed by Okazaki [7.140] (Fig. 7.18), who follows the change from one to the other, mentioning a transition zone between 240 and 298°C [7.141]. Hirakawa [7.142], working with Fe_7Se_8 monocrystals, has observed an easy magnetization direction parallel to axis c at low temperatures, and gradually becoming perpendicular as the temperature rises. The resulting moment increases along with x [7.143]. Ferrimagnetism is closely bound up with the order of the vacancies, and $T_C = 187°C$ for Fe_7Se_8 [7.141, 7.144]; $T_C = 170°C$ [7.145]. The magnetic anisotropy has been studied [7.90]. The neutron diffraction method has enabled Andresen and Leciejewicz [7.146] to define the direction of the spins in the 3c overstructure of Fe_7Se_8, perpendicular to axis c above 130°K, and almost parallel below 120°K. The distribution of Fe(II) and Fe(III) corresponds to case a in Fig. 7.3. For $0.125 < x < 0.25$, Serre and Druilhe [7.147] have observed a flattening in the conductivity slope below T_C, and then semiconductor behavior with an activation energy of 0.1 eV. A change in the slope at high temperatures probably reflects an alteration in the order of vacancies (Fig. 7.19).

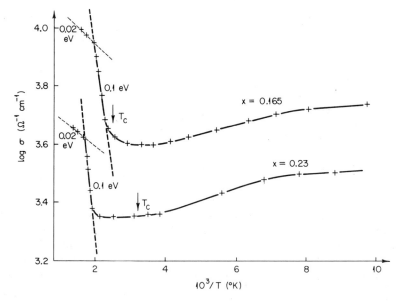

Fig. 7.19. Logarithmic variation in the conductivity of $B8_1$ iron selenides $Fe_{1-x}Se$ in relation to the temperature reciprocal (based on Serre and Druilhe [7.147]).

The diselenide $FeSe_2$ has been found to be semiconducting by Fischer [7.148]. The resistivity and Hall effect curves show an excitation mechanism across a forbidden band of 0.3 eV. Dudkin and Vaidanich [7.149] indicate $\sigma = 2\ \Omega^{-1}cm^{-1}$, $\alpha = 62\ \mu V\ deg^{-1}$ and $E_G = 0.9$ eV. Cobalt selenides form a $B8_1$ phase for $0.02 < x < 0.27$ [7.6, 7.22]. The diselenide is metallic [7.149].

7.4. Tellurides

The rhodium selenide Rh_2Se_5 is supraconducting [7.150]. Nickel monoselenides, like sulfides, have $B1_3$ and $B8_1$ phases. An electrical investigation of them has been done by Kuznetsov et al. [7.101]. For the diselenide, $2S \sim 2$ [7.63].

The same comments can be made on the conduction mechanism for selenides as for sulfides, although the more covalent character of the T—X bonds involves a more marked delocalization of the d electrons. The effect of the parameter c on conductivity along the axis is illustrated by the irregularity in chromium selenides.

7.4. TELLURIDES

All the monotellurides in the first transition series have the $B8_1$ structure, and the phase has a very wide homogeneity domain (Fig. 7.20). The T—X bonds are fully covalent for formulas T_2Te_3. In the other series, this structure is found for ZrTe, HfTe, NbTe, TaTe, RhTe, PdTe, and PtTe. A classification of ditellurides was proposed in Section 7.1. Two higher tellurides have been

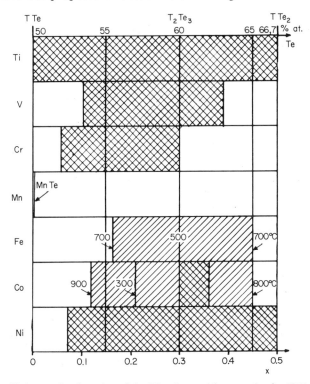

Fig. 7.20. Homogeneity domains of the $B8_1$ phase with vacancies for $\square_x T_{1-x}$ tellurides in the first transition series. The single hatchings refer to high temperatures.

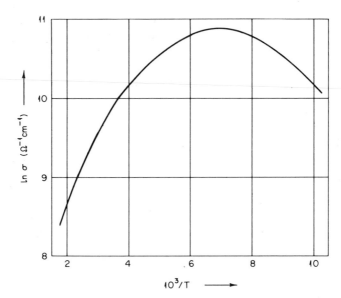

Fig. 7.21. Logarithmic variation in the conductivity of $TiTe_2$ in relation to the temperature reciprocal (based on Grimmeiss et al. [7.25]).

identified: monoclinal $ZrTe_3$ and quadratic $TaTe_3$. Titanium tellurides form a single $B8_1$–C6 phase for $0 < x < 0.5$, the susceptibility of which is 2×10^{-4} [7.6]. Grimmeiss et al. [7.25] considered monocrystal TiTe and $TiTe_2$ as impure semiconductors (Fig. 7.21). Chevreton and Bertaut [7.111] have studied the distribution of vacancies in Ti_3Te_4. Greenaway and Nitsche [7.26] have studied the optical properties of $TiTe_2$. Vanadium monotelluride is paramagnetic, with $2S = 1$ [7.151]. Niobium and tantalum ditellurides have a layer structure [7.152].

Chromium tellurides form a $B8_1$ phase for $0 < x < \frac{1}{3}$, with monoclinal distortion for $0.16 < x < 0.25$. CrTe is ferromagnetic with $T_C = 66°C$, but the specific magnetization is reported to correspond to $2S = 2.4$ [7.153], whereas its law of decrease with temperatures agrees with Weiss' theory if $2S = 4$ [7.154], as is logical for Cr(II) (cf. theory [7.155]). There are irregularities in calorific capacity at 60°C, and in resistivity at 58°C (5×10^{-4} Ω cm at room temperature) [7.156–7.158]. $T_C = 72°C$ and $2S = 2.8$ [7.159]. The longitudinal and transverse magnetoresistances are both negative (the former is usually positive, cf. Chapter 9) [7.157, 7.160, 7.161]. Guillaud [7.162] has suggested the existence of negative interactions between the chromium atoms in different crystallographic sites. Lotgering and Gorter [7.118] have found $T_C \sim 80$–$85°C$ and $2S = 2.45$. Kikoin et al. [7.163] have studied the Hall effect in the magnetic domain: $R_1 < 0$ is 100 times higher than in iron or

7.4. Tellurides

nickel; $R_0 < 0$ is low, varying little with the temperature. Gaidukov et al. [7.164] believe that they have observed two magnetic phases for $x > 0.20$. Kikoin et al. [7.165, 7.166] have established a law linking magnetoresistance and magnetization. Men'shikov and Nemnonov [7.14] have shown that the d electrons do not play any significant part in the Cr—Te bonds. Grazhdankina et al. [7.167, 7.168] have observed a drop in T_C from 58 to 31°C and a 5% increase in resistivity, at a pressure of 4.6 Torr cm^{-2}. Chevreton et al. [7.169] and Kieu Van Con and Suchet [7.170] have studied the crystallographic structure, and they deny the existence of stoichiometric CrTe. Kieu Van Con and Suchet, working with slowly cooled polycrystal ingots of Cr_3Te_4, have observed a slight exponential decrease in resistivity above 115°C, corresponding to $E_A \sim 0.02$ eV (Fig. 7.22). Andresen [7.171], using neutron diffraction, has found ferromagnetic structures with spins parallel to the c axis for Cr_5Te_6 and Cr_2Te_3, and moments of 2.7 and 2.1 μ_B.

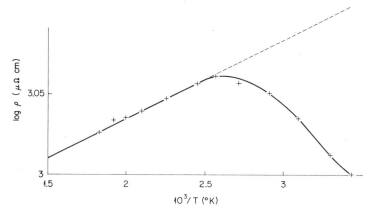

Fig. 7.22. Logarithmic variation in the resistivity of Cr_3Te_4 in relation to the temperature reciprocal; $E_A \sim 0.02$ eV (based on Kieu Van Con and Suchet [7.170]).

The exact conduction mechanism has recently been the subject of various investigations. Suchet [7.172] interprets variation in magnetization in relation to x as the change from ferrimagnetism to partly balanced antiferromagnetism (double hexagonal cell):

$$x = 0.08: \quad [\square_1 Cr(II)_5]_A \; [Cr(II)_4 Cr(III)_2]_B \; Te_{12} \tag{7.3}$$

$$x = 0.25: \quad [\square_3 Cr(II)_3]_A \; [Cr(III)_4 Cr(III)_2]_B \; Te_{12} \tag{7.4}$$

Albers and Haas [7.173] have observed a resistivity maximum and a sign-change in the Seebeck effect for Cr_3Te_4 (Fig. 7.23). Bertaut et al. [7.41, 7.43], using neutron diffraction, have found a ferromagnetic structure ($T_C = 329°K$),

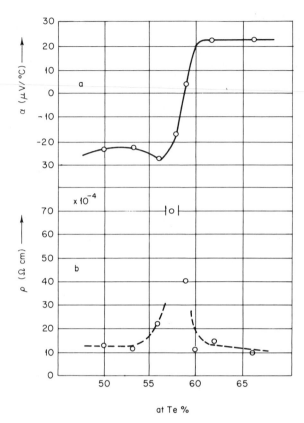

Fig. 7.23. Linear variation in the Seebeck coefficient (a) and resistivity (b) of $Cr_{1-x}Te$ in terms of at. %Te. $x = 0$: 50%; $x = 0.25$: 57.1%; $x = 0.33$: 60%. (Based on Albers and Haas [7.173]).

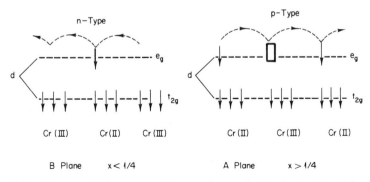

Fig. 7.24. Diagram of the two possible transfer mechanism in Cr_3Te_4, with a slight defect (n) or excess (p) of Te (based on Suchet and Imbert [7.174]).

7.4. Tellurides

on which slight antiferromagnetism is superimposed below 80°K. Suchet and Imbert [7.174] and Suchet and Druilhe [7.175] have suggested that there is a transfer mechanism in planes A (with vacancies) and B (full) (Fig. 7.24), which matches formulas (7.3) and (7.4), the results found by Albers et al., and the case b in Fig. 7.3. Nogami [7.176] [7.177] has returned to the Hall effect for CrTe, and found $R_0 \sim +10^{-3}$ cm^3 C^{-1} and $R_1 \sim -0.05$ to -0.1 cm^3 C^{-1}. Serre and Suchet [7.178] have measured R_0 in terms of x, and they confirm the sign change of the carriers for 0.25. Magnetic measurements [7.118] and neutron diffraction [7.179, 7.180] have revealed noncollinear spin configurations in CrTe–CrSb solutions. The magnetic [7.117, 7.118] and electrical [7.167] properties of CrTe–CrSe solutions are also being investigated.

Molybdenum ditelluride is diamagnetic, and decomposes at 815°C [7.6]. According to Brixner [7.125] it is an n-type semiconductor with high Seebeck effect. Its optical absorption threshold is at 6600 Å ($E_G = 1.9$ eV) [7.128]. Revolinsky and Beerntsen [7.129] have given the $\rho(T)$ and $\alpha(T)$ curves. $\rho = 25$ Ω cm and $\alpha = -360$ μV deg^{-1} at 100°C and $\mu = 12$ cm^2 V^{-1}sec^{-1} at room temperature [7.127]. Lepetit [7.181] has measured ρ and R_H in n and p monocrystals (Fig. 7.25). $\rho(T)$ gives $E_G = 1$ eV, agreeing with the optical

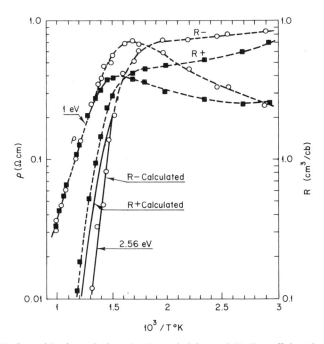

Fig. 7.25. Logarithmic variations in the resistivity and Hall coefficient in MoTe$_2$ in relation to the temperature reciprocal. Calculation shows that minority carriers intervene in R_H (based on Lepetit [7.181]).

spectrum. $\mu \sim 25$ (holes) to 50 (electrons) cm^2 V^{-1}sec^{-1}. Brown [7.182] has discovered a high-temperature β phase (linked with C6). Revolinsky and Beerntsen [7.183] have found that the resistivity here is 100 times lower than that of α(C7). For the effect of various impurities, see [7.127]. Tungsten ditelluride is also diamagnetic. According to Brixner [7.125], it is an n-type semiconductor, changing to p-type above 100°C. Revolinsky et al. [7.129] have given the $\rho(T)$ and $\alpha(T)$ curves: 2×10^{-3} Ω cm at 25°C and 30 μV deg^{-1} at 130°C. The structure has been described by Brown [7.182]. Kabashima [7.184] has measured ρ, α, and R_H in monocrystals (Fig. 7.26). WTe$_2$–MoTe$_2$ solid solutions have been studied by Revolinsky and Beerntsen [7.129], and the quaternary system W–Mo–Se–Te by Champion [7.130].

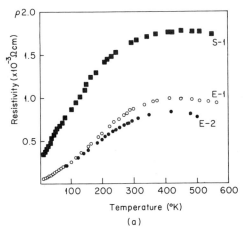

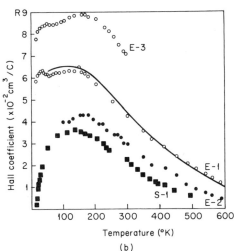

Fig. 7.26. Linear variation in the resistivity (a) and Hall coefficient (b) of WTe$_2$ in relation to temperature: E = monocrystals, S = sintered samples (based on Kabashima [7.184]).

7.4. Tellurides

The $B8_1$ phase domain in manganese tellurides is narrow; MnTe is antiferromagnetic with $2S = 5$ [7.6]. Palmer [7.133] has indicated its semiconductor character; $\rho \sim 1\,\Omega$ cm with thermal hysteresis, $R_H = +0.6$ (room temperature) and $+2.6$ cm^3 C^{-1} ($-196°$C). Uchida et al. [7.185] have confirmed that $T_N = 55°$C and the variation in susceptibility at $37°$C in polycrystals. The resistivity and Seebeck effect show a maximum ($\sim 2.2\,\Omega$ cm and 0.6 mV deg^{-1}), and the Hall effect a bend (Fig. 7.27) at T_N. Pressure reduces ρ and raises T_N [7.186]. Johansen [7.187] shows that the compound contains vacancies, and exists for $0.002 < x < 0.013$ at $800°$C. The neutron analysis has been done by Doroshenko et al. [7.188]. Yadaka et al. [7.189], working with monocrystals, have observed a p-type and an intrinsic excitation mechanism above $200°$C. Komatsubara et al. [7.190] discuss the magnetic properties. Wasscher et al. [7.191, 7.192] have studied the particularly marked irregularity in the Seebeck effect around T_N (Fig. 7.28), explaining it as the effect of dragging of the carriers by the spin periodicity ("magnon drag"). In this connection, it might be mentioned that the same irregularity is found in other substances (cf. for instance, CrO_2, Section 5.3 [5.137], Fe_2O_3–TiO_2, Section 6.4 [6.152], Cr_2Te_3, this section [7.175], etc.). Kumitomi et al. [7.193], followed by Sirota and Makovetskii [7.194], have pursued the analysis done by Doroshenko et al., and have shown that the directions of spins are perpendicular to axis c, and each (001) plane ferromagnetic (cf. [7.195] for thermal conductivity). Suchet et al. [7.174, 7.175] have suggested a transfer mechanism explaining the p-type. Wasscher [7.196] points out very slight ferromagnetism below $260°$K. The resistivity of Cr_xMn_{1-x}Te solutions ($x \leqslant 0.05$) increases rapidly with x, and

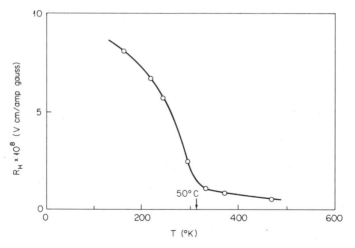

Fig. 7.27. Linear variation in the Hall coefficient of MnTe in relation to temperature (based on Uchida et al. [7.185]).

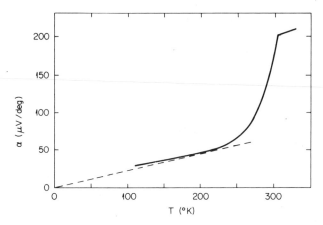

Fig. 7.28. Linear variation in the Seebeck effect in MnTe in relation to temperature (based on Wasscher and Haas [7.191]).

there is no maximum [7.189]. $Ge_xMn_{1-x}Te$ solutions ($x > 0.9$), with the B1 structure, are ferromagnetic [7.197]: Rodot et al. [7.198] have studied their properties ($T_C = 143°K$ for $x = 0.95$) and believe that they might be semiconducting, although ρ increases with T. Manganese ditelluride is antiferromagnetic, with an order of the first kind [7 64, 7 65] and a special bond [7.21]. The electrical study has been done by Sawaoka et al. [7.199, 7.200]. Resistivity shows an irregularity at $T_N = 87°K$. Pressure reduces ρ and raises T_N.

Iron tellurides form only two stable phases at room temperature: $FeTe_{0.8}$ to $FeTe_{0.9}$ (B10) and $FeTe_{1.9}$ to $FeTe_{2.1}$ (C18). A $B8_1$ phase with monoclinal distortion is stable above approximately 500°C [7.201]. The B10 phase is antiferromagnetic with $2S = 2$ [7.202, 7.203]. Its semiconductor character has been attributed by Finlayson et al. [7.204] to a transfer mechanism. According to Tsubokawa and Chiba [7.205], $T_N = 63°K$, but varies with the composition [7.206]. Fujime et al. [7.84] have measured $R_H = +2.1 \times 10^{-3}$ cm^3 C^{-1} and $\mu = 3.2$ cm^2 V^{-1}sec^{-1}. Leciejewicz [7.207], using neutron diffraction, has found that the spins are parallel to the tetragonal axis for $T > T_N$, and perpendicular to it below this. The $B8_1$ phase was first thought to be antiferromagnetic [7.202, 7.203], but Suchet and Serre [7.207] have discovered a Pauli paramagnetism. Aramu and Manca [7.208] mentioned the semiconductor character of Fe_2Te_3, with $E_G = 2 \times 0.34$ eV, $\sigma = 450$ Ω^{-1} cm^{-1} and $\mu = 11.5$ cm^2 V^{-1} sec^{-1} (holes) (Fig. 7.29). Suchet [7.209] has found $E_A = 0.08$ eV. Suchet and Imbert [7.174] have tried unsuccessfully to observe the Mössbauer resonances of Fe(II) and Fe(III) in Fe_3Te_4 separately, and have suggested a transfer mechanism in which the d electrons are localized for a period of time short in relation to the transfer time. The Fe(II) configuration would then be too

7.4. Tellurides

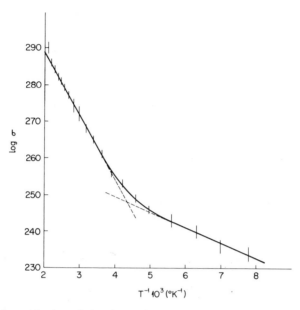

Fig. 7.29. Logarithmic variation in conductivity of Fe_2Te_3 in relation to reciprocal temperature (based on Aramu and Manca [7.208]).

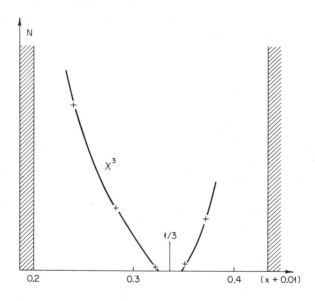

Fig. 7.30. Variation in the cube of the susceptibility or the number of carriers in $Fe_{1-x}Te$ in terms of x (based on Suchet and Druilhe [7.175]).

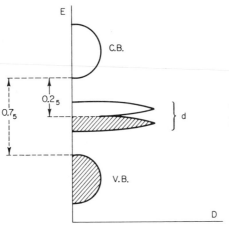

Fig. 7.31. Approximate band structure of Fe_2Te_3 at 90°K (based on Manca et al. [7.212]).

transient to be observed. The digest [7.175] reviews these results (Fig. 7.30). Manca and Saut [7.210] have found a forbidden band. Chappert and Fatseas [7.211] show that there is still no Zeeman effect in the Mössbauer spectrum at 1.5°K. Finally, Manca et al. [7.212], in a thorough investigation which goes back over earlier work, adding an optical investigation, have observed a minimum number of carriers for $x = \frac{1}{3}$, using three different methods (magnetic susceptibility, isomeric shift, electronic specific heat), and have proposed the band structure in Fig. 7.31. The effect of the carrier density on the shift δ in the Mössbauer resonance frequency is mentioned here for the first time.

Cobalt tellurides form a $B8_1$ phase with a wide domain. The mention of slight ferromagnetism [7.213, 7.214] was due to the presence of free Co, since the composition CoTe does not exist [7.7, 7.215–7.217]. The phase diagram is given by [7.218]. Saut [7.219] has found resistivity/temperature curves of metallic type, and a Pauli paramagnetism where x is low. The number of carriers, worked out from the susceptibility, tends to approach zero for $x = 0.5$. Manca and Saut [7.210] have found a forbidden band. The ditelluride is reported to have the C6 structure, but to change to C18 by annealing at 250°C [7.6]. Zhuze and Regel [7.220] have suggested that it is semiconducting, and this seems to be confirmed for the C6 structure phase by findings of Saut and Manca et al. Rhodium ditelluride is reported to be supraconducting below 1.5°K [7.150]. Nickel tellurides form a single $B8_1$–C6 phase for $0 < x < 0.5$, the susceptibility of which is independent of temperature [7.221]. Zhuze and Regel [7.220] have observed metallic conductivity. Fujime et al. [7.84] have measured $R_H = 3.5 \times 10^{-5}$ cm^3 C^{-1} and $\mu = 0.49$ cm^2 V^{-1}sec^{-1}. Saut [7.222] has confirmed the Pauli paramagnetism, and has found two minimum carrier points (Fig. 7.32). The ditelluride is reported to be semiconducting [7.220]. Palladium ditelluride is supraconducting below 1.7°K

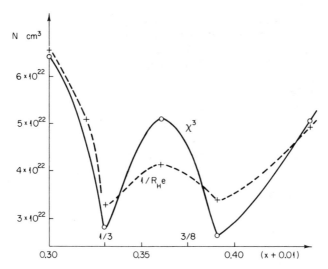

Fig. 7.32. Variation in the number of carriers in $Ni_{1-x}Te$, calculated from the susceptibility (solid line) or Hall effect (broken line) (based on Saut [7.222]).

[7.223]. Kjekshus and Pearson [7.224] have studied the (metallic) electrical properties of PdTe and $PdTe_2$, and confirmed the supraconductivity of $PdTe_2$.

The semiconduction mechanism of tellurides with $B8_1$ structure generally involves transfers. It should be noted, however, that $x = 0.002$ would correspond to approximately 10^{20} transfers cm^{-3}, which is relatively low. It is therefore possible that both mechanisms are superimposed in MnTe. In the final terms of the series (Fe, Co, Ni), pseudometals appear, in which semiconducting properties may be observed for particular compositions. One should note the great stability of the configuration d^5 (MnTe, Fe_2Te_3, and $CoTe_2$ in the $B8_1$–C6 structure).

The characteristic feature of the chalcogenides considered in this chapter is probably the transformation of an antiferromagnetism to slight ferrimagnetism through the presence of ordered vacancies in the $B8_1$ structure. This is frequently accompanied by transfer semiconduction.

REFERENCES

[7.1] A. F. Wells, "Structural Inorganic Chemistry," 3rd ed. Oxford Univ. Press (Clarendon), London and New York, 1962.
[7.2] R. W. G. Wyckoff, "Crystal Structures," 2nd ed., Vol. 1. Wiley (Interscience), New York, 1963.
[7.3] E. F. Bertaut, *Acta Crystallogr.* **6**, 557 (1953).
[7.4] F. Jellinek, *Acta Crystallogr.* **10**, 620 (1957).
[7.5] M. Chevreton, E. F. Bertaut, and S. Brunie, *Bull. Soc. Sci. Bretagne* **39**, 77 (1964).

[7.6] P. Pascal, "Nouveau traité de chimie minérale." Masson, Paris, 1956–1963.
[7.7] H. Haraldsen, *Congr. Int. Chim. Pure Appl. 16th Paris 1957*. Birkhäuser, Bâle 1957.
[7.8] A. Kjekshus and W. B. Pearson, *Progr. Solid State Chem.* **1**, 83 (1964).
[7.9] H. Krebs, "Grundzüge der anorganischen Kristallchemie." Enke, Stuttgart, 1968 [*English transl.:* McGraw-Hill, Maidenhead, 1968].
[7.10] W. B. Pearson, *Can. J. Phys.* **35**, 886 (1957).
[7.11] A. T. Shuvayev, *Izv. Akad. Nauk SSSR, Ser. Fiz.* **25**, 896 (1961).
[7.12] A. Z. Men'shikov and S. A. Nemnonov, *Fiz. Metal. Metallov.* **14**, 187 (1962).
[7.13] A. Z. Men'shikov, S. A. Nemnonov, and L. B. Mishchenko, *Fiz. Metal. Metallov.* **14**, 383 (1962).
[7.14] A. Z. Men'shikov and S. A. Nemnonov, *Fiz. Metal. Metallov.* **10**, 390 (1960).
[7.15] J. Barstad, F. Grønvold, E. Røst, and E. Vestersjø, *Acta Chem. Scand.* **20**, 2865 (1966).
[7.16] J. P. Suchet, *Mater. Res. Bull.* **2**, 547 (1967).
[7.17] B. E. Brown and D. J. Beerntsen, *Acta Crystall.* **18** (Part I), 31 (1965).
[7.18] L. Pauling, "The Nature of the Chemical Bond." Cornell Univ. Press, Ithaca, New York, 1950.
[7.19] F. Hulliger and E. Mooser, *J. Phys. Chem. Solids* **26**, 429 (1965); F. Hulliger, *ibid.* **26**, 639 (1965).
[7.20] F. Hulliger and E. Mooser, *Progr. Solid State Chem.* **1**, 330 (1965).
[7.21] N. Elliott, *J. Chem. Phys.* **33**, 903 (1960).
[7.22] M. Hansen, "Constitution of Binary Alloys." McGraw-Hill, New York 1958; R. P. Elliott, idem. 1961.
[7.23] H. Hahn and B. Harder, *Z. Anorg. Chem.* **288**, 239 (1956).
[7.24] F. K. McTaggart, *Aust. J. chem.* **2**, 471 (1958).
[7.25] H. G. Grimmeiss, A. Rabenau, H. Hahn, and P. Ness, *Z. Elektrochem.* **65**, 776 (1961).
[7.26] D. L. Greenaway and R. Nitsche, *J. Phys. Chem. Solids* **26**, 1445 (1965).
[7.27] V. P. Zhuze and S. M. Ryvkin, *Dokl. Akad. Nauk SSSR* **62**, 55 (1948).
[7.28] S. Brunie and M. Chevreton, *C.R. Acad. Sci. Paris* **258**, 5847 (1964).
[7.29] M. Chevreton and A. Sapet, *C.R. Acad. Sci. Paris* **261**, 928 (1965).
[7.30] I. G. Fakidov and N. P. Grazhdankina, *Dokl. Akad. Nauk SSSR* **63**, 27 (1948); **75**, 19 (1950).
[7.31] I. G. Fakidov and N. P. Grazhdankina, *Tr. Inst. Fiz. Metal, Akad. Nauk SSSR (Ural Filial)* **15**, 60, 65 (1954).
[7.32] I. H. Watanabe and N. Tsuya, *Sci. Rep. Res. Inst. Tôhoku Univ.* **A2**, 503 (1950).
[7.33] K. Yosida, *Physica* **17**, 794 (1951).
[7.34] F. S. Smirnov, *Zh. Tekh. Fiz.* **23**, 50 (1953).
[7.35] M. Yuzuri, T. Hirone, H. Watanabe, S. Nagasaki, and S. Maeda, *J. Phys. Soc. Japan* **12**, 385 (1957).
[7.36] T. Kamigaichi, K. Masumoto, and T. Hirahara, *J. Phys. Soc. Japan* **15**, 1355 (1960).
[7.37] T. Kamigaichi, *J. Sci. Hiroshima Univ.* **A24**, 371 (1960).
[7.38] K. Dwight, R. W. Germann, N. Menyuk, and A. Wold, *J. Appl. Phys. suppl.* **33**, 1341 (1962).
[7.39] M. Yuzuri, Y. H. Kang, and Y. Goto, *J. Phys. Soc. Japan suppl.* **17** (B-I), 253 (1962).
[7.40] M. Yuzuri and Y. Nakamura, *J. Phys. Soc. Japan* **19**, 1350 (1964).
[7.41] E. F. Bertaut, G. Roult, A. Delapalme, G. Bassi, M. Mercier, A. Murasik, Vu Van Qui, R. Aleonard, R. Pauthenet, M. Chevreton, and R. Jansen, *J. Appl. Phys.* **35** (Pt. 2), 952 (1964).
[7.42] J. T. Sparks and T. Komoto, *J. Phys. Paris* **25**, 567 (1964).
[7.43] E. F. Bertaut, G. Roult, R. Aleonard, R. Pauthenet, M. Chevreton, and R. Jansen,

J. Phys. **25**, 582 (1964); Coll. Int. Diffract. Diffus. Neutrons, p. 158. C.N.R.S., Paris, 1964.
[7.44] A. K. Dutta, Nature **156**, 240 (1945).
[7.45] A. K. Dutta, Nature **159**, 477 (1947).
[7.46] J. Lagrenaudie, J. Phys. Radium **13**, 311 (1952).
[7.47] J. Lagrenaudie, French Patent 1.087.879, March 1, 1955.
[7.48] P. W. Bridgman, Amer. Acad. Arts Sci. **81**, 265 (1952).
[7.49] F. Regnault, P. Aigrain, C. Dugas, and B. Jancovici, C.R. Acad. Sci. Paris **235**, 31 (1952).
[7.50] R. Mansfield and S. A. Salam, Proc. Phys. Soc. **66B**, 377 (1953).
[7.51] J. Lagrenaudie, J. Phys. Radium **15**, 299 (1954).
[7.52] V. I. Lyashenko and O. V. Snitko, Radiotekhn. Elektron. (URSS) **2**, 269 (1957).
[7.53] I. Nakada, J. Phys. Soc. Japan **13**, 1547 (1958).
[7.54] S. A. Salam, Proc. Math. Phys. Soc. UAR (Egypt) **24**, 41 (1960).
[7.55] R. F. Frindt and A. D. Yoffe, Proc. Roy. Soc. **A273**, 69 (1963).
[7.56] S. S. Bhatnagar, B. Prakash and J. Singh, J. Indian Chem. Soc. **16**, 313 (1939).
[7.57] H. Bizette, Ann. Phys. Paris **1**, 233 (1946).
[7.58] W. S. Carter and K. W. Stevens, Proc. Phys. Soc. **69B**, 1006 (1956).
[7.59] P. J. Wojtowicz, Phys. Rev. **107**, 429 (1957).
[7.60] L. M. Corliss, N. Elliott, and J. M. Hastings, Phys. Rev. **104**, 924 (1956).
[7.61] S. S. Batsanov and V. V. Kopytina, Vest. Moscow Univ. Ser. Mat. Mekh. Astron. Fiz. Khim. **12**, 227 (1957).
[7.62] L. Neel and R. Benoit, C.R. Acad. Sci. Paris **237**, 444 (1953).
[7.63] F. Hulliger, Helv. Phys. Acta **32**, 615 (1959).
[7.64] L. M. Corliss, N. Elliott, and J. M. Hastings, J. Appl. Phys. **29**, 391 (1958).
[7.65] J. M. Hastings, N. Elliott, and L. M. Corliss, Phys. Rev. **115**, 13 (1959).
[7.66] R. Benoit, J. Chim. Phys. **52**, 119 (1955).
[7.67] S. Miyahara, Proc. Phys. Math. Soc. Japan **22**, 358 (1940).
[7.68] T. Hirone and N. Tsuya, Phys. Rev. **83**, 1063 (1951).
[7.69] K. Yosida, Progr. Theor. Phys. **6**, 356 (1951).
[7.70] E. F. Bertaut, C.R. Acad. Sci. Paris **234**, 1295 (1952).
[7.71] E. F. Bertaut, J. Phys. Radium **13**, 372 (1952); **16**, 425 (1955).
[7.72] R. Benoit, C.R. Acad. Sci. Paris **234**, 2175 (1952).
[7.73] R. Pauthenet, C.R. Acad. Sci. Paris **234**, 2261 (1952).
[7.74] L. Neel, Proc. Phys. Soc. **A65**, 869 (1952).
[7.75] L. Neel, Rev. Mod. Phys. **25**, 58 (1953).
[7.76] R. Benoit, State thesis, Grenoble (1954).
[7.77] T. Hirone, S. Maeda, S. Chiba, and N. Tsuya, J. Phys. Soc. Japan **9**, 500 (1954).
[7.78] T. Hirone, S. Maeda, and N. Tsuya, J. Phys. Soc. Japan **9**, 503 (1954).
[7.79] T. Kamigaichi, J. Sci. Hiroshima Univ. **A19**, 499 (1956).
[7.80] T. Kamigaichi, T. Hihara, H. Tazaki, and E. Hirahara, J. Phys. Soc. Japan **11**, 606 (1956).
[7.81] T. Hihara, M. Murakami, and E. Hirahara, J. Phys. Soc. Japan **12**, 743 (1957).
[7.82] E. Hirahara and M. Murakami, J. Phys. Chem. Solids **7**, 281 (1958).
[7.83] T. Hihara, J. Sci. Hiroshima Univ. **A22**, 215 (1958); **24**, 31 (1960).
[7.84] S. Fujime, M. Murakami, and E. Hirahara, J. Phys. Soc. Japan **16**, 183 (1961).
[7.85] M. Murakami, J. Phys. Soc. Japan **16**, 187 (1961).
[7.86] J. B. Goodenough, J. Appl. Phys. suppl. **33**, 1197 (1962).
[7.87] J. T. Sparks, W. Mead, A. J. Kirschbaum, and W. Marshall, J. Appl. Phys. **31**, 356S (1960).
[7.88] J. T. Sparks, W. Mead, and T. Komoto, J. Phys. Soc. Japan suppl. **17** (B-I), 249 (1962).
[7.89] M. Bin and R. Pauthenet, C.R. Acad. Sci. Paris **254**, 3078 (1962).

[7.90] T. Hirone, K. Adachi, M. Yamada, S. Chiba, and S. Tazawa, *J. Phys. Soc. Japan suppl.* **17** (B-I), 257 (1962).
[7.91] M. Bin and R. Pauthenet, *J. Appl. Phys.* **34**, 1161 (1963).
[7.92] K. Ôno, Y. Ishikawa, A. Ito, and E. Hirahara, *J. Phys. Soc. Japan suppl.* **17** (B-I), 125 (1962).
[7.93] H. Horita and E. Hirahara, *J. Phys. Soc. Japan* **21**, 1447 (1966).
[7.94] S. S. Hafner, B. J. Evans, and G. M. Kalvius, *Solid State Commun.* **5**, 17 (1967).
[7.95] A. Theodossiou, *Phys. Rev.* **137A**, 1321 (1965).
[7.96] J. C. Marinace, *Phys. Rev.* **96**, 593 (1954).
[7.97] G. Borgonovi and G. Caglioti, *Nuovo Cimento* **24**, 1174 (1962).
[7.98] T. Rosenqvist, *J. Iron Steel Inst.* **176**, 37 (1954).
[7.99] M. Laffitte, State thesis, Paris (1958).
[7.100] R. Benoit, *J. Chim. Phys.* **52**, 201 (1955).
[7.101] V. G. Kuznetsov *et al.*, *Vop. Met. Fiz. Poluprov., Akad. Nauk SSSR* **4**, 159 (1961).
[7.102] E. A. Dancy and G. J. Derge, *Trans. Met. Soc. AIME* **227**, 1034 (1963).
[7.103] R. Perthel, *Ann. Phys. Leipzig.* **5**, 273 (1960).
[7.104] K. Hauffe and H. G. Flindt, *Z. Phys. Chem.* **200**, 199 (1952).
[7.105] K. Shimomura, *J. Sci. Hiroshima Univ.* **A16**, 319 (1952).
[7.106] J. T. Sparks and T. Komoto, *J. Appl. Phys.* **34**, 1191 (1963).
[7.107] I. Tsubokawa, *J. Phys. Soc. Japan* **13**, 1432 (1958).
[7.108] J. Rey and M. Laffitte, *Acta Crystall.* **21**, 686 (1966).
[7.109] H. Fischmeister, *Acta Chem. Scand.* **13**, 852 (1959).
[7.110] R. A. Munson, *Inorgan. Chem.* **5**, 1296 (1966).
[7.111] M. Chevreton and F. Bertaut, *C.R. Acad. Sci. Paris* **255**, 1275 (1962).
[7.112] M. Chevreton and S. Brunie, *Bull. Soc. Fr. Minéral. Cristallog.* **87**, 277 (1964).
[7.113] S. M. Ryvkin, *Zh. Tekh. Fiz.* **18**, 1951 (1948).
[7.114] F. K. Putseiko, *Dokl. Akad. Nauk SSSR* **67**, 1009 (1949).
[7.115] R. K. Quinn, R. Simmons, and J. J. Banewicz, *J. Phys. Chem. Solids* **70**, 230 (1966).
[7.116] L. H. Brixner, *J. Electrochem. Soc.* **110**, 289 (1963).
[7.117] I. Tsubokawa, *J. Phys. Soc. Japan* **11**, 662 (1956).
[7.118] F. K. Lotgering and E. W. Gorter, *J. Phys. Chem. Solids* **3**, 238 (1957).
[7.119] L. M. Corliss, N. Elliott, J. M. Hastings, and R. L. Sass, *Phys. Rev.* **122**, 1402 (1961).
[7.120] K. Masumoto, T. Hihara, and T. Kamigaichi, *J. Phys. Soc. Japan* **17**, 1209 (1962).
[7.121] M. Chevreton and F. Bertaut, *C.R. Acad. Sci. Paris* **253**, 145 (1961).
[7.122] E. F. Bertaut, A. Delapalme, F. Forrat, G. Roult, F. de Bergevin, and R. Pauthenet, *J. Appl. Phys. suppl.* **33**, 1123 (1962).
[7.123] M. Chevreton, M. Murat, C. Eyraud, and E. F. Bertaut, *J. Phys.* **24**, 443 (1963).
[7.124] K. Masumoto, *J. Sci. Hiroshima Univ.* **A27**, 87 (1964).
[7.125] L. H. Brixner, *J. Inorg. Nucl. Chem.* **24**, 257 (1962).
[7.126] L. H. Brixner and G. Teufer, *Inorg. Chem.* **2**, 992 (1963).
[7.127] W. T. Hicks, *J. Electrochem. Soc.* **111**, 1058 (1964).
[7.128] R. F. Frindt, *J. Phys. Chem. Solids* **24**, 1107 (1963).
[7.129] E. Revolinsky and D. Beerntsen, *J. Appl. Phys.* **35**, 2086 (1964).
[7.130] J. A. Champion, *Brit. J. Appl. Phys.* **16**, 1035 (1965).
[7.131] C. G. Shull, W. A. Strauser, and E. O. Wollan, *Phys. Rev.* **83**, 333 (1951).
[7.132] E. Uchida and H. Kondo, *Busseiron Kenkyu* **59**, 88 (1953).
[7.133] W. Palmer, *J. Appl. Phys.* **25**, 125 (1954).
[7.134] T. R. McGuire and R. R. Heikes, *J. Appl. Phys. suppl.* **31**, 276 (1960).
[7.135] B. V. Karpenko and A. A. Berdyshev, *Fiz. Tverd. Tela* **5**, 3397 (1963).
[7.136] T. Hirone, S. Maeda, and N. Tsuya, *J. Phys. Soc. Japan* **9**, 496 (1954).
[7.137] T. Hirone and S. Chiba, *J. Phys. Soc. Japan* **11**, 666 (1956).
[7.138] A. Okazaki and K. Hirakawa, *J. Phys. Soc. Japan* **11**, 930 (1956).

[7.139] N. Kh. Abrikosov, *Izv. Akad. Nauk SSSR, Sekt. Fiz. Khim. Analiza* **27**, 157 (1956).
[7.140] A. Okazaki, *J. Phys. Soc. Japan* **14**, 112 (1959).
[7.141] A. Okazaki, *J. Phys. Soc. Japan* **16**, 1162 (1961).
[7.142] K. Hirakawa, *J. Phys. Soc. Japan* **12**, 929 (1957).
[7.143] M. S. Maxim, *Stud. Cercet. Fiz. Romin.* **9**, 323 (1958).
[7.144] F. A. Sidorenko, P. V. Gel'd, and L. B. Dubrovskaya, *Fiz. Metal. Metalloved.* **8**, 465 (1959).
[7.145] F. Lihl and H. Ebel, *Arch. Eisenhüttenw.* **32**, 489 (1961).
[7.146] A. F. Andresen and J. Leciejewicz, *J. Phys.* **25**, 574 (1964); *Coll. Int. Diffract. Diffus. Neutrons*, p. 150. C.N.R.S., Paris (1964).
[7.147] J. Serre and R. Druilhe, *C.R. Acad. Sci. Paris* **262B**, 639 (1966).
[7.148] G. Fischer, *Can. J. Phys.* **36**, 1435 (1958).
[7.149] L. D. Dudkin and V. I. Vaidanich, *Fiz. Tverd. Tela* **2**, 1526 (1960).
[7.150] B. T. Matthias, E. Corenzwit, and E. C. Miller, *Phys. Rev.* **93**, 1415 (1954).
[7.151] F. M. Gal'perin and T. M. Perekalina, *Dokl. Akad. Nauk SSSR* **69** 19 (1949).
[7.152] B. E. Brown, *Acta Crystallog.* **20** (Pt 2), 264 (1966).
[7.153] C. Guillaud and S. Barbezat, *C.R. Acad. Sci. Paris* **222**, 386 (1946).
[7.154] C. Guillaud, *C.R. Acad. Sci. Paris* **222**, 1224 (1946).
[7.155] S. V. Vonsovskii, *Zh. Tekh. Fiz.* **18**, 131 (1948).
[7.156] I. K. Kikoin, *Dokl. Akad. Nauk SSSR* **68**, 481 (1949).
[7.157] I. G. Fakidov, N. P. Grazhdankina, and I. K. Kikoin, *Dokl. Akad. Nauk SSSR* **68**, 491 (1949).
[7.158] K. P. Belov, *Dokl. Akad. Nauk SSSR* **71**, 261 (1950).
[7.159] F. M. Gal'perin and T. M. Perekalina, *Dokl. Akad. Nauk SSSR* **69**, 19 (1949).
[7.160] I. G. Fakidov and N. P. Grazhdankina, *Dokl. Akad. Nauk SSSR* **66**, 847 (1949).
[7.161] T. D. Zotov and Ya. S. Shur, *Dokl. Akad. Nauk SSSR* **86**, 267 (1952).
[7.162] C. Guillaud, *C.R. Acad. Sci. Paris* **235**, 468 (1952).
[7.163] I. K. Kikoin, E. M. Buryak, and Yu. A. Muromkin, *Dokl. Akad. Nauk SSSR* **125**, 1011 (1959).
[7.164] L. G. Gaidukov, V. N. Novogrudskii, and I. G. Fakidov, *Fiz. Metal. Metalloved.* **9**, 152 (1960).
[7.165] I. K. Kikoin, N. A. Babushkina, and T. N. Igosheva, *Zh. Eksp. Teor. Fiz.* **39**, 1172 (1960).
[7.166] I. K. Kikoin and T. N. Igosheva, *Zh. Eksp. Teor. Fiz.* **46**, 1923 (1964).
[7.167] N. P. Grazhdankina, L. G. Gaidukov, K. P. Rodionov, M. I. Oleinik, and V. A. Shchipanov, *Zh. Eksp. Teor. Fiz.* **40**, 433 (1961).
[7.168] N. P. Grazhdankina, *Zh. Eksp. Teor. Fiz.* **48**, 1257 (1965).
[7.169] M. Chevreton, E. F. Bertaut, and F. Jellinek, *Acta Crystallog.* **16** (Pt. 5), 431 (1963).
[7.170] Kieu van Con and J. Suchet, *C.R. Acad. Sci. Paris* **256**, 2823 (1963).
[7.171] A. F. Andresen, *Acta. Chem. Scand.* **17**, 1335 (1963).
[7.172] J. Suchet, *C.R. Acad. Sci. Paris* **257**, 1756 (1963).
[7.173] W. Albers and C. Haas, *Phys. Semicond.* 1261 (Conf. Paris 1964), Dunod, Paris 1964.
[7.174] J. Suchet and P. Imbert, *C.R. Acad. Sci. Paris* **260**, 5239 (1965).
[7.175] J. Suchet and R. Druilhe, *Coll. Int. Dérivés Semimétalliques*, p. 307 (Orsay 1965), C.N.R.S., Paris (1967).
[7.176] M. Nogami, *Rep. Fac. Eng. Shizuoka Univ.* **16**, 27 (1965).
[7.177] M. Nogami, *Jap. J. Appl. Phys.* **5**, 134 (1966).
[7.178] J. Serre and J. Suchet, *C.R. Acad. Sci. Paris* **264B**, 1412 (1967).
[7.179] D. E. Cox, G. Shirane, and W. J. Takei, *Proc. Int. Conf. Magn.* p. 291. Inst. Phys. & Phys. Soc., London, 1965.
[7.180] W. J. Takei, D. E. Cox, and G. Shirane, *J. Appl. Phys.* **37**, 973 (1966).

[7.181] A. Lepetit, *J. Phys.* **26**, 175 (1965).
[7.182] B. E. Brown, *Acta Crystallogr.* **20** (Pt. 2), 268 (1966).
[7.183] E. Revolinsky and D. J. Beerntsen, *J. Phys. Chem. Solids* **27**, 523 (1966).
[7.184] S. Kabashima, *J. Phys. Soc. Japan* **21**, 945 (1966).
[7.185] E. Uchida, H. Kondo, and N. Fukuoka, *J. Phys. Soc. Japan* **11**, 27 (1956).
[7.186] N. P. Grazhdankina, *Zh. Eksp. Teor. Fiz.* **33**, 1524 (1957); **48**, 1257 (1965).
[7.187] H. A. Johansen, *J. Inorg. Nucl. Chem.* **6**, 344 (1958).
[7.188] A. V. Doroshenko, V. V. Klyuskin, A. A. Loshmanov, and V. I. Gomankov, *Fiz. Metal. Metalloved.* **12**, 911 (1961).
[7.189] H. Yadaka, T. Harada, and E. Hirahara, *J. Phys. Soc. Japan* **17**, 875 (1962).
[7.190] T. Komatsubara, M. Murakami, and E. Hirahara, *J. Phys. Soc. Japan* **18**, 356 (1963).
[7.191] J. D. Wasscher and C. Haas, *Phys. Lett.* **8**, 302 (1964); **10**, 160 (1964).
[7.192] J. D. Wasscher, A. M. Seuter, and C. Haas, *Phys. Semicond. Conf. Paris, 1964* p. 1269. Dunod, Paris (1964).
[7.193] N. Kumitomi, Y. Hamaguchi, and S. Anzai, *J. Phys.* **25**, 568 (1964).
[7.194] N. N. Sirota and G. I. Makovetskii, *Dokl. Akad. Nauk. SSSR* **170**, 1300 (1966).
[7.195] E. D. Devyatkova, A. V. Golubkov, E. K. Kudinov, and I. A. Smirnov, *Fiz. Tverd. Tela* **6**, 1813 (1964).
[7.196] J. D. Wasscher, *Solid State Commun.* **3**, 169 (1965).
[7.197] M. Chomentowski, H. Rodot, G. Villers, and M. Rodot, *C.R. Acad. Sci. Paris* **261**, 2198 (1965).
[7.198] M. Rodot, J. Lewis, H. Rodot, G. Villers, J. Cohen, and P. Mollard, *Proc. Int. Conf. Phys. Semicond. Kyoto, 1966, J. Phys. Soc. Japan suppl.* **21**, 627 (1966).
[7.199] A. Sawaoka and S. Miyahara, *J. Phys. Soc. Japan* **20**, 2087 (1965).
[7.200] A. Sawaoka, S. Miyahara, and S. Minomura, *J. Phys. Soc. Japan* **21**, 1017 (1966).
[7.201] F. Grønvold, H. Haraldsen, and J. Vihovde, *Acta Chem. Scand.* **8**, 1927 (1954).
[7.202] E. Uchida and H. Kondo, *J. Phys. Soc. Japan* **10**, 357 (1955).
[7.203] S. Chiba, *J. Phys. Soc. Japan* **10**, 837 (1955).
[7.204] D. M. Finlayson, D. Greig, J. P. Llewellyn, and T. Smith, *Proc. Phys. Soc.* **B69** (Pt. 8), 860 (1956).
[7.205] I. Tsubokawa and S. Chiba, *J. Phys. Soc. Japan* **14**, 1120 (1959).
[7.206] R. Naya, M. Murakami, and E. Hirahara, *J. Phys. Soc. Japan* **15**, 360 (1960).
[7.207] J. Suchet and J. Serre, *C.R. Acad. Sci. Paris* **260**, 3890 (1965).
[7.208] F. Aramu and P. Manca, *Nuovo Cimento* **33**, 1025 (1964).
[7.209] J. Suchet, *C.R. Acad. Sci. Paris* **259**, 3219 (1964).
[7.210] P. Manca and G. Saut, *C.R. Acad. Sci. Paris* **262B**, 1621 (1966).
[7.211] J. Chappert and G. Fatseas, *C.R. Acad. Sci. Paris* **262B**, 242 (1966).
[7.212] P. Manca, J. P. Suchet, and G. A. Fatseas, *Ann. Phys. Paris* **1**, 621 (1966).
[7.213] E. Uchida and H. Kondo, *Busseiron Kenkyu* **79**, 92 (1954).
[7.214] E. Uchida, *J. Phys. Soc. Japan* **10**, 517 (1955).
[7.215] J. Haraldsen, F. Grønvold, and T. Hurlen, *Z. Anorg. Allg. Chem.* **283**, 143 (1956).
[7.216] E. Uchida, *J. Phys. Soc. Japan* **11**, 465 (1956).
[7.217] S. M. Ariya, E. M. Kolina, and M. S. Apurina, *Zh. Neorg. Khim.* **2**, 23 (1957).
[7.218] L. D. Dudkin and K. A. Dyuldina, *Zh. Neorg. Khim.* **4**, 2313 (1959).
[7.219] G. Saut, *C.R. Acad. Sci. Paris* **261**, 3339 (1965).
[7.220] V. P. Zhuze and A. R. Regel, *Zh. Tekh. Fiz.* **25**, 978 (1955).
[7.221] E. Uchida and H. Kondo, *J. Phys. Soc. Japan* **11**, 21 (1956).
[7.222] G. Saut, *C.R. Acad. Sci. Paris* **263B**, 1174 (1966).
[7.223] J. Guggenheim, F. Hulliger, and J. Müller, *Helv. Phys. Acta* **34**, 408 (1961).
[7.224] A. Kjekshus and W. B. Pearson, *Can. J. Phys.* **43**, 438 (1965).

Chapter 8 ||| **Compounds of Rare Earths and Similar Elements**

8.1. MAIN STRUCTURES

In the same way as the letter T was used to symbolize any transition metal, the letter L will be used to symbolize the elements Sc, Y, or any lanthanide Ln, and the letter A to symbolize any actinide. Although this letter has already been used elsewhere, there should be no confusion.

The main crystallographic structures met with in this class of compounds have the symbols B1 (rock salt), C1 (fluorite), C11 (CaC_2), C32 (AlB_2), C38 (Cu_2Sb), $D2_1$ (CaB_6), $D5_2$ (La_2O_3), $D5_3$ (α-Mn_2O_3) and $D7_3$ (Th_3P_4). The familiar B1 structure is very common among binary compounds of rare earths and similar elements, including nearly all LV, AV, LVI, AVI compounds, and AC carbides. The simple C1 structure includes all the AO_2 oxides and a few LnO_2 oxides, where the electron formula of the Ln element is particularly stable: $Ce(IV)(f^0)$, $Pr(IV)(f^1)$, $Tb(IV)(f^7)$. It produces a structure with vacancies, $LO_{2-x}\square_x$, the homogeneity domain of which seems to be $0 < x < \frac{1}{4}$ [8.1]. In contrast to T atoms, L and A atoms can therefore be met with in cubic environments involving resonating sp^3 orbital functions. Another example will be found in the $D7_3$ structure. Finally, the C32 structure (AlB_2), referred to in Section 4.1, includes the borides LB_2 and AB_2 and a few silicides (USi_2). In this section we shall deal with the new structures $D5_2$ and $D5_3$ for L_2O_3 oxides and particularly $D7_3$, with or without vacancies, which assumes great importance for L_3VI_4 and A_3VI_4 chalcogenides and A_3V_4 compounds. The structures $D2_1$ (CaB_6), C11 (CaC_2) and C38 (Cu_2Sb), involving the compounds LB_6, LC_2 and AV_2, respectively, will then be mentioned briefly. Readers wishing for more detailed information on all these structures, as well as on rarer structures such as C23 (β-US_2), C222 (U_3O_8), $D5_8$ (Sb_2S_3), $L1_2$ (Cu_3Au) may refer to the general works by Wells [8.2] and Wyckoff [8.3].

L_2O_3 oxides, in which the L element is trivalent, can crystallize in three different structures, the first (A) of which is hexagonal, with the symbol $D5_2$ and La_2O_3 as its prototype. Figure 8.1 shows that the environment of the L atom is quasi octahedral, but that a seventh oxygen atom is added above the center of one of the faces of the octahedron, and that it repels the three oxygen atoms occupying the corners of this face. This seven coordination, which is quite rare, has been confirmed by neutron diffraction [8.4]. The second structure (B) is very complex, and contains three sorts of nonequivalent L sites, each of which has six almost equidistant oxygen neighbors, with the seventh oxygen atom slightly farther away [8.5]. It has been described as monoclinic for Sm_2O_3, Gd_2O_3 [8.3] and Dy_2O_3 [8.6]. The third structure (C) is cubic, with the symbol $D5_3$ and α-Mn_2O_3 as its prototype (cf. Section 5.1). Figure 8.2 shows that it is derived from the C1 structure, of which it is a distorted overstructure with vacancies $L_2O_3\square$. What is more, there is a continuous transition with C1 in LO_2–$L_2'O_3$ solid solutions. Each L atom has a quasi-octahedral environment, but two kinds of nonequivalent sites can be distinguished. Each has six neighboring oxygen atoms at the corners of a cube, but in the first kind the missing oxygen atoms are at the ends of the face diagonal, while in the second kind they are at the ends of a cube diagonal. Structure (A) appears to be stable at higher temperatures, while (C) is stable at lower temperatures [8.2]. The distribution of the three structures has been studied by Iandelli [8.7]. It is at present as follows [8.2, 8.3, 8.6]:

Sc	Y	La	Ce	Pr	Nd	Pm	Sm	Eu	Gd	Tb	Dy	Ho	Er	Tm	Yb	Lu
		A	A	A	A		A									
					B	B		B	B	B		B				
	C	C	C	C	C	C	C	C	C	C	C	C	C	C	C	C

The $D7_3$ cubic structure of Th_3P_4 is very common among compounds of rare earths and similar elements [8.8, 8.9]. In it, the L or A element occurs in a trivalent state in a roughly cubic environment, similar to that of Cs in CsCl, while the metalloid has a roughly octahedral environment. As first approximation, one can therefore assume the intervention of sp^3 orbital functions on the metal and p^3 orbital functions on the metalloid, which resonance for each of them. Figure 8.3 shows, in the absence of resonance, the tetrahedron of X atoms which would surround each L or A atom, and the three tetrahedra with a common corner occupied by an X atom, providing it with three L or A neighbors. The resonance by permutation of the sp^3 bonds reestablishes the cubic environment of the metal, as in the structure B2 (CsCl) or C1 (CaF_2) [8.10]. This structure produces a structure with vacancies $\square_x L_{3-x} VI_4$ or $\square_x A_{3-x} VI_4$, with $0 < x < \frac{1}{3}$, for a large number of chalcogenides of rare earths and similar elements prepared at high temperature (for the rare earths, cf. [8.10a]. For instance, the density drops from 5.675 for Ce_3S_4 ($x = 0$) to 5.186

8.1. Main Structures

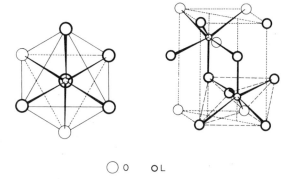

Fig. 8.1. Structure (A) of L_2O_3 oxides. On the left, in projection, the 7 oxygen atoms surrounding the L atom: octahedron plus extra oxygen atom (thick circle) above one of the faces (based on Wells [8.2]).

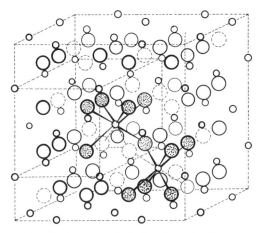

Fig. 8.2. Structure (C) of L_2O_3 oxides. It derives from the C1 structure of fluorite, with the withdrawal of a quarter of the oxygen atoms (dotted circles). The two kinds of non-equivalent L sites are indicated (based on Wells [8.2]).

for Ce_2S_3 ($x = \frac{1}{3}$) [8.11, 8.12]. On the other hand, no significant deviation from stoichiometry has been recorded for the A_3V_4 compounds.

LB_6 borides and a few AB_6 borides (Th, Pu) crystallize in the $D2_1$ structure of CaB_6. In it one finds B_6 octahedra linked by their corners, and these B—B covalent bonds provide the crystal with a highly rigid skeleton, and explain the stability, hardness and refractoriness of the borides, comparable with those of the element boron (Fig. 8.4). The L atoms form a separate interstitial sub-lattice. The elementary cell is of the B2 type [8.13, 8.14]. LC_2 and AC_2 carbides crystallize in the C11 structure of CaC_2. This is in fact a B1 structure in which the metalloid sites are occupied by C_2 pairs along [001]. The carbon atoms in

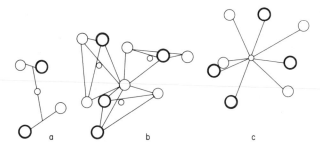

Fig. 8.3. $D7_3$ structure of Th_3P_4. ThP_4 tetrahedron (a), set of 3 tetrahedra surrounding one P atom (b) and total coordination of Th (c) (based on Obolonchik and Lashkarev [8.10]).

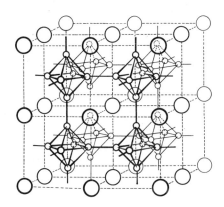

Fig. 8.4. $D2_1$ structure of CaB_6. The small circles represent the boron atoms (based on Wells [8.2]).

these pairs are probably linked by the triple bond found in acetylene [8.2]. Many AV_2 compounds crystallize in the C38 structure of Cu_2Sb or Fe_2As (Fig. 8.5). There are two different A atom sites, with complex coordinations [8.3].

Lanthanides and actinides are mainly of valency 3. The other possible valencies are shown below:

		4	4	?					4						
La	Ce	Pr	Nd	Pm	Sm	Eu	Gd	Tb	Dy	Ho	Er	Tm	Yb	Lu
					?	2	2						?	2	

		4	4	4	4	4	4					
Ac	Th	Pa	U	Np	Pu	Am	Cm	Bk	Cf	...	No	
				2	2	2					2	

They illustrate the special stability of the electron configurations $4f^0$(La(III), Ce(IV)), $4f^7$(Eu(II), Gd(III), Tb(IV)), $4f^{14}$Yb((II), Lu(III)), $5f^0$(Ac(III), Th(IV)), $5f^7$(Am(II), Cm(III)), and $5f^{14}$(No(II)). Among the lanthanides, the development of properties show a clear break for gadolinium. Among

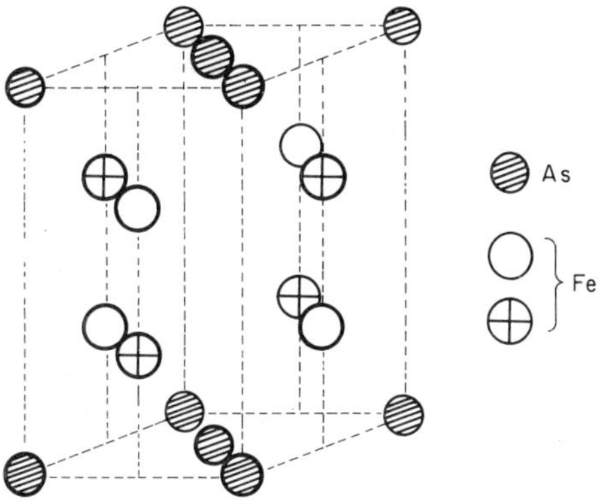

Fig. 8.5. C38 structure of Fe_2As (or Cu_2Sb) (based on Wells [8.2]).

actinides, valency 4 is most common, and valencies 5, 6, and 7 can be found. This is due to the possibility of transitions between the 5f and 6d levels [8.2]. It should be noted that transitions from 4f to 5d also exist in lanthanides, but they are less probable and therefore less obvious.

The levels occupied by the f electrons in the atom are too deep for them to be affected much by neighboring atoms. The effect of the crystal field, although not quite nonexistent, does not cause any division of the f level (cf. Section 3.1). The break in properties observed for the configuration f^7, however, shows that it is convenient to consider the levels $f\alpha$ and $f\beta$ as separate, in the same way as the $d\alpha$ and the $d\beta$ levels of transition metal compounds. The orbital moments are not blocked and, except in particular cases, the magnetic properties depend on the resultant J of the orbital and spin moments L and S (cf. Section 3.2). The magnetic moments of the trivalent ions were calculated by Hund from theoretical values for L, S, and J, on the assumption that all the ions occupy the ground state. Van Vleck's [8.15] more general theory modifies the values for Sm and Eu (Fig. 8.6), matching experimental results.

Synthesis of compounds of rare earths and similar elements is generally recent. Problems of purity (lanthanides) or radioactivity (actinides) make it very difficult. For these reasons, the research described in this chapter is less advanced than that dealt with in earlier chapters.

8.2. IIIB, IVB, AND VB COMPOUNDS

As already mentioned, diborides crystallize in the C32 structure, but the (quadratic) structure of LB_4 and AB_4 borides already has a boron skeleton [8.16], like that of LB_6 and LB_{12} borides. Eick's [8.17] digest gives the cell parameters for these different compounds. The only ones with useful electrical properties are hexaborides. Lafferty [8.18, 8.19] shows that the valency

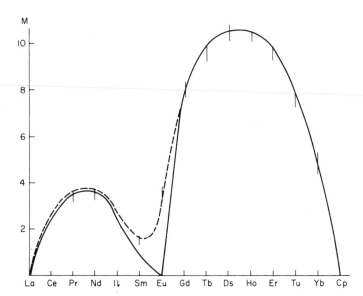

Fig. 8.6. Magnetic moments M of lanthanide ions. Comparison between (——) Hund's and (– – –) Van Vleck's theories and (|) experimental results. [Based on Pascal, "Nouveau Traité de Chimie Minérale." Masson, Paris, 1956–1963.]

electrons in the rare earth atom are not accepted by the B_6 complex, and behave like free electrons. LaB_6 has a resistivity of $57 \times 10^{-6}\ \Omega$ cm, follows the law of metals, and has a very marked thermionic emission. It is used as a surface layer on a cathode of carburized refractory metal. According to Benoit [8.20], LaB_6 and YB_6 are diamagnetic and the others paramagnetic. Measurement of their resistivity [8.21], and special investigations of YB_6 [8.22], SmB_6 [8.23] and LaB_6 [8.24] confirm their metallic character and thermionic properties. Flodmark [8.25] has investigated the nature of the bonds, concluding that there is metallic conduction. Binder *et al.* [8.26] reached this conclusion for SmB_6, EuB_6, and YbB_6 only, on the basis of crystal parameters.

Samsonov *et al.* have summarized this work [8.27] and discussed the effect of porosity on ρ and R_H [8.28]. The number of free electrons per L atom, which is 0.05 for CaB_6, varies between 0.9 and 1.1 for LB_6 borides, except for EuB_6 (0.09) and YbB_6 (0.05). It is 1.99 per Th atom in ThB_6. The authors referred to suggest that the L atom loses three electrons, two of which are transferred to the boron atom, within the framework of the L—B bonds, the other one remaining free. It will be noted that the stable configurations $4f^7$ and $4f^{14}$ correspond at Eu(II) and Yb(II), thus explaining the irregularity in their borides, and that Th, after transferring two electrons to the boron, still possesses two 6d electrons, probably free in this band. Coles *et al.* [8.29] mention the intervention of free electrons in the antiferromagnetic interaction

8.2. IIIB, IVB, and VB Compounds

of GdB_6, and the strong magnetic scattering. The EPR shows that the configuration $4f^7$ exists in the paramagnetic domain. Tsarev et al. [8.30] believe that their optical measurements of LaB_6 can be explained by the existence of a forbidden band of 0.08 eV, but Kauer [8.31] thinks that a metal is involved (10^{22} electrons cm^{-3}, $\mu = 32$ cm^2 $V^{-1}sec^{-1}$, $\varepsilon = 15.6$ and $m^*/m = 0.32$). In conclusion, the electrons of LB_6 borides seem to move in a band (5d?), almost empty for Eu and Yb. It is hard to say whether metals or pseudometals are involved. Bilz' model (cf. Section 4.2) may perhaps be applied to them.

Among monocarbides, ScC has a resistivity of 0.26 Ω cm, compared with 0.05 for TC compounds of the first transition series. Its metallic character is therefore less marked [8.30]. The resistivity temperature coefficient is positive for UC [8.31, 8.32], ThC and PuC [8.32]. ThC is supraconducting below 4°K, and PuC presents an irregularity around $T_N = 100°K$. The resistivity of UC rises from 0.05 Ω cm (300°K) to 0.2 Ω cm (2000°K) [8.32, 8.33]. The magnetic susceptibility of PuC varies considerably with the stoichiometry [8.34]. Auskern et al. [8.35] have measured ρ, R_H, and α, and found 0.16 electrons per Th atom in $ThC_{0.96}$ (0.025 Ω cm), the electronic structure of which they claim to be similar to that of TiC, ZrC, and HfC. Kruger and Moser [8.36] have found a resistivity of 0.26 Ω cm for PuC, almost unaffected by temperature.

Dicarbides have the C11 structure, except for YC_2 (hexagonal). Gadolinium is trivalent ($4f^7$) in GdC_2 [8.1]. The resistivity of lanthanon carbides has been studied [8.37]. Auskern et al. [8.35] have found 1.6 electrons per Th atom in ThC_2 (0.03 Ω cm). TbC_2 and HoC_2 have a complex antiferromagnetic order [8.38]. Ln_2C_3 compounds are centered cubic. Atoji et al. [8.39–8.41] have used neutron diffraction to show that the L atoms in the carbides generally have valency 3, except for YbC_2 (2.8) and Ce_2C_3 (3.4) The uranium in UC_2 seems to be tetravalent. The d band makes an important contribution to conduction in dicarbides A digest has been produced on the structure of rare earth silicides [8.42]. The disilicides LSi_2 and ASi_2, with a structure β-$ThSi_2$, are the most familiar [8.43]. Certain digermanides such as $ScGe_2$ [8.44] and YGe_2 [8.45] are supraconductors.

The commonest nitrides are LN and AN with the B1 structure. There exist a few higher actinide nitrides, $Th_2N_3(D5_2)$, UN_2, and $PaN_2(C1)$. Mononitrides are refractory (ScN, 2900°C). Their cell parameter varies lineally with the trivalent ionic radius (Fig. 8.7). The L atoms therefore probably have valency 3 (4 for Ce) [8.46]. For the values of cell parameters, see [8.17]. ScN has a resistivity of 3×10^{-4} Ω cm at room temperature [8.1]. Wilkinson et al. [8.47] and Child et al. [8.48] have confirmed valency 3 by neutron diffraction and have found ferromagnetism at low temperature, with the exception of TmN ($T_C = 43°K$ for TbN). Interactions with the crystal field reduce the moments and bring about a complex magnetic order. Costa et al. [8.49] have found $\rho = 0.45$ Ω cm for PuN.

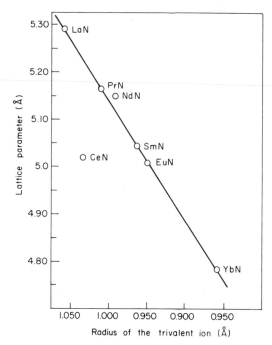

Fig. 8.7. Linear variation in the lattice parameter of LN nitrides, in terms of the radius of the trivalent ion (based on Eick et al. [8.46]).

Sclar [8.50] estimates a forbidden band E_G of around 2 eV for LN nitrides. Didchenko and Gortsema [8.51] have studied the paramagnetism of sintered powders and found θ parameters of 0°K (PrN) to 60 (GdN) and 0 (TmN). The constant paramagnetism of CeN indicates delocalization of the 4f electron of Ce(III), which therefore changes to Ce(IV). YbN is antiferromagnetic. The resistivity at room temperature, given in Fig. 8.8, shows the peaks for EuN and YbN. $\alpha \sim +10$ to $60\ \mu\mathrm{V}\ \mathrm{deg}^{-1}$ for CeN. YbN is the only one to have semiconductor properties, and the Hall effect for it shows $N = 2 \times 10^{21}\ \mathrm{cm}^{-3}$ and $\mu = 0.3\ \mathrm{cm}^2\ \mathrm{V}^{-1}\mathrm{sec}^{-1}$. For the theory of interactions with the crystal field, cf. [8.52]. Sclar [8.53] has studied the optical absorption of films evaporated in a vacuum, confirming, except for ScN and YN, the estimated value of $E_G \sim 2$ eV, and explaining the apparent metallic properties by a metal excess or slight oxidation leading to the appearance of valency 2 and a conduction in the 5d band. For GdN, one has $T_C = 69°\mathrm{K}$ and $7\mu_B$ [8.54]. Busch et al. [8.55, 8.56], finally, have done a complete study of the magnetic properties of the series, and have shown that the terms in which L has an uneven number of 4f electrons all show a magnetic order at sufficiently low temperatures.

Phosphides have the formulas LP and AP (B1), plus Th_3P_4 and U_3P_4 ($D7_3$)

8.2. IIIB, IVB, and VB Compounds

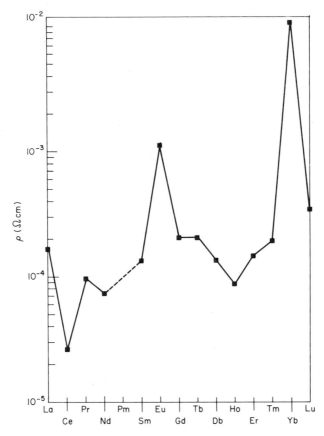

Fig. 8.8. Logarithmic variation in the resistivity at room temperature in the series of nitrides (based on Didchenko and Gortsema [8.51]).

and UP_2 (C38). Monophosphides, except for CeP, are fairly unstable [8.1]. Iandelli's [8.57] digest gives the values for the cell parameters. The fact that they are proportional to the trivalent ionic radii (Fig. 8.9) shows that the metal has valency 3. These are paramagnetic materials, but the Curie–Weiss law sometimes suggests ferromagnetism at low temperatures (GdP, $\theta = 40°K$). Child et al. [8.48] have found, by neutron diffraction, antiferromagnetism at low temperatures (complex order for HoP) as well as marked interactions with the crystal field. For the theory of these, see [8.52]. According to Busch et al., GdP [8.58], TbP, DyP, HoP, and ErP [8.59] are antiferromagnetic materials, and induced ferromagnetism appears only in high fields (metamagnetism). Tsuchida and Wallace [8.60] have found a magnetic order for CeP and a Van Vleck paramagnetism for PrP at low temperatures. The order is confirmed,

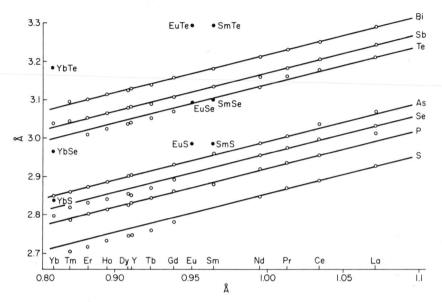

Fig. 8.9. Linear variation in the lattice parameter of LX compounds (X = P, As, Sb, Bi, S, Se, Te) in terms of the radius of the trivalent ion (based on Iandelli [8.57]).

but the valency of Ce changes with temperature [8.61]. For GdP, TbP, and DyP, Yaguchi [8.62] has found resistivities of around 5×10^{-3} Ω cm, with metallic properties. For PuP, Kruger and Moser [8.36] have found $\rho = 0.75$ Ω cm at room temperature and $\alpha < 0$ above 740°K. The resistivity increase from PuC to PuN and PuP occurs in other actinides, and is probably bound up with the filling of the 6d band and the 5f–6d transitions.

Arsenides have the formulas LAs and AAs (B1), plus Th_3As_4 and U_3As_4 ($D7_3$) and $ThAs_2$ and UAs_2 (C38). Iandelli's [8.56] digest gives the values of the cell parameters. Their surprising degree of proportionality to the trivalent ionic radii (Fig. 8.9) makes valency 3 probable. These are paramagnetics. Child et al. [8.48], using neutron diffraction, have found antiferromagnetism at low temperatures, as well as marked interactions with the crystal field. Reid et al. [8.63], using polycrystal ingots of NdAs, GdAs, and SmAs (arsenic deficiency) have found 10^{20} to 10^{21} electrons cm^{-3}, $\rho \sim 0.1$–0.2 Ω cm, $\mu \sim 50$–200 cm^2 V^{-1}sec^{-1}, $\alpha \sim -10 \,\mu$V deg^{-1}. Despite the variation in mobility shown in Fig. 8.10, the uniformity in the number of carriers, regardless of temperature, rules out any excitation mechanism. Busch et al. [8.64], working with monocrystals, have found that GdAs, TbAs, and ErAs are metamagnetic at low temperatures ($T_N = 25$°K for GdAs). HoAs is antiferro- ($H < 1.8$ kOe), ferri- ($1.8 < H < 6$ kOe), and then ferromagnetic ($H = 6$ kOe). Similarly, DyAs moves from ferri- to ferromagnetism ($T_C = 8.5$°K). These results

8.2. IIIB, IVB, and VB Compounds

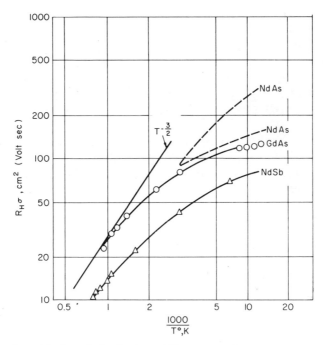

Fig. 8.10. Logarithmic variation in the mobility $\mu = R_H \sigma$ in relation to the temperature reciprocal for NdAs, GdAs, and NdSb of n-type (based on Reid et al. [8.63]).

correspond with the theory in [8.52]. Tsuchida and Wallace [8.60], working at low temperatures, have found a magnetic order for CeAs and Van Vleck paramagnetism for PrAs. The order is confirmed, but the valency of Ce changes with the temperature [8.61].

Antimonides have the formulas LSb and ASb (B1), plus Th_3Sb_4 and U_3Sb_4 ($D7_3$) and $ThSb_2$ and USb_2 (C38). Iandelli's [8.56] digest gives the value of the cell parameters. Their very surprising degree of proportionality to the trivalent ionic radii (Fig. 8.9) may indicate valency 3. Miller and Himes [8.65] mention resistivities of around 10^{-4} Ω cm, and an n-type for LaSb, ErSb, and YSb. Busch et al. [8.66] mention LaSb as being dia-, PrSb, NdSb, SmSb, HoSb, and EuSb as being para-, TbSb as being antiferro-, and DySb as being metamagnetic at low temperatures. Child et al. [8.48], using neutron diffraction, have found antiferromagnetism at low temperatures, as well as marked interactions with the crystal field. Reid et al. [8.63], using polycrystal ingots of NdSb, have found 10^{21} electrons cm^{-3}, $\rho \sim 0.07$ Ω cm, $\mu \sim 50$ cm^2 $V^{-1} sec^{-1}$. Despite the mobility variation shown in Fig. 8.10, the uniformity of the number of carriers, regardless of temperature, rules out any excitation mechanism. Busch et al. [8.67], using monocrystals, have found antiferromagnetism for

Gd ($T_N = 28°K$), and metamagnetism for Nd, Tb, Dy, Ho, and Er. Tsuchida and Wallace [8.60] have found several successive magnetic orders for CeSb, and Van Vleck paramagnetism for PrSb. The order is confirmed, but the valency of Ce changes with the temperature [8.61]. The concentration of free electrons per Ho(III) atom varies from 0 (HoSb) to 1 (HoTe) in solid solutions [8.68]. Holtzberg et al. [8.68a] mentioned that Gd_4Sb_3 (anti $D7_3$) is ferromagnetic ($T_C = 260°K$) and metallic (220 $\mu\Omega$ cm). Its resistivity is maximum near 180°K.

In conclusion, one could extend to this whole section the suggestion of a transfer mechanism in a 5d (L) or 6d (A) band, but only the overlapping of these with a wide band would explain the high mobilities of certain LAs and LSb compounds.

8.3. Oxides

EuO is the only rare earth monoxide of interest to us. Matthias et al. [8.69] mention its ferromagnetism below $T_C = 77°K$, with a moment of $7\mu_B$, confirmed by neutron diffraction [8.70] and explained [8.71], as well as a resistance of $10^9 \Omega$ on pressed powders, indicating a semiconductor character. Wachter [8.72] and Busch et al. [8.73, 8.74] have found an optical absorption corresponding to $E_G = 1.115$ eV at room temperature, with $dE_G/dT = -10^{-4}$ eV deg^{-1}, but a reverse shift below 90°K (Fig. 8.11). This phenomenon had been predicted by Irkhin and Turov [8.75] and Irkhin [8.76]. For the Mössbauer resonance of ^{151}Eu, cf. [8.77]. Samokhvalov et al. [8.78] have shown that the magnetization intensity follows a law in $T^{3/2}$ between 1.7 and 43°K. Smit [8.79] suggests that the magnetism is due to the direct overlapping of the 5d orbitals, partly occupied. Suits et al. [8.80] have studied the Faraday and Cotton–Mouton magnetooptical effects, finding $T_C = 69°K$ and suggesting a $4f^7$–$4f^6 5d^1$ transition. Greiner and Fan [8.81] have studied the Kerr longitudinal magnetooptical effect, and confirmed the preceding hypothesis. These three effects are very marked in EuO. An oxide with the formula Eu_3O_4 has an orthorhombic structure [8.82]. Samokhvalov et al. [8.83] have found in it ferromagnetism and a magnetization intensity varying in T^2 with $T_C = 7.8°K$, while Wang [8.84] has found ferrimagnetism with $T_C = 77°K$ and Holmes and Schiebet [8.85] metamagnetism with $T_N = 5°K$.

Research into L_2O_3 semiconductors goes back much further. Foëx [8.86, 8.87] has studied the variation in resistivity of sintered powders when cooling in different atmospheres. Figures 8.12 [(a) La_2O_3, (b) Nd_2O_3, (c) Sm_2O_3] and 8.13 [(a) Ce_2O_3, (b) Pr_2O_3] show that the resistivity of cerium and praseodymium oxides, which can reach a higher valency state, depends very much on the composition. On PrO_x, Martin [8.87a] measures α as a function of x, then Honig et al. [8.87b] measure σ and α (monocrystals) and claim a semiconduction by transfers between Pr(II) and Pr(III). See also [8.87c] (PrO_x—O_2

8.3. Oxides

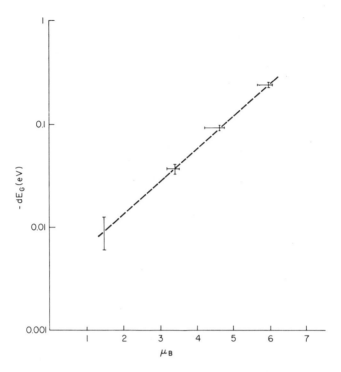

Fig. 8.11. Logarithmic variation in the forbidden band in EuO, in relation to spontaneous magnetization (based on Busch *et al.* [8.73]).

phase diagram) and [8.87d] (general discussion of LO_x—O_2 phase diagrams). La_2O_3 has dielectric properties [8.88], and Eu_2O_3 activates the emission spectra of alkaline-earth oxides [8.89]. Sc_2O_3, Y_2O_3, La_2O_3 ($4f^0$), and Lu_2O_3 ($4f^{14}$) are of course diamagnetic. The others follow the Curie–Weiss law, with the exception of Sm_2O_3 and Eu_2O_3, in which L atoms of different valencies are present, in a proportion varying with temperature [8.1, 8.90]. Y_2O_3 is antiferromagnetic below 3°K [8.91]. Noddack and Walch [8.92] have confirmed semiconductor properties and the increase in conductivity resulting from the addition of LO_2 dioxides. Solid solutions of different oxides present phase changes [8.93]. For the neutron investigation, cf. [8.94]. Schab and Bohla [8.95] state that Eu_2O_3 is a *p*-type semiconductor with a resistivity of 10^6 Ω cm and an activation energy of 1.1 eV. Tare and Schmalzried [8.96] have studied the Seebeck effect (which is positive) of Gd_2O_3, Dy_2O_3, Sm_2O_3, and Y_2O_3, in terms of the oxygen pressure P between 650 and 900°C. Ionic conductivity appears below 10^{-5} atm. For the paramagnetism of Eu_2O_3, cf. [8.97]. Bogoroditskii *et al.* [8.98] have given the resistivity/temperature

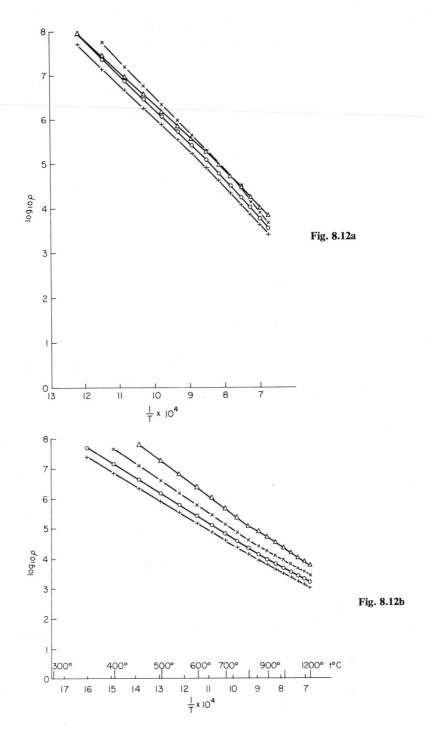

Fig. 8.12a

Fig. 8.12b

8.3. Oxides

curves and the dielectric constants for numerous oxides. The following activation energies have been observed between 300 and 1300°C approximately:

La_2O_3	Nd_2O_3	Sm_2O_3	Eu_2O_3	Gd_2O_3	Tb_2O_3	Dy_2O_3
2.86	2.24	2.27	1.84	2.9	0.86	3.08 eV

Ho_2O_3	Er_2O_3	Tm_2O_3	Yb_2O_3	Lu_2O_3
2.84	3.26	3.17	2.99	3.94 eV.

Tallman and Vest [8.99] have shown that Y_2O_3 is an *n*-type or *p*-type semiconductor, depending on T and P. For Y_2O_3–ThO_2 solid solutions, cf. [8.100]. For $PrO_{1.5+x}$—CeO_2, see the digest by Leroy-Eyring [8.100a].

LO_2 peroxides usually have a fairly wide homogeneity range resulting from the presence of oxygen vacancies. The most stable is CeO_2 (melting point 2600°C) [8.1]. Croatto [8.101] shows that the conductivity increases as one

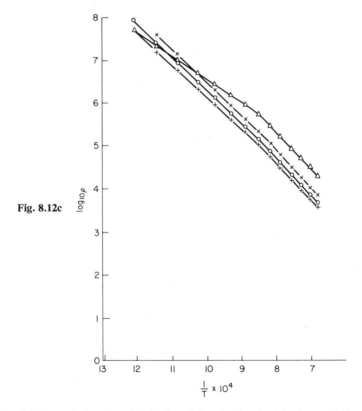

Fig. 8.12c

Fig. 8.12. Logarithmic variation in resistivity in relation to the absolute temperature reciprocal: (a) La_2O_3, (b) Nd_2O_3, (c) Sm_2O_3 (based on Foëx [8.87]). Key: (+) oxygen, (O) air, (×) nitrogen, (△) hydrogen.

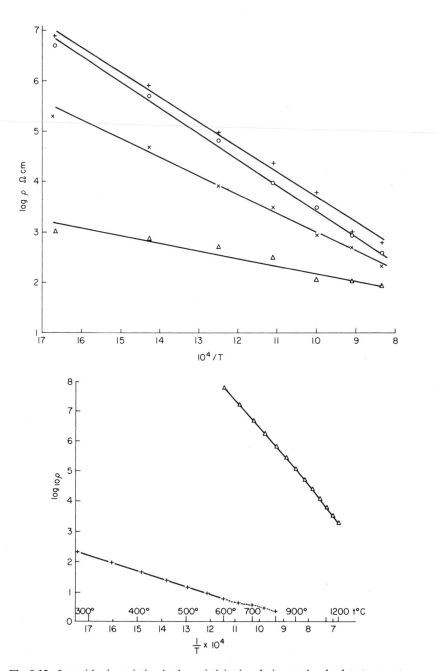

Fig. 8.13. Logarithmic variation in the resistivity in relation to the absolute temperature reciprocal: (a) Ce_2O_3, (b) Pr_2O_3 (based on Foëx [8.86, 8.87]). Key: (a) (top) Ce_2O_3 (△) hydrogen, CeO_2 (+) oxygen, (O) air, (×) nitrogen; (b) (bottom) Pr_2O_3 (△) hydrogen, Pr_6O_{11} (+) oxygen.

8.3. Oxides

moves away from the formula CeO_2. Foëx [8.86, 8.87] gives the resistivity/temperature curves for CeO_2 and PrO_2 (Fig. 8.13) and shows that the magnetic susceptibility of PrO_2 increases up to 15×10^{-6} for Pr_6O_{11}, while its conductivity is multiplied by 10^7. The presence of oxygen vacancies in CeO_2 is studied in [8.102–8.104], and in TbO_2 in [8.105]. The dielectric constant of CeO_2 is 26 [8.106] and its susceptibility is 0.1×10^{-6} [8.107]. Czanderna and Honig [8.108] link the resistivity and Seebeck effect of CeO_2 with the surface loss of oxygen; Greener et al. [8.108a] find that σ varies as $P^{-1/5}$ above 1100°C. McChesney et al. [8.109] have found that stoichiometric PrO_2 ($P > 200$ atm) is antiferromagnetic, with $T_N = 14°K$. A very large number of LO_2–$L_2'O_3$ solid solutions exists, with continuous variation from the C1 structure to the "C" structure of L_2O_3. There are also LO_2–AO_2 solutions like ThO_2–CeO_2, which show a resistivity maximum for 1% of CeO_2 [8.1].

The electrical properties of actinide oxides have had hardly any work done on them, except for thorium and uranium. Thoria ThO_2 is one of the best refractory materials known (3200°C) and its high power of emission with 1% CeO_2 has led to its use in gas incandescent lighting (Auer gas-mantle). In addition, it has been used as a thermionic cathode in vacuum lamps. Its dielectric constant is 16.5, and its susceptibility 0.1×10^{-6}. The resistivity/temperature curve is that of a semiconductor [8.1]. Foëx [8.110] has shown that maximum resistivity occurs in a reducing medium around 1000°C, and in an oxidized medium above 1400°C. The curve $\log \rho(1/T)$ besides shows a break [8.111, 8.112] (Fig. 8.14). The activated state, used in thermionic cathodes (0.5 A cm^{-2}), appears around 1900°K and increases up to 2650°K. It reflects the departure of oxygen [8.113]. The semiconducting properties of activated thoria have been described by various authors, who disagree on whether there is n- or p-type semiconduction [8.114–8.118]. One of them proposes a band structure with $E_G = 6.5$ eV and an impurity level [8.119]. In fact, research by Danforth [8.120–8.122] and Danforth and Bodine [8.123] has revealed a polarization phenomenon resulting from partly ionic conductivity, and Goldwater [8.124] has even carried out an electrolysis experiment. Takahashi [8.125] has found that the total conductivity, which he believes to be basically electronic, varies in accordance with $P^{1/2}$. See also Danforth [8.126] and Mesnard [8.127]. Gammage and Young [8.128] have added Ta_2O_5 and observed a resistivity drop, probably due to valency induction. Lasker and Rapp [8.129] and Bauerle [8.100] have studied the mixed conductivity of ThO_2–Y_2O_3 solutions, in relation to P. No clear conclusion can be drawn from all this research, concerning the electron conduction mechanism. McNeilly [8.130] has studied PuO_{2-x} for $0 < x < 0.3$, and found $E_G = 1.8$ eV, and an activation energy of 0.52 eV.

Uranium oxides have been the subject of a great deal of research in the field of nuclear energy. UO_2, heated in the air to less than 300°C, produces UO_{2+x}

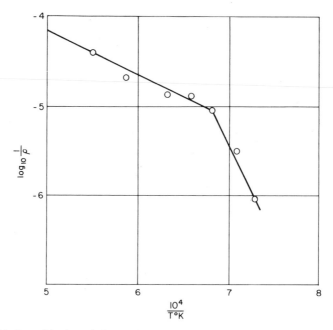

Fig. 8.14. Logarithmic variation in the conductivity of ThO_2 deposited on W in relation to the absolute temperature reciprocal (based on Hanley [8.111]).

(notably U_3O_7 and U_4O_9) with structures derived from C1, while above 300°C it more or less quickly produces U_3O_8 with a C222 orthorhombic structure. With oxygen under pressure, one could reach UO_3. Makarov [8.131] has suggested the existence of a series of terms of the general formula U_nO_{2n+2}. The U–O phase diagram has been considered by many writers [8.132–8.136] (Fig. 8.15). Neth [8.137] gives the development of resistivity in relation to the oxygen content; there is a maximum for $UO_{2.37}$. Zhukovskii et al. [8.138] have observed this maximum for $UO_{2.07}$, with a rise from $UO_{2.56}$ to U_3O_8. The dioxide melts at around 2500°C, and its p-type semiconductor character has long been known [8.1]. Early electrical studies have been summarized by Meyer [8.139]. See also Day et al. [8.140] and Gruen [8.141]. Dawson and Lister [8.142, 8.143] and Trzebiatowski and Selwood [8.144] have found a susceptibility of 9×10^{-6} and a paramagnetic moment of 3.2 μ_B. These results suggest a $5f^2$ or $5f^16d^1$ formula for U(IV), rather than $6d^2$. The increase in susceptibility as a result of oxidation would correspond with the appearance of U(VI) atoms. Arrott and Goldman [8.145] have found antiferromagnetism with $T_N = 29$°K, and they attribute departures from the Curie–Weiss law to interstitial oxygen.

The semiconductor character of UO_2 has been confirmed by Willardson

8.3. Oxides

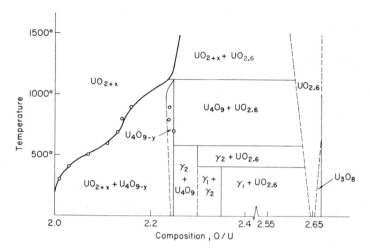

Fig. 8.15. Phase diagram for the U–O system. The circles indicate the limits given by Grønvold [8.132] (based on Roberts [8.135]).

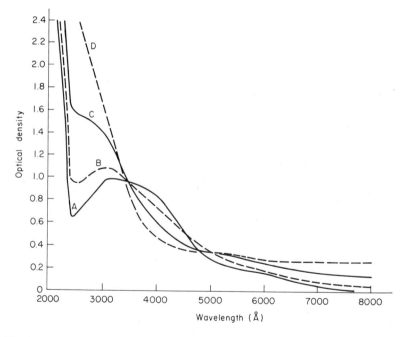

Fig. 8.16. Optical absorption of thin films of UO_{2+x} (based on Ackermann et al. [8.146]): Phase UO_{2+x}: (A) O/U = 2.00, (B) O/U = 2.14, (C) O/U = 2.18. Phase U_4O_9 (D) O/U = 2.23.

et al. [8.133], who have observed an excitation mechanism, extrinsic below 800°C (*p*-type resulting from the interstitial oxygen) and intrinsic above, with $E_G = 3$ eV. U_4O_9 is reported to be *n*-type. Ackermann *et al.* [8.146] have studied the optical absorption of thin films of UO_{2+x} ($0 < x < 0.25$) deposited on silica. Figure 8.16 shows a principal threshold at 2400 Å (5.2 eV) and two neighboring wide lines varying with x (3 and 4 eV). Companion and Winslow [8.147] have also recorded a peak at 1.86 eV. Aronson *et al.* [8.148] have studied the resistivity and the Seebeck effect in UO_{2+x} and have proposed the following equation, corresponding to the theoretical model of Heikes and Johnston [8.149] (cf. Section 3.3):

$$\sigma = (A/T)2x(1-2x)\exp(-E_A/KT)$$

where $A = 3.8 \times 10^6$, $E_A = 0.30$ eV, and μ is calculated at 0.02 cm² V⁻¹sec⁻¹

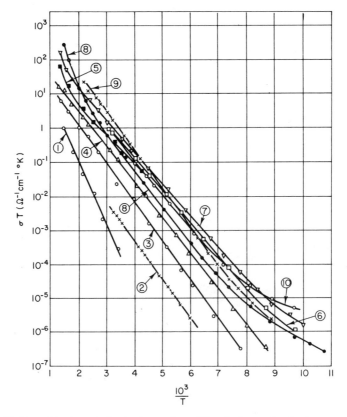

Fig. 8.17. Logarithmic variation in the product σT in relation to the absolute temperature reciprocal for UO_{2+x} (based on Nagels *et al.* [8.155]): U/O: ① < 2.001, ② < 2.001, ③ < 2.001, ④ < 2.001 ⑤ 2.001, ⑥ 2.001, ⑦ 2.006, ⑧ 2.007, ⑨ 2.008, ⑩ 2.016.

8.3. Oxides

at 600°C for a hopping mechanism between U(IV) and U(V). Andersen's [8.150] theory applies to the interstitial oxygen atoms [8.151]. Leask et al. [8.152] have observed a $T_N = 30°K$ for UO_2, but with a simple susceptibility peak at 6.4°K for UO_{2+x}, as for U_3O_8. The adsorption effects the resistivity of sintered samples [8.153]. The first research into monocrystals of UO_{2+x} was done by Nagels et al. [8.154, 8.155]. Figure 8.17 shows the variation of σT between 90 and 800°K for $0 < x < 0.007$. E_A varies between 0.34 and 0.19 eV, and $\mu \leqslant 0.015 \text{ cm}^2 \text{ V}^{-1}\text{sec}^{-1}$ at room temperature. This suggests transport by small polarons. For the effect of radiations, cf. [8.156]. The conductivity rise between UO_2 and UO_{2+x} is proportional to x [8.157]. $E_G = 1.3 \text{ eV}$ (n-type) between 1100 and 2000°C [8.158]. Iida [8.159, 8.160] has studied UO_{2-x} reduced oxides, finding 0.2 to 0.4 eV below 700°C and 1.3 above it. Willis and Taylor [8.161] have used neutron diffraction to show that the moments $(1.82\mu_B)$ are in the (200) planes. For the origin of the departures from stoichiometry, cf. [8.162]. Devreese et al. [8.163] have confirmed the transfer mechanism.

The oxide U_3O_8 seems to be able to accept oxygen up to UO_3, with an ill-defined structure but known semiconductor properties [8.1]. Haraldsen and Bakken [8.164] have explained the variation in magnetic susceptibility by the

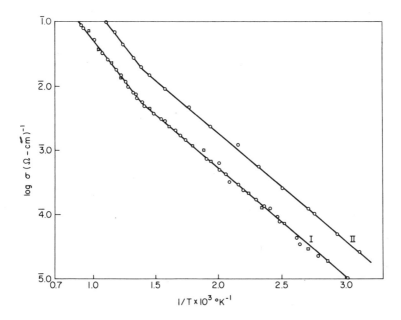

Fig. 8.18. Logarithmic variation in the conductivity of U_3O_8 in air in relation to the absolute temperature reciprocal (based on George and Kharkhanavala [8.171]). Key: U_3O_8 (I) from ammodiuranate and (II) pure nuclear; (O) heating, (□) cooling.

presence of U(V) atoms, and Dawson and Lister [8.142] have found a paramagnetic moment of 1.39 μ_B. Andresen [8.165] has studied the structure with neutrons. The conductivity is of n-type [8.132, 8.166]. A heterojunction, $n(U_3O_8)$–p(other oxide), supplies a voltage under radiation [8.167]. Maurat and Eyraud [8.168] have measured the conductivity between 300 and 900°K, and between U_3O_8 and $UO_{2.9}$. Zhukovskii *et al.* have done the same for the ranges U_3O_8–UO_3 between 25 and 200°C [8.169], and U_3O_8–UO_2 between 25 and 600°C [8.138]. Leask *et al.* [8.152] have found a susceptibility maximum at 6.4°K in the region U_3O_8–$UO_{2.10}$. Tkachenko *et al.* [8.170] have studied the conductivity of U_3O_8 and UO_3 between 80 and 900°C and their dissociation. George and Kharkhanavala [8.171] have found $E_A = 0.64$ eV up to 450°C and 1.09 eV above this, the break being linked to the orthorhombic–hexagonal transformation of the second order (Fig. 8.18). Conductivity is proportional to $P^{-1/6}$. Nirmal Singh and Kharkhanavala [8.172] have measured the Hall effect, finding 1.3×10^{14} electrons cm^{-3}.

In conclusion, it may be said that conduction in oxides of rare earths and actinides generally appears to be due to d-level transfers.

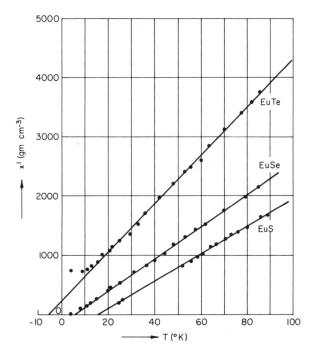

Fig. 8.19. Variation in the magnetic susceptibility reciprocal of EuS, EuSe, and EuTe in relation to temperature (based on Van Houten [8.177]).

8.4. CHALCOGENIDES

Apart from EuS and CeS, LS monosulfides have been prepared only recently [8.173]. Apart from EuS, which is not very stable, they are refractory (2000°C) and CeS is used as a crucible [8.174]. Valency L(II) exists in SmS and EuS, and the other terms contain a high density of free electrons [8.1, 8.173]. The ferromagnetism of EuS was discovered simultaneously by McGuire et al. [8.175], Busch et al. [8.176] and Van Houten [8.177], who indicate $T_C = 18$, 15, and 16°K, and 6.8, 6.85, and 6.5 μ_B, respectively (Fig. 8.19). A resistivity of 10^7 Ω cm at room temperature on pressed powder samples indicates semiconducting properties [8.176, 8.177]. The activation energy is 2.1 eV (resistivity/temperature at 1000°C) or higher than 3 eV (optical absorption at room

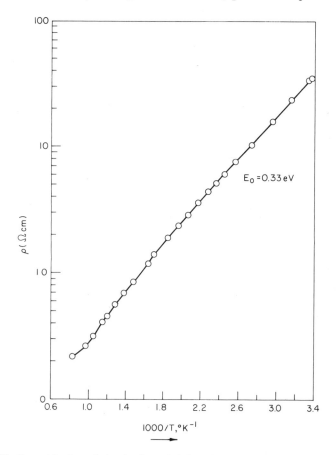

Fig. 8.20. Logarithmic variation in the resistivity of $YbS_{1.05}$ in relation to the absolute temperature reciprocal (based on Didchenko and Gortsema [8.51]).

temperature [8.176]. The magnetic interactions are discussed in [8.71, 8.178–8.180]. There is positive direct interaction among the Eu atoms, less marked than in EuO, and negative indirect Eu–S–Eu interaction [8.180]. Semiconductor properties are mentioned by Houston [8.181] for SmS, and by Didchenko and Gortsema [8.51] for SmS, EuS, and YbS [L(II), high cell parameter]. YbS is insulating, but for $YbS_{1.05}$, $\rho \sim 30$ Ω cm at room temperature (Fig. 8.20), $p > 10^{17}$ cm^{-3}, $e_a = 0.33$ eV and $\mu < 1.3$ cm^2 $V^{-1}sec^{-1}$. For SmS, $\rho \sim 0.5$ Ω cm at room temperature, $N = 4.3 \times 10^{18}$ cm^{-3}, $E_G = 0.24$ eV and $\mu = 22.4$ cm^2 $V^{-1}sec^{-1}$. For CeS, cf. [8.182]. McClure [8.183], using the MO method, has proposed a model for LS crystals involving excitation from the 4f level to the 6s band of the metal. The p-type of YbS is said to be the result of an excess of sulfur.

According to Zhuze et al. [8.184], LaS, CeS, PrS, and NdS are metallic (one electron in 5d). Zhuze, Golubkov et al. [8.185, 8.186] have done a comprehensive study of the properties of SmS, and have suggested that the 4f level is separated from the 5d band by 0.22 eV (Figs. 8.21 and 8.22). We feel that it is more likely to be an Sm(II)–Sm(III) transfer semiconductibility mechanism below approximately 500°K ($E_A \sim 0.1$ eV), with the 4f–5d excitation mechanism becoming predominant only above this temperature. Busch et al. [8.73, 8.74] have found an optical absorption threshold in EuS corresponding to $E_G = 1.645$ eV at room temperature, with $dE_G/dT = -1.7 \times 10^{-4}$ eV deg^{-1},

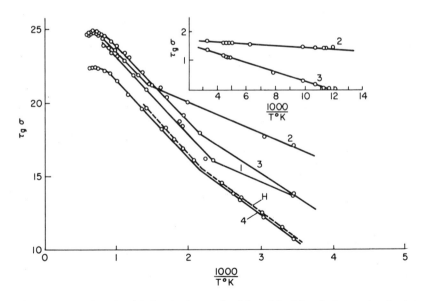

Fig. 8.21. Logarithmic variation in the conductivity of SmS in relation to the absolute temperature reciprocal (based on Zhuze et al. [8.185]). Curve H based on Houston [8.181].

8.4. Chalcogenides

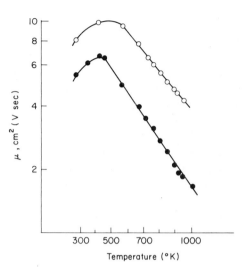

Fig. 8.22. Logarithmic variation in the mobility of SmS in relation to temperature (based on Golubkov et al. [8.186] (two samples).

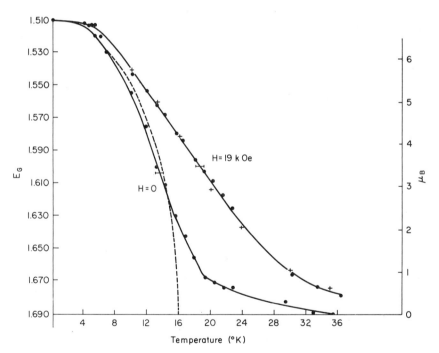

Fig. 8.23. Variation in E_G, worked out from the absorption threshhold, and in the magnetization of EuS in relation to temperature (based on Busch et al. [8.74]). Key: (●—●) experiment, $H=0$ and $H=19$ kOe, (+—+) Brillouin, $B_{7/2}$ $\theta = 16°K$, $H_i = 11.2$ kOe. (---) Brillouin, $B_{7/2}$ $\theta = 16°K$, $H_i = 0$.

but a reverse shift below 20°K. As with EuO, E_G drops by 0.2 to 0.3 eV below T_C, and the phenomenon is proportional to the magnetization (Fig. 8.23). Smit [8.79] suggests magnetism resulting from the overlapping of the 5d orbitals. Greiner et al. [8.81] have studied the Kerr longitudinal magneto-optical effect, and mention the 4f–5d transition. The solid solutions EuS–LaS [8.187, 8.188], EuS–GdS [8.189, 8.190], and EuS–EuSe [8.191] have been studied. In the first two systems, T_C increases and ρ decreases as the La (Fig. 8.24) or Gd content rises, and a maximum of ρ is noted at T_C.

AS monosulfides have only recently been discovered. US melts at more than 2000°C, has low susceptibility [8.192, 8.193], and is reported to be an electrical conductor [8.1]. Ferromagnetism is mentioned [8.194]. Tetenbaum [8.195] has studied the resistivity and Seebeck effect in US, ThS and their solid solutions (metallic). Kruger and Moser [8.36] have found that PuS is an intrinsic semiconductor between 425 and 700°K, with $E_G = 0.24$ eV and $\alpha = +115\ \mu V\ deg^{-1}$, and they speculate on its band structure.

In L_5S_7 sulfides, with monoclinic structure, Flahaut and Guittard [8.196] have found that Y_5S_7 has metallic properties, as the valency Y(III) would have suggested.

The sulfides $\square_x L_{3-x} S_4$ have a homogeneity domain of $0 < x < \frac{1}{3}$. According to Eastman et al. [8.197], Ce_3S_4 melts at about 2000°C, and $\rho = 4 \times 10^{-4}\ \Omega\ cm$ at room temperature. Appel and Kurnick [8.198] consider that it has semiconductor properties, which they explain by the small polaron theory, while Ryan et al. [8.199] observe insulating properties (Ce_2S_3) or apparent metallic properties (Ce_3S_4). For $x = 0.08$, $\rho = 1.5 \times 10^{-3}\ \Omega\ cm$, $\alpha = -70\ \mu V\ deg^{-1}$, $\mu = 3\ cm^2\ V^{-1}sec^{-1}$ and $N = 2 \times 10^{21}\ cm^{-3}$. Solid solutions with BaS have a much higher thermoelectric factor of merit f [8.199a]. Cutler et al. [8.200] have confirmed this, and concluded that there is a transfer mechanism (overlapping of the 5d orbitals). Yb_2S_3 is reported to have not a $D7_3$ structure, but another [8.201]. Marchenko and Samsonov [8.182, 8.202] have measured ρ and α on powder samples sintered under H_2S and densified to 85% and have found $e_d = 0.25$ (Ce_2S_3) and 0.32 (La_2S_3) and $E_G = 1.12$ (Ce_2S_3) and 1.32 eV (La_2S_3) (Fig. 8.25). Cutler et al. have speculated on the role of the vacancies in $Ce_{3-x}S_4$ ($x < 0.3$) [8.203], and have concluded that there is a transfer mechanism ($x > 0.3$) [8.204] Figure 8.26 shows the exponential law for μ. See [8.205] for electrical and thermal conductivities in La_2S_3, Ce_2S_3, Pr_2S_3 and Nd_2S_3. Methfessel [8.189] and Holtzberg et al. [8.206] have shown that the magnetic interactions in Gd_2S_3 depend on the density of carriers and that T_C increases with N (Ruderman–Kittel–Yosida indirect exchange). Lashkar'ev et al. [8.207] have shown that Sm_2S_3 is a refractory semiconductor with $E_G = 2.96$

Fig. 8.24. Logarithmic variation in the quotient ρ/T of $La_xEu_{1-x}S$ solutions in relation to the absolute temperature reciprocal (based on Heikes et al. [8.188]).

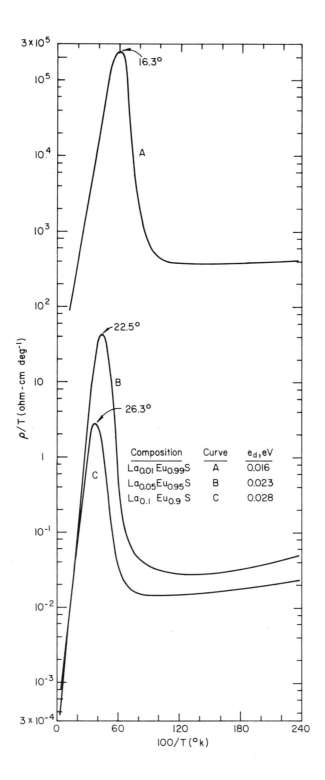

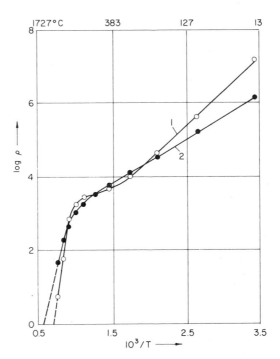

Fig. 8.25. Logarithmic variation in the resistivity of La_2S_3 (1) and Ce_2S_3 (2) in relation to the absolute temperature reciprocal (based on Samsonov and Marchenko [8.202]). Porosity of samples: 13–16%.

eV; they speculate on the band structure and emphasize the covalent aspect of the bonds. The reflection has been interpreted [8.208]. LS_2 disulfides have been very little studied (L(III)).

The composition of the sulfides U_2S_3 and US_2 has been discussed [8.209], and their susceptibility is low [8.193].

Selenides and tellurides of rare earths and actinides have recently been dealt with in a short book by Obolonchik et al. [8.10], which is very straightforward and to which readers may refer.

LSe monoselenides are refractory (1700–2100°C). Miller et al. [8.210] have mentioned $\rho = 1.7 \times 10^{-4}$ Ω cm at room temperature for ErSe. The ferromagnetism of EuSe was discovered simultaneously by McGuire et al. [8.175], Busch et al. [8.176] and Van Houten [8.177], who indicate $T_C = 8 \pm 3$, 5.5, and 6°K, and ?, 6.75, and 7 μ_B, respectively (Fig. 8.19). A resistivity of 10^7 Ω cm at room temperature on pressed powder samples indicates semiconducting properties [8.176, 8.177]. The activation energy is 1.6 eV (resistivity/temperature at 1000°C) or 2 eV (optical absorption at room temperature) [8.176]. The magnetic interactions are discussed in [8.71, 8.178–8.180]. There is positive

8.4. Chalcogenides

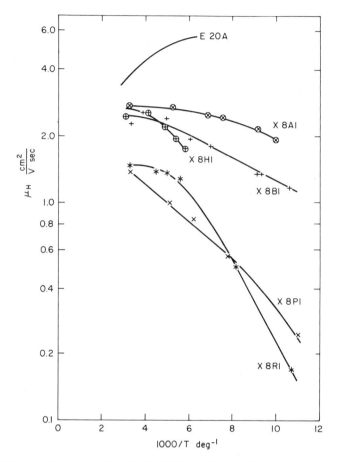

Fig. 8.26. Logarithmic variation in the Hall mobility in $Ce_{3-x}S_4$ ($0.3 < x < 0.33$) in relation to the absolute temperature reciprocal (based on Cutler and Leavy [8.204]). Carrier density $\times 10^{18}$ cm^{-3}: X8R1 = 0.50 X8P1 = 0.87; X8H1 = 4.3; X8B1 = 5.2; X8A1 = 14; E20A = 83.

direct interaction among the Eu atoms, less marked than in EuS, and a negative Eu–Se–Eu indirect interaction [8.180]. According to Zhuze et al. [8.184], LaSe, CeSe, PrSe, and NdSe are metallic (one electron in 5d). Reid et al. [8.63, 8.211] have found SmSe and YbSe (L(II)) to be semiconducting ($\rho = 2000$ and 100 Ω cm) and NdSe, GdSe, and ErSe (L(III)) to be metallic ($\rho \sim 10^{-4}$ Ω cm); 4f→5d or 6s excitation has been suggested in the first case [8.183], with $E_G = 0.72$ (SmSe, n-type) and 0.34 eV (YbSe, p-type). Busch et al. [8.73] have found an optical absorption threshold in EuSe corresponding to $E_G = 1.78$ eV at room temperature, with $dE_G/dT = -3.2 \times 10^{-4}$ eV deg^{-1}. ErSe is metallic [8.212]. Argyle et al. [8.213] have studied the magnetic dichroism of EuSe, and concluded that there is metamagnetism at 4.2°K. The wavelength of the absorption threshold increases in a magnetic field (ferromagnetic region). This would correspond to a $4f^7 \rightarrow 4f^6 5d$ transition [8.80]. Overlapping of the 5d orbitals has been suggested [8.79]. Busch et al. [8.74] have confirmed the metamagnetism of EuSe above 3°K ($T_N = 4.6°K$). As with EuO and EuS, E_G drops by 0.2 to 0.3 eV below T_C (Fig. 8.27, cf. 8.23). The reflection spectrum [8.208] and luminescence [8.214] have been explained. The solid solutions EuSe–LaSe [8.215], EuSe–GdSe [8.189, 8.190, 8.215], SmSe–NdSe [8.211], EuSe–EuS, and EuSe–EuTe [8.191] have been studied. Figure 8.28 refers to $Eu_{1-x}Gd_xSe$, in which T_C attains 45°K for $0.07 < x < 0.25$.

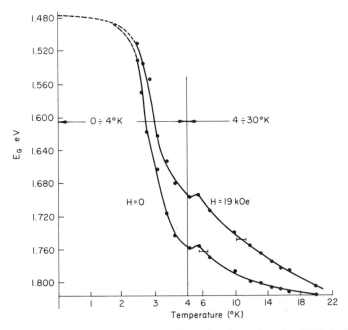

Fig. 8.27. Variation in E_G, worked out from the absorption threshold, in EuSe, in relation to temperature (based on Busch et al. [8.74]).

Selenides with the $D7_3$ structure have the same homogeneity domain as sulphides, but other structures are possible [8.201]. Miller, Reid et al. [8.63, 8.210, 8.215a] indicate a high value for ρ (Y_2Se_3), 10^{-3} (Ce_2Se_3, Gd_2Se_3) and 292 Ω cm (CeSe), $\mu = 3.7$ (Ce_2Se_3) and 0.4 cm² V⁻¹sec⁻¹ ($CeSe_2$), and Holtzberg et al. [8.206] have mentioned the semiconducting character of Gd_2Se_3 (3 Ω cm at room temperature, $T_N = 6°K$). Gd in the vacancies of $D7_3$ increases σ and produces ferromagnetism ($Gd_{2.1}Se_{2.9}$: $T_C = 80°K$, $\rho = 1.4 \times 10^{-3}$ Ω cm). Polarization of the conduction electrons by the 4f–5d exchanges is said to

8.4. Chalcogenides

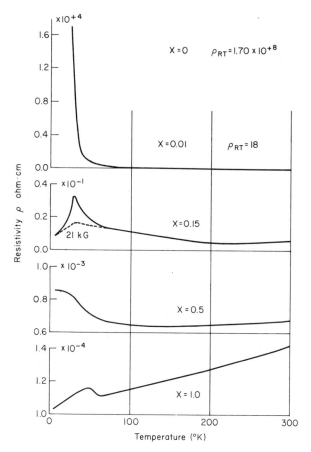

Fig. 8.28. Linear variation in the resistivity of $Eu_{1-x}Gd_xSe$ solid solutions in relation to absolute temperature (based on Holtzberg et al. [8.190]).

explain the relation between σ and T_C [8.189, 8.190]. Semiconductor properties have been mentioned by Haase and Steinfink [8.212] for Er_2Se_3 and $ErSe_2$ ($E_G = 1.66$ and 1.06 eV) and by Lashkar'ev and Paderno [8.216] for Pr_2Se_3 and Nd_2Se_3 ($E_G = 1.81$ and 1.6 eV, $\mu_P > \mu_N$). For L_2Se_3 reflection spectra, cf [8.208]. The solid solutions Gd_2Se_3–EuSe (paramagnetic) and Gd_2Se_3–YSe (ferromagnetic) have been studied [8.206].

The compounds U_3Se_4 ($D7_3$ structure) and U_2Se_3 (Sb_2S_3 structure) are reported to be electrical conductors. USe_2 is polymorphous and USe_3 contains U(IV) [8.1].

LTe monotellurides are refractory (1700–2100°C). The antiferromagnetism of EuTe was discovered simultaneously by McGuire et al. [8.175], Busch et al. [8.176] and Van Houten [8.177], who indicate T_N = ?, 8, and 6°K, respectively

(Fig. 8.19). A resistivity of 10^7 Ω cm at room temperature on pressed powder samples indicates semiconductor properties [8.176, 8.177]. The activation energy is 1.1 eV (resistivity/temperature at 1000°C), and 2.2 eV (optical absorption at room temperature) [8.176]. The magnetic interactions are discussed in [8.178, 8.180]. The negative Eu–Te–Eu indirect interaction predominates over direct interaction. Neutron investigation indicates an irregularity in the magnetization of the sublattices [8.180], and antiferromagnetism of the second order with spins in the (111) planes and $T_N = 7.8°K$ [8.217]. Busch et al. [8.58] have made the spins parallel under 75 kOe, and measured 6.1 μ_B (metamagnetism). According to Zhuze et al. [8.184], LaTe, CeTe, PrTe, and NdTe are metallic (one electron in 5d). Reid et al. [8.63, 8.211] have found SmTe and YbTe (L(II)) to be semiconducting (ρ = 2000 and 7000 Ω cm), and NdTe, GdTe, and ErTe (L(III)) to be metallic ($\rho \sim 10^{-4}$ Ω cm). 4f→5d or 6s excitation has been suggested in the first case [8.183] with E_G = 0.6 eV (YbTe,

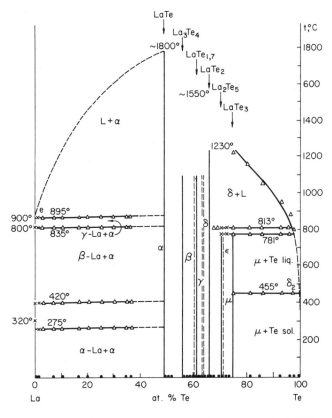

Fig. 8.29. La–Te phase diagram (based on Eliseev et al. [8.221]).

8.4. Chalcogenides

p-type). Busch *et al.* [8.73, 8.74] have found an optical absorption threshold in EuTe corresponding to $E_G = 1.05$ eV at room temperature, with $dE_G/dT = -2 \times 10^{-4}$ eV deg^{-1}, but with no reverse shift below $T_N = 9.8°$K. However, the cell is contracted [8.218]. Overlapping of the 5d orbitals has been suggested [8.79]. LaTe [8.219] and ErTe [8.220] are metallic. The solid solutions HoSb–HoTe [8.68] and EuSe–EuTe [8.191] have been studied.

The polytellurides often have a complex phase diagram [8.221] (Fig. 8.29). Miller, Reid *et al.* [8.63, 8.210] have mentioned the high Seebeck effects, of around 200 μV deg^{-1}, in Nd_2Te_3 and Gd_2Te_3 ($E_G = 0.7$ eV, $\mu = 55$ cm^2 V^{-1}sec^{-1}) (Fig. 8.30); see [8.221a] for the other compounds. Semiconductor properties have been mentioned by Andrellos and Bro [8.222] for $NdTe_2$ ($E_G = 0.48$ eV), by [8.223] for Ce_2Te_3, Pr_2Te_3, and Nd_2Te_3 ($E_G = 1.2, 1.3$, and 0.8 eV, respectively), by Ramsey *et al.* [8.218, 8.224] for La_2Te_3 and $LaTe_2$ and by Haase and Steinfink [8.220] for Er_2Te_3.

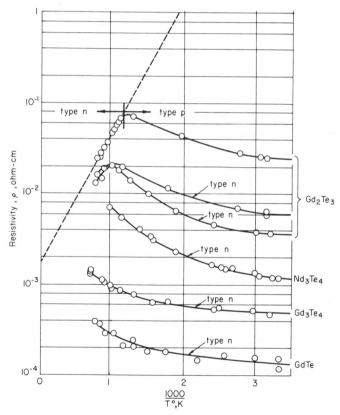

Fig. 8.30. Logarithmic variation in the resistivity of Nd_2Te_3 and Gd_2Te_3 in relation to the absolute temperature reciprocal (based on Reid *et al.* [8.63]).

Excitation of a 4f electron to 5d seems to be the characteristic feature of these chalcogenides. If it involves all the L atoms [stable L(III) valency], it involves a transfer mechanism the metallic character of which is a result of their density and of the overlapping of the 5d orbitals. If it remains an exception, and is bound up with thermal agitation [stable L(II) valency], it involves an excitation mechanism, and E_G measures the gap 4f–5d. Nonstoichiometry in relation to LX (B1) or $\Box L_2 X_3$ (D7$_3$) causes L(II)–L(III) transfers, with metallic properties beyond a certain density. The case of EuX ferromagnetic semiconductors is of great interest. Mössbauer resonance parameters depend on the covalent aspect of the Eu—X bond [8.225] and probably also on the 4f–5d transitions, which can alter the screen effect of the d electrons.

REFERENCES

[8.1] P. Pascal, "Nouveau traité de chimie minérale." Masson, Paris, 1956–1963.
[8.2] A. F. Wells, "Structural Inorganic Chemistry," 3rd ed. Oxford Univ. Press (Clarendon), London and New York, 1962.
[8.3] R. W. G. Wyckoff, "Crystal Structures," 2nd ed. Vol. 1, 2. Wiley (Interscience), New York, 1963–64.
[8.4] W. C. Koehler and E. O. Wollan, Acta Crystallog. **6**, 741 (1953).
[8.5] D. T. Cromer, J. Phys. Chem. **61**, 753 (1957).
[8.6] M. Hansen, "Constitution of Binary Alloys." McGraw-Hill, New York, 1958; R. P. Elliott, idem, 1961.
[8.7] A. Iandelli, Gazz. Chim. Ital. **77**, 312 (1947).
[8.8] J. Flahaut, M. Guittard, M. Patrie, M. P. Pardo, S. M. Golabi, and L. Domange, Acta Crystallog. **19** (Pt. 1), 14 (1965).
[8.9] F. Holtzberg and S. Methfessel, J. Appl. Phys. **37**, 1433 (1966).
[8.10] V. A. Obolonchik and G. V. Lashkar'ev, "Selenidy i telluridy redkozemel'nykh metallov i aktinoidov." Naukova Dumka, Kiev, 1966.
[8.10a] J. F. Miller, L. K. Matson, and R. C. Himes, in "Rare Earth Research" (K. S. Vorres, ed.), Vol. 2, p. 135. Gordon and Breach, New York, 1964.
[8.11] W. H. Zachariasen, Acta Crystallogr. **1**, 265 (1948); **2**, 57 (1949).
[8.12] A. D. Wadsley, "Non-stoichiometric Compounds" (L. Mandelcorn, ed.), p. 98. Academic Press, New York, 1964.
[8.13] E. F. Bertaut and P. Blum, C.R. Acad. Sci. Paris **234**, 2621 (1952).
[8.14] R. Kiessling, Acta Chem. Scand. **4**, 209 (1950).
[8.15] J. H. van Vleck, "The Theory of Electric and Magnetic Susceptibilities." Oxford Univ. Press, London and New York, 1932.
[8.16] A. Zalkin and D. H. Templeton, Acta Crystallogr. **6**, 269 (1953).
[8.17] H. A. Eick, "Rare Earth Research" (E. V. Kleber, ed.), p. 297. Macmillan, New York, 1961.
[8.18] J. M. Lafferty, Phys. Rev. **79**, 1012 (1950).
[8.19] J. M. Lafferty, J. Appl. Phys. **22**, 299 (1951).
[8.20] R. Benoit, J. Chim. Phys. **52**, 119 (1955).
[8.21] G. V. Samsonov and E. Grodshtein, Zh. Fiz. Khim. **30**, 379 (1956).
[8.22] G. A. Kudintseva, M. D. Polyakova, G. V. Samsonov, and B. M. Tsarev, Fiz. Metal. Metalloved. **6**, 272 (1958).

[8.23] G. V. Samsonov, N. N. Zhuravlev, Yu. B. Paderno, and V. R. Melikadamyan, *Kristallografiya* **4**, 538 (1959).
[8.24] Yu. B. Paderno, G. V. Samsonov, and V. S. Fomenko, *Fiz. Metal. Metalloved.* **10**, 633 (1960).
[8.25] S. Flodmark, *Ark. Fys.* **14**, 513 (1959).
[8.26] I. Binder, S. Laplaca, and B. Post, *Conf. Boron, Ashbury Park, New Jersey, 1959* USASRDL (1959).
[8.27] G. V. Samsonov and Yu. B. Paderno, "Boridy redkozemel'nykh metallov." Izd. Akad. Nauk Ukr. SSR, Kiev, 1961.
[8.28] G. V. Samsonov, E. E. Vainshtein, and Yu. B. Paderno, *Fiz. Metal. Metalloved.* **13**, 744 (1962).
[8.29] B. R. Coles, T. Cole, J. Lambe, and N. Laurance, *Proc. Phys. Soc.* **79** (Pt. 1), 84 (1962).
[8.29a] B. M. Tsarev and S. V. Illarionov, *Poroshkovaya Met. Akad. Yauk Ukr. SSR* **2**, 85 (1962).
[8.29b] E. Kauer, *Phys. Lett.* **7**, 171 (1963).
[8.30] S. N. L'vov, V. F. Nemchenko, and G. V. Samsonov, *Dokl. Akad. Nauk SSSR* **135**, 3 (1960).
[8.31] L. B. Griffiths, *Nature* **193**, 362 (1962).
[8.32] P. Costa and R. Lallement, *Phys. Lett.* **7**, 21 (1963).
[8.33] J. J. Norreys and M. J. Wheeler, *Trans. Brit. Ceram. Soc.* **161**, 183 (1963).
[8.34] R. Lallement, P. Costa, and R. Pascard, *J. Phys. Chem. Solids* **26**, 1255 (1965).
[8.35] A. B. Auskern, S. Aronson, J. Sadofsky, and F. J. Salzano, *J. Phys. Chem. Solids* **27**, 613 (1966).
[8.36] O. L. Kruger and J. B. Moser, *J. Chem. Phys.* **46**, 891 (1967).
[8.37] F. H. Spedding, K. Gschneider, Jr., and A. H. Daane, *Trans. AIME* **215**, 192 (1959).
[8.38] M. Atoji, *Phys. Lett.* **23**, 208 (1966).
[8.39] M. Atoji, *J. Chem. Phys.* **35**, 1950 (1961).
[8.40] M. Atoji and D. E. Williams, *J. Chem. Phys.* **35**, 1960 (1961).
[8.41] M. Atoji, *J. Phys. Soc. Japan suppl.* **17** (B-II), 395 (1962).
[8.42] C. E. Lundin, "Rare Earth Research" (E. V. Kleber, ed.), p. 306. Macmillan, New York, 1961.
[8.43] E. F. Bertaut and P. Blum, *Acta Crystallogr.* **3**, 319 (1950).
[8.44] B. T. Matthias, E. Corenzwit, and W. H. Zachariasen, *Phys. Rev.* **112**, 89 (1958).
[8.45] K. A. Gschneider, Jr. and B. T. Matthias, "Rare Earth Research" (E. V. Kleber, ed.), p. 158. Macmillan, New York, 1961.
[8.46] H. A. Eick, N. C. Baenziger, and L. Eyring, *J. Amer. Chem. Soc.* **78**, 5987 (1956).
[8.47] M. K. Wilkinson, H. R. Child, J. W. Cable, E. O. Wollan, and W. C. Koehler, *J. Appl. Phys. suppl.* **31**, 358S (1960).
[8.48] H. R. Child, M. K. Wilkinson, J. W. Cable, W. C. Koehler, and E. O. Wollan, *Phys. Rev.* **131**, 922 (1963).
[8.49] P. Costa, R. Lallement, F. Anselin, and D. Rossignol, "*Compounds of Interest in Nuclear Reactor Technology*," 10, 83. Edward Brothers, Ann Arbor, Michigan, 1964.
[8.50] N. Sclar, *J. Appl. Phys.* **33**, 2999 (1962).
[8.51] R. Didchenko and F. P. Gortsema, *J. Phys. Chem. Solids* **24**, 863 (1963).
[8.52] G. T. Trammell, *Phys. Rev.* **131**, 932 (1963).
[8.53] N. Sclar, *J. Appl. Phys.* **35**, 1534 (1964).
[8.54] J. P. Rebouillat and J. J. Veyssie, *C.R. Acad. Sci. Paris* **259**, 4239 (1964).
[8.55] G. Busch, P. Junod, F. Levy, and O. Vogt, *Proc. Int. Conf. Magn. Nottingham, 1964* p. 472, Inst. Phys. & Phys. Soc., London, 1964.

[8.56] G. Busch, P. Junod, F. Levy, A. Menth, and O. Vogt, *Phys. Lett.* **14**, 264 (1965).
[8.57] A. Iandelli, "Rare Earth Research" (E. V. Kleber, ed.), p. 135. Macmillan, New York, 1961.
[8.58] G. Busch, P. Junod, P. Schwob, O. Vogt, and F. Hulliger, *Phys. Lett.* **9**, 7 (1964).
[8.59] G. Busch, P. Schwob, O. Vogt, and F. Hulliger, *Phys. Lett.* **11**, 100 (1964).
[8.60] T. Tsuchida and W. E. Wallace, *J. Chem. Phys.* **43**, 2885 (1965).
[8.61] G. Busch and O. Vogt, *Phys. Lett.* **20**, 152 (1966).
[8.62] K. Yaguchi, *J. Phys. Soc. Japan* **21** 1226 (1966).
[8.63] F. J. Reid, L. K. Matson, J. F. Miller, and R. C. Himes, *J. Electrochem. Soc.* **111**, 943 (1964).
[8.64] G. Busch, O. Vogt, and F. Hulliger, *Phys. Lett.* **15**, 301 (1965).
[8.65] J. F. Miller and R. C. Himes, "Rare Earth Research" (E. V. Kleber, ed.), p. 232. Macmillan, New York, 1961.
[8.66] G. Busch, P. Junod, M. Risi, and O. Vogt, *Proc. Int. Conf. Phys. Semicond., Exeter, 1962* Inst. Phys. & Phys. Soc., London 1962.
[8.67] G. Busch, O. Marincek, A. Menth, and O. Vogt, *Phys. Lett.* **14**, 262 (1965).
[8.68] G. Busch and O. Vogt, *Helv. Phys. Acta* **39**, 199 (1966); *Phys. Lett.* **22**, 388 (1966).
[8.68a] F. Holtzberg, S. Methfessel, and J. C. Suits, *in* "Rare Earth Research" (K. S. Vorres, ed.), Vol. 2, p. 385. Gordon and Breach, New York, 1964.
[8.69] B. T. Matthias, R. M. Bozorth, and J. H. van Vleck, *Phys. Rev. Lett.* **7**, 160 (1961).
[8.70] N. G. Nereson, C. E. Olsen, and G. P. Arnold, *Phys. Rev.* **127**, 2101 (1962).
[8.71] J. Callaway, *Nuovo Cimento* **26**, 625 (1962).
[8.72] P. Wachter, *Helv. Phys. Acta* **37**, 637 (1964).
[8.73] G. Busch, P. Junod, and P. Wachter, *Phys. Lett.* **12**, 11 (1964).
[8.74] G. Busch and P. Wachter, *Phys. Kondens. Mater.* **5**, 232 (1966).
[8.75] Iu. P. Irkhin and E. A. Turov, *Fiz. Metal. Metalloved.* **4**, 9 (1957).
[8.76] Iu. P. Irkhin, *Tr. Inst. Fiz. Met. Ural, Fil. Akad. Nauk* **20**, 95 (1958).
[8.77] P. Brix, S. Hufner, P. Kienle, and D. Quitmann, *Phys. Lett.* **13**, 140 (1964).
[8.78] A. A. Samokhvalov, V. G. Bamburov, N. V. Volkenshtein, T. D. Zotov, A. A. Ivakin, Yu. N. Morozov, and M. I. Simonova, *Fiz. Metal. Metalloved.* **20**, 309 (1965).
[8.79] J. Smit, *J. Appl. Phys.* **37**, 1455 (1966).
[8.80] J. C. Suits, B. E. Argyle, and M. J. Freiser, *J. Appl. Phys.* **37**, 1396 (1966).
[8.81] J. H. Greiner and G. J. Fan, *Appl. Phys. Lett.* **9**, 27 (1966).
[8.82] H. Bärnigshausen and G. Brauer, *Acta Crystallogr.* **15** (Pt. 10), 1059 (1962).
[8.83] A. A. Samokhvalov, V. G. Bamburov, N. V. Volkenshtein, T. D. Zotov, A. A. Ivakin, Yu. N. Morozov, and M. I. Simonova, *Fiz. Metal. Metalloved.* **20**, 308 (1965).
[8.84] F. F. Y. Wang, *Phys. Status Solidi* **14**, 189 (1966).
[8.85] L. Holmes and M. Schieber, *J. Appl. Phys.* **37**, 968 (1966).
[8.86] M. Foëx, *Bull. Soc. Chim.* **11**, 17 (1944).
[8.87] M. Foëx, *C.R. Acad. Sci. Paris* **220**, 359 (1945).
[8.87a] R. L. Martin, *Nature* **165**, 202 (1950).
[8.87b] J. M. Honig, A. A. Cella, and J. C. Cornwell, *in* "Rare Earth Research" (K. S. Vorres, ed.), Vol. 2, p. 555. Gordon and Breach, New York, 1964.
[8.87c] B. G. Hyde, D. J. M. Bevan, and Leroy-Eyring, *in* "Rare Earth Research" (K. S. Vorres, ed.), Vol. 2, p. 277. Gordon and Breach, New York, 1964.
[8.87d] B. G. Hyde and Leroy-Eyring, *in* "Rare Earth Research" (Leroy-Eyring, ed.), Vol. 3, p. 623. Gordon and Breach, New York, 1965.
[8.88] S. Nagasawa, *J. Electrochem. Soc. Japan* **18**, 158 (1950).
[8.89] G. Brauer, *Z. Naturforsch.* **6a**, 561 (1951).

[8.90] C. Henry la Blanchetais, *J. Rech. CNRS Paris* **6**, 103 (1954).
[8.91] W. E. Henry, *J. Appl. Phys.* **29**, 524 (1958).
[8.92] W. Noddack and H. Walch, *Z. Elektrochem.* **63**, 269 (1959).
[8.93] R. L. Mozzi and O. J. Guentert, *J. Chem. Phys.* **36**, 298 (1962).
[8.94] A. Fert, *Bull. Soc. Fr. Minéral. Cristallogr.* **85**, 267 (1962).
[8.95] G. M. Schab and F. Bohla, *Naturwissenschaften* **50**, 567 (1963).
[8.96] V. B. Tare and H. Schmalzried, *Z. Phys. Chem. (Frankfurt am Main)* **43**, 30 (1964).
[8.97] M. Schieber and L. Holmes, Magnetic Ceramics Meeting (London 1963), *Proc. Brit. Ceram. Soc.* 139 (1964).
[8.98] N. P. Bogoroditskii, V. V. Pasynkov, R. R. Basili, and Yu. M. Volokobrinskii, *Dokl. Akad. Nauk SSSR* **160**, 578 (1965).
[8.99] N. M. Tallman and R. W. Vest, *J. Amer. Ceram. Soc.* **49**, 401 (1966).
[8.100] J. E. Bauerle, *J. Chem. Phys.* **45**, 4162 (1966).
[8.100a] Leroy-Eyring *in* "Rare Earth Research" (J. F. Nachman and C. E. Lundin, eds.), Vol. 1, p. 339. Gordon and Breach, New York, 1962.
[8.101] U. Croatto, *Ric. Sci.* **13**, 830 (1942).
[8.102] U. Croatto and M. Bruno, *Ric. Sci.* **15**, 578 (1949).
[8.103] M. Bruno, *Ric. Sci.* **20**, 645 (1950).
[8.104] G. Rienäcker and M. Birckenstaedt, *Z. Anorg. Chem.* **265**, 99 (1951).
[8.105] D. M. Gruen, W. C. Koehler, and J. J. Katz, *J. Amer. Chem. Soc.* **73**, 1475 (1951).
[8.106] H. Gränicher, *Helv. Phys. Acta* **24**, 619 (1951).
[8.107] C. Henry la Blanchetais, State thesis, Paris 1954.
[8.108] A. W. Czanderna and J. M. Honig, *J. Phys. Chem. Solids* **6**, 96 (1958).
[8.108a] E. H. Greener, J. M. Wimmer, and W. M. Hirthe, *in* "Rare Earth Research" (K. S. Vorres, ed.), Vol. 2, p. 539. Gordon and Breach, New York, 1964.
[8.109] J. B. McChesney, H. J. Williams, R. C. Sherwood, and J. F. Potter, *J. Chem. Phys.* **41**, 3177 (1964).
[8.110] M. Foëx, *C.R. Acad. Sci. Paris* **215**, 534 (1942).
[8.111] T. E. Hanley, *J. Appl. Phys.* **19**, 583 (1948).
[8.112] G. Mesnard, *C.R. Acad. Sci. Paris* **232**, 1744 (1951).
[8.113] G. Mesnard, *C.R. Acad. Sci. Paris* **231**, 768, 833 (1950).
[8.114] D. A. Wright, *Proc. Phys. Soc.* **62B**, 188 (1949).
[8.115] W. E. Danforth and F. H. Morgan, *Phys. Rev.* **79**, 142 (1950).
[8.116] T. Arizumi and L. Esaki, *J. Phys. Soc. Japan* **5**, 163, 174 (1950).
[8.117] G. Mesnard and R. Uzan, *Le Vide Paris* **6**, 1052, 1091 (1951).
[8.118] O. A. Weinreich and W. E. Danforth, *J. Franklin Inst.* **255**, 569 (1953).
[8.119] G. Mesnard, *J. Phys. Radium* **14**, 179 (1953).
[8.120] W. E. Danforth, *J. Franklin Inst.* **251**, 515 (1951); **258**, 233 (1954).
[8.121] W. E. Danforth, *Phys. Rev.* **86**, 416 (1952).
[8.122] W. E. Danforth, *J. Chem. Phys.* **23**, 591 (1955).
[8.123] W. E. Danforth and J. H. Bodine, *J. Franklin Inst.* **260**, 467 (1955).
[8.124] D. L. Goldwater, *J. Franklin Inst.* **261**, 331 (1956).
[8.125] S. Takahashi, *Le Vide Paris* **10**, 352 (1955).
[8.126] W. E. Danforth, *J. Franklin Inst.* **266**, 483 (1958).
[8.127] G. Mesnard, *C.R. Acad. Sci. Paris* **236**, 904 (1958).
[8.128] R. B. Gammage and D. A. Young, *Nature* **207**, 74 (1965).
[8.129] M. F. Lasker and R. A. Rapp, *Z. Phys. Chem. (Frankfurt am Main)* **49**, 198 (1966).
[8.130] C. E. McNeilly, *J. Nucl. Mater.* **11**, 53 (1964).
[8.131] E. S. Makarov, *Dokl. Akad. Nauk SSSR* **139**, 612 (1961).
[8.132] F. Grønvold, *J. Inorg. Nucl. Chem.* **1**, 357 (1955).
[8.133] R. K. Willardson, J. W. Moody, and H. L. Goering, *J. Inorg. Nucl. Chem.* **6**, 19 (1958).

[8.134] A. Burdese, *Gazz. Chim. Ital.* **89**, 718 (1959).
[8.135] L. E. J. Roberts, *Quart. Rev.* **15**, No. 4 (1961).
[8.136] C. Alexander, Thesis, Ohio State Univ. (1961).
[8.137] F. Neth, Thesis, Tübingen (1955).
[8.138] V. M. Zhukovskii, V. G. Vlasov, and A. G. Lebedev, *Fiz. Metal. Metallov.* **14**, 319 (1962).
[8.139] W. Meyer, *Z. Elektrochem.* **50**, 274 (1944).
[8.140] J. Day, M. Freymann, and R. Freymann, *C.R. Acad. Sci. Paris* **229**, 1013 (1949).
[8.141] D. M. Gruen, *J. Amer. Chem. Soc.* **76**, 2117 (1954).
[8.142] J. K. Dawson and M. W. Lister, *J. Chem. Soc.* 2181 (1950).
[8.143] J. K. Dawson and B. A. J. Lister, *J. Chem. Soc.* 5041 (1952).
[8.144] W. Trzebiatowski and P. W. Selwood, *J. Amer. Chem. Soc.* **72**, 4504 (1950).
[8.145] A. Arrott and J. E. Goldman, *Phys. Rev.* **108**, 948 (1957).
[8.146] R. J. Ackermann, R. J. Thorn, and G. H. Winslow, *J. Opt. Soc. Amer.* **49**, 1107 (1959)
[8.147] A. Companion and G. H. Winslow, *J. Opt. Soc. Amer.* **50**, 1043 (1960).
[8.148] S. Aronson, J. E. Rulli, and B. E. Schaner, *J. Chem. Phys.* **35**, 1382 (1961).
[8.149] R. R. Heikes and W. D. Johnston, *J. Chem. Phys.* **26**, 582 (1957).
[8.150] L. A. Andersen, *Proc. Roy. Soc. London* **185**, 69 (1946).
[8.151] P. Gerdanian and M. Dode, *C.R. Acad. Sci. Paris* **254**, 1005 (1962).
[8.152] M. J. M. Leask, L. E. J. Roberts, A. J. Walter, and W. P. Wolf, *J. Chem. Soc.* 4788 (1963).
[8.153] H. E. Schmidt and B. Raz, *J. Nucl. Mater.* **8**, 265 (1963).
[8.154] P. Nagels, M. Denayer, and J. Devreese, *Solid State Commun.* **1**, 35 (1963).
[8.155] P. Nagels, J. Devreese, and M. Denayer, *J. Appl. Phys.* **35**, 1175 (1964).
[8.156] P. Nagels and M. Denayer, *Bull. Soc. Belge Phys.* **4**, 35 (1964).
[8.157] A. Duquesnoy and F. Marion, *C.R. Acad. Sci. Paris* **258**, 4072, 4550 (1964).
[8.158] H. P. Myers, T. Jonsson, and R. Westin, *Solid State Commun.* **2**, 321 (1964).
[8.159] S. Iida, *J. Phys. Soc. Japan* **20**, 291 (1965).
[8.160] S. Iida, *Jap. J. Appl. Phys.* **4**, 833 (1965).
[8.161] B. T. M. Willis and R. I. Taylor, A.E.R.E., R.4999 (1965).
[8.162] R. J. Thorn and G. H. Winslow, *J. Chem. Phys.* **44**, 2632 (1966).
[8.163] J. Devreese, R. de Coninck and H. Pollak, *Phys. Status Solidi* **17**, 825 (1966).
[8.164] H. Haraldsen and R. Bakken, *Naturwissenschaften* **28**, 127 (1940).
[8.165] A. F. Andresen, *Acta Crystallogr.* **11**, 612 (1958).
[8.166] A. Duquesnoy and F. Marion, *C.R. Acad. Sci. Paris* **258**, 5657 (1964).
[8.167] Yu. K. Gu'kov, A. V. Zvonarev, and V. P. Klichkova, *At. Energ.* **8**, 72 (1960).
[8.168] M. Maurat and C. Eyraud, *C.R. Acad. Sci. Paris* **254**, 3084 (1962).
[8.169] V. M. Zhukovskii, V. G. Vlasov, and A. G. Lebedev, *Fiz. Metal. Metalloved.* **14**, 475 (1962).
[8.170] E. V. Tkachenko, A. D. Neuimin, V. G. Vlasov, and V. N. Strekalovskii, *Fiz. Metal. Metalloved.* **16**, 193 (1963).
[8.171] A. M. George and M. D. Kharkhanavala, A.E.E.T./CD/13, Bombay (1963); *J. Phys. Chem. Solids* **24**, 1207 (1963).
[8.172] Nirmal Singh and M. D. Kharkhanavala, A.E.E.T./CD/27, Bombay (1964).
[8.173] M. Picon and M. Patrie, *C.R. Acad. Sci. Paris* **242**, 1321 (1956).
[8.174] L. Brewer, N. R. LeRoy, L. A. Bromley, and E. D. Eastman, *J. Amer. Ceram. Soc.* **34**, 128 (1951).
[8.175] T. R. McGuire, B. E. Argyle, M. W. Schafer, and J. S. Smart, *Appl. Phys. Lett.* **1**, 17 (1962).

[8.176] G. Busch, P. Junod, M. Risi, and O. Vogt, *Proc. Int. Conf. Phys. Semicond., Exeter, 1962* p. 727. Inst. Phys. & Phys. Soc., London, 1962.
[8.177] S. van Houten, *Phys. Lett.* **2**, 215 (1962).
[8.178] U. Enz, J. F. Fast, S. van Houten, and J. Smit, *Philips Res. Rep.* **17**, 451 (1962).
[8.179] T. R. McGuire, B. E. Argyle, M. W. Schafer, and J. S. Smart, *J. Appl. Phys.* **34**, 1345 (1963).
[8.180] T. R. McGuire and M. W. Schafer, *J. Appl. Phys.* **35**, 984 (1964).
[8.181] M. D. Houston, *in* "Rare Earth Research" (E. V. Kleber, ed.), p. 255, Macmillan, New York 1961.
[8.182] V. I. Marchenko and G. V. Samsonov, *Ukr. Fiz. Zh.* **8**, 140 (1963).
[8.183] J. W. McClure, *J. Phys. Chem. Solids* **24**, 871 (1963).
[8.184] V. P. Zhuze, A. V. Golubkov, E. V. Goncharova, and V. M. Sergeeva, *Fiz. Tverd. Tela* **6**, 257 (1964).
[8.185] V. P. Zhuze, A. V. Golubkov, E. V. Goncharova, T. I. Komarova, and V. M. Sergeeva, *Fiz. Tverd. Tela* **6**, 268 (1964).
[8.186] A. V. Golubkov, E. V. Goncharova, V. P. Zhuze, and I. G. Manoilova, *Fiz. Tverd. Tela* **7**, 2430 (1965).
[8.188] R. R. Heikes and C. W. Chen, Paper 64-9F3-442-P4, October 6, 1964, Westinghouse Research Labs.; see also C. W. Chen, F. Carter, and R. R. Heikes, Abstract in *J. Appl. Phys.* **36** (Pt. 2), 1160 (1965).
[8.189] S. Methfessel, Ferromagn. Conf. (Düsseldorf 1964), *Z. Angew. Phys.* **18**, 414 (1965).
[8.190] F. Holtzberg, T. R. McGuire, S. Methfessel, and J. C. Suits, *Proc. Int. Conf. Magn., Nottingham, 1964* p. 470, Inst. Phys. and Phys. Soc., London, 1964.
[8.191] G. Busch, P. Schwob, and O. Vogt, *Helv. Phys. Acta* **39**, 591 (1966).
[8.192] E. D. Eastman, L. Brewer, L. A. Bromley, P. W. Gilles, and N. L. Lofgren, *J. Amer. Chem. Soc.* **72**, 4019 (1950).
[8.193] M. Picon and J. Flahaut, *C.R. Acad. Sci. Paris* **241**, 655 (1955).
[8.194] M. A. Kanter and C. W. Kazmierowicz, *J. Appl. Phys.* **35** (Pt. 2), 1053 (1964).
[8.195] M. Tetenbaum, *J. Appl. Phys.* **35**, 2468 (1964).
[8.196] J. Flahaut and M. Guittard, *C.R. Acad. Sci. Paris* **243**, 1210 (1956).
[8.197] E. D. Eastman, L. Brewer, and N. R. LeRoy, *J. Amer. Chem. Soc.* **74**, 835 (1952).
[8.198] J. Appel and S. W. Kurnick, *J. Appl. Phys. suppl.* **32**, 2206 (1961).
[8.199] F. M. Ryan, I. N. Greenberg, F. L. Carter, and R. C. Miller, *J. Appl. Phys.* **33**, 864 (1962).
[8.199a] S. W. Kurnick, R. L. Fitzpatrick, and M. F. Merriam *in* "Rare Earth Research" (J. F. Nachman and C. E. Lundin, eds.), Vol. 1, p. 249, Gordon and Breach, New York, 1962.
[8.200] M. Cutler, R. L. Fitzpatrick, and J. F. Leavy, *J. Phys. Chem. Solids* **24**, 319 (1963).
[8.201] J. Flahaut, L. Domange, M. Guittard, M. P. Pardo, and M. Patrie, *C.R. Acad. Sci. Paris* **257**, 1530 (1963).
[8.202] G. V. Samsonov and V. I. Marchenko, *Dokl. Akad. Nauk SSSR* **152**, 671 (1963).
[8.203] M. Cutler, J. F. Leavy, and R. L. Fitzpatrick, *Phys. Rev.* **133**, A1143 (1964).
[8.204] M. Cutler and J. F. Leavy, *Phys. Rev.* **133**, A1153 (1964).
[8.205] V. I. Marchenko and I. G. Barantseva, *Inzhen. Fiz. Zh.* **7**, 120 (1964).
[8.206] F. Holtzberg, T. R. McGuire, S. Methfessel, and J. C. Suits, *J. Appl. Phys.* **35** (Pt. 2), 1033 (1964).
[8.207] G. V. Lashkar'ev, Yu. B. Paderno, S. V. Radzikivs'ka, and V. P. Fedorchenko, *Ukr. Fiz. Zh.* **10**, 520 (1965).
[8.208] B. K. Zalevskii, G. V. Lashkar'ev, V. V. Sobolyev, and N. N. Syrbu, *Ukr. Fiz. Zh.* **11**, 638 (1966).

[8.209] J. Flahaut, *Bull. Soc. Chim. Fr.* 772 (1958).
[8.210] J. F. Miller, F. J. Reid, and R. C. Himes, *J. Electrochem. Soc.* **106**, 1043 (1959).
[8.211] F. J. Reid, L. K. Matson, J. F. Miller, and R. C. Himes, *J. Phys. Chem. Solids* **25**, 969 (1964).
[8.212] D. J. Haase and H. Steinfink, *J. Appl. Phys.* **36**, 3490 (1965).
[8.213] B. E. Argyle, J. C. Suits, and M. J. Freiser, *Phys. Rev. Lett.* **15**, 822 (1965).
[8.214] G. Busch and P. Wachter, *Helv. Phys. Acta* **39**, 197 (1966).
[8.215] F. Holtzberg, T. R. McGuire, S. Methfessel, and J. C. Suits, *Phys. Rev. Lett.* **13**, 18 (1964).
[8.215a] J. F. Miller, L. K. Matson, and R. C. Himes, *in* "Rare Earth Research" (J. F. Nachman and C. E. Lundin, eds.), Vol. 1, p. 233, Gordon and Breach, New York, 1962.
[8.216] G. V. Lashkar'ev and Yu. B. Paderno, *Ukr. Fiz. Zh.* **10**, 566 (1965).
[8.217] G. Will, S. J. Pickart, H. A. Alperin, and R. Nathans, *J. Phys. Chem. Solids* **24**, 1679 (1963).
[8.218] D. S. Rodbell, L. M. Osika, and P. E. Lawrence, *J. Appl. Phys.* **36**, 666 (1965).
[8.219] T. H. Ramsey, H. Steinfink, and E. J. Weiss, *J. Appl. Phys.* **36**, 548 (1965).
[8.220] D. J. Haase and H. Steinfink, *J. Appl. Phys.* **137**, 2246 (1966).
[8.221] A. A. Eliseev, E. I. Yarembash, V. G. Kuznetsov, I. I. Antonova, and Z. P. Stoyantsova, *Izv. Akad. Nauk SSSR, Neorg. Mat.* **1**, 1027 (1965).
[8.221a] W. L. Mularz and S. J. Wolnik, *in* "Rare Earth Research" (K. S. Vorres, ed.), Vol. 2, p. 473. Gordon and Breach, New York, 1964.
[8.222] J. C. Andrellos and P. Bro, *Solid State Electron.* **5**, 414 (1962).
[8.223] E. I. Yarembash, E. S. Vigileva, A. A. Eliseev, and V. I. Kalitin, *Izv. Akad. Nauk, Ser. Fiz.* **28**, 1306 (1964).
[8.224] T. H. Ramsey, H. Steinfink, and E. J. Weiss, *J. Appl. Phys.* **34**, 2917 (1963).
[8.225] J. Danon and A. M. de Graaf, *J. Phys. Chem. Solids* **27**, 1953 (1966).

Part Three SOME PRESENT PROBLEMS AND POSSIBLE APPLICATIONS

Chapter 9 | Various Magnetoelectric Effects

9.1. MAGNETIC SCATTERING

The resistivity of nonmagnetic crystals results from the scattering of the charge carriers by impurities, lattice vibrations, and electrical dipoles. In magnetic crystals, scattering resulting from the disorder of the atomic magnetic moments is also possible. The electron is a particle with a magnetic moment known as the "spin moment." It is therefore to be expected that neither the electron moving freely in an AB band, nor the corresponding "hole," which moves in the B band, will be unaffected by the presence of atomic magnetic moments. The importance of the scattering is shown by an abrupt change in the slope or a maximum at the Curie temperature in the resistivity versus temperature curves.

It has long been known that these curves show a change at T_C for ferromagnetic metals. If, as Becker and Döring [9.1] have done, one makes the experimental curves for Ni and Pd, placed one below the other in the periodic table, coincide at this point, the lower values for the former in its ferromagnetic region are clearly apparent (Fig. 9.1). Mott [9.2, 9.3] has supplied a satisfactory quantitative explanation of this irregularity, by using a model with two energy bands in which the electrons in the 4s band are responsible for the electrical conductibility, while the contribution from those in the 3d band is negligible because of their high effective mass. The 4s electrons can undergo s–s or s–d transitions, during which their spin is assumed to remain unchanged. The resistivity becomes higher as the s–d transitions increase in frequency. The probability of these s–d transitions is greatest in the paramagnetic state and lesser in the ferromagnetic state, in which the moments are ordered. For this

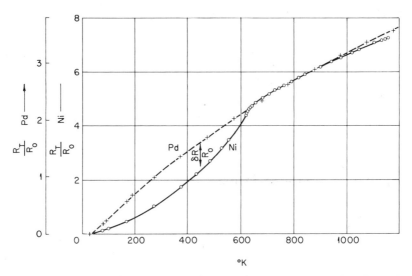

Fig. 9.1. Comparison of the resistivities of Ni and Pd (based on Becker and Döring [9.1]).

theoretical aspect, readers may also refer to [9.4–9.7] as well as the excellent digest by Coles [9.8], for a more general view of magnetic scattering in transition metals and their alloys.

If one ignores the effect of impurities at low temperatures, the resistivity at medium temperatures may be taken to contain, in addition to the usual term resulting from the thermal agitation of the atoms, another term resulting from the effect of the disorder of the atomic moments on the spins of the free electrons:

$$\rho = \rho_{\text{atom}} + \rho_{\text{spin}} \qquad (9.1)$$

The first term is a linear function of T. The second is nil at low temperatures, presents a positive temperature coefficient in the domain in which thermal agitation of the atoms gradually reduces the ferromagnetic (or antiferromagnetic) order, and then, in the paramagnetic domain, above T_C, assumes a value calculated by Friedel and de Gennes [9.5]. One can simplify the expression given by these authors by combining the influence of various factors that are independent of temperature; S, as usual, refers to the component of the spin quantum numbers, and N to the number of free carriers per cubic centimeter:

$$\rho_{\text{spin}} = (1/N)\, S(S+1) \times C^{te} \qquad (9.2)$$

Weiss and Marotta [9.9] have calculated the contribution of the two terms, at different temperatures, to the resistivity of α-Fe (Fig. 9.2).

For semiconductor compounds, there is no doubt that one finds the two terms of the Eq. (9.1) for the charge carriers entirely delocalized in the crystal,

9.1. Magnetic Scattering

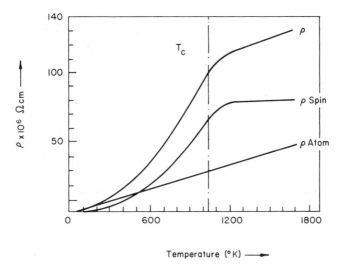

Fig. 9.2. Contributions by the thermal agitation of atoms and the disorder of their magnetic moments to the resistivity of α-Fe (based on Weiss and Marotta [9.9]).

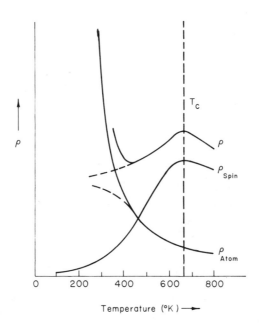

Fig. 9.3. Contributions by the thermal agitation of atoms and the disorder of their magnetic moments to the resistivity of a semiconductor (based on Suchet [9.10]).

in other words those created in the B and AB bands by excitation across the forbidden energy band. Here, however, contrary to what holds for metals and alloys, N varies exponentially in relation to temperature, and the rapid variation resulting from this in the first term is usually greater than for the second term. It can happen, however, when E_G is low and S high, that the positive temperature coefficient of the term ρ_{spin} momentarily predominates over the negative coefficient of ρ_{atom}, for temperatures immediately below T_C. Figure 9.3 shows the shape of the resulting curve, and the resistivity maximum which may be observed close to T_C. Similar curves have been met with earlier in this book, for example for the oxide CrO_2 (cf. Fig. 5.20a, Section 5.3). Some authors have also observed them for the telluride MnTe (cf. Section 7.4).

It should be pointed out, however, that a curve of this shape does not reciprocally prove the existence of magnetic scattering in a semiconductor. It was recalled, in Section 3.4, that the internal field bound up with the magnetic order involved a marked energy difference between electrons with spins α and β. The effect of this phenomenon on the reduction of E_G has been mentioned by theoretical scientists [9.11]. It can result in an interaction between the upper sublevel and the AB band, and metallic behavior below T_C. Such behavior may also have other causes, such as a variation in the cell parameter, or even a phase change around T_C. In addition, it should not be forgotten that in transition metal compounds, the excitation and transfer mechanisms are always superimposed to some degree (cf. Section 3.3 [9.12]) with

$$\sigma = N_1 e\mu_1 \text{ (excitation)} + N_2 e\mu_2 \text{ (transfers)} \qquad (9.3)$$

When the densities of carriers created by each of the mechanisms are comparable, those resulting from transfers (2) predominate at low temperatures and those resulting from excitations (1) at high temperatures.

As was written in Section 3.4 [Eq. (3.15)], when the carriers created by excitation predominate,

$$\rho = 1/\sigma = (1/Ne)(1/\mu_1 + 1/\mu_1') \qquad (9.4)$$

where $\mu_1 =$ the mobility of the carriers created by the first mechanism. The second term, nil at absolute zero, represents the magnetic scattering, and in the paramagnetic domain it reaches the value:

$$1/\mu_1' \sim KS(S+1) \qquad (9.5)$$

where K expresses the influence of various factors independent of the temperature. Table 9.1 summarizes the probable electrical behavior of a magnetic crystal [9.12]. For crystals not involving any interactions between the AB band and d sublevels, transfer (2) carriers generally predominate at low temperatures if the composition is not strictly stoichiometric. As the temperature rises, the excitation (1) carriers excited in the AB band are more

TABLE 9.1

	$T \ll T_C$	$T < T_C$	$T > T_C$
sc$_2$/sc$_1$	$N_2 \gg N_1$ $N_2 = C^{te}$ $\mu_2 \sim \exp(-E_A/kT)$ $d\sigma/dT > 0$	$\begin{cases} N_1 > N_2 \\ d\mu_1'/dT \ll 0 \\ d\sigma/dT < 0 \end{cases}$ $\begin{cases} N_1 < N_2 \\ d\mu_2/dT > 0 \\ d\sigma/dT > 0 \end{cases}$	$N_1 \gg N_2$ $N_1 \sim \exp(-E_G/2kT)$ $\mu_1' \ll \mu_1$ $d\sigma/dT > 0$
metal/sc$_1$		$N \sim 10^{22}$ cm^{-3} $d\sigma/dT \sim -C^{te}$	

numerous, and at medium temperatures they equal the transfer (2) carriers. But as one approaches T_C (or T_N), a magnetic scattering term $1/\mu_1'$ appears, reducing the mobility of the excitation (1) carriers considerably. It is therefore possible that in the Eq. (9.3) for conductivity, the contribution of the second term may not always be negligible compared with that of the first term for $T > T_C$. This very summary description is in practice complicated by the presence of extrinsic semiconduction resulting from impurities and defects, which has been ignored here.

So far as we know, attempts made to broach the theory of magnetic scattering in semiconductors from a more rigorous viewpoint have been very few in number. Haas [9.13] has recently investigated this point, and he has reported on the main experimental results concerning the resistivity, magnetoresistance and mobility of magnetic semiconductors in terms of the coupling between carriers and moments. When this coupling is very marked, he admits the existence of a "magnetic polaron."

9.2. MAGNETORESISTANCE

Readers may refer to the works by Vonsovskii and Shur [9.14], Vonsovskii Bates [9.15], [9.16], Belov [9.17], and Bozorth [9.18]. The main aim of Jan's digest [9.19] is to set the phenomenon in the general context of galvanomagnetic effects, and Belov's [9.20] work deals with the effect in the immediate neighborhood of T_C.

Magnetoresistance is the relative increase of resistivity $\Delta\rho/\rho$. In diamagnetic crystals, it results, like the Hall effect, from the effect of the Lorentz forces on a group of electrons endowed with an overall movement (these forces are those exerted by two fields, electric and magnetic, on a moving charge). It is lower when the field is parallel to the current (longitudinal magnetoresistance) than when it is perpendicular (transverse magnetoresistance), and is proportional to H^2 in weak fields and to H in strong fields. In the latter case, the slope is less marked for semiconductors than metal alloys [9.21]. The intensity of the phenomenon decreases as the temperature rises and as the thermal agitation reduces the mean free path of the carriers. In paramagnetic materials, scattering of the electrons by the disordered atomic moments introduces an additional component of the resistivity, as mentioned in the previous section. The effect of the field, which tends to order the moments, thus reduces ρ, and the slope $\Delta\rho/\rho$ (H) becomes slight (cerium $+0.3 \times 10^{-6}$) or even negative (transition metal alloys [9.19], diamagnetic semiconductors which high doping has made paramagnetic [9.22]).

In ferromagnetics, the magnetoresistance is bound up with the existence of an induction B, but it increases more slowly with the field than this induction (Fig. 9.4). One should distinguish the terms H_{ext} and B_{ext}, which refer to the ambient medium, from H_{int} and B_{int}, which refer to the sample:

$$\begin{aligned} B_{ext} &= H_{ext} \\ B_{int} &= H_{int} + 4\pi I \\ H_{int} &= B_{ext} - F_d I \\ B_{int} &= H_{ext} + (4\pi - F_d) I \end{aligned} \qquad (9.6)$$

where F_d is the demagnetizing factor. The influence of H_{ext}, which is always masked in weak fields, would tend to raise the resistance slightly, as in non-magnetic crystals. The effect of the second term, on the other hand, depends on the angle θ between I and the measurement current. It is often positive (in metals and alloys) for longitudinal magnetoresistance, and always negative for transverse magnetoresistance (Fig. 9.5). The resistivity of a ferromagnetic domain may be represented by a development into series in terms of $\cos\theta$ [9.24]. Uneven powers are absent because of the symmetry of the effect, and so one finds that $\Delta\rho/\rho$ is proportional to $(B-H)^2$. This law is generally followed, in the absence of mechanical tension, for $60° < \theta < 90°$. At the beginning of the magnetization curve it is better to replace it by a law in $(I^2 - I_0^2)$ where I_0 reflects the variations in magnetization resulting from wall displacements, which do not affect the magnetoresistance since they involve rotations of 180°, keeping $\cos\theta$ at the same value. A law in $H^{2/3}$ is observed in the immediate neighborhood of T_C [9.20].

The greatest slope in the magnetoresistance curve in terms of the field reaches a maximum at T_C, the temperature at which $(B-H)$ depends most on H [9.25, 9.26]. Figure 9.6 shows this phenomenon for nickel. When saturation is reached, both magnetoresistances decrease lineally when the field continues to increase. This is due, as in paramagnetics, to the effect of the field, which tends to improve the order of the moments further, and thus reduce ρ. This

9.2. Magnetoresistance

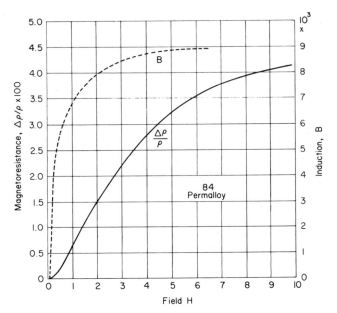

Fig. 9.4. Linear variation in the magnetoresistance and induction of 84 permalloy in relation to the external field (based on McKeehan [9.23]).

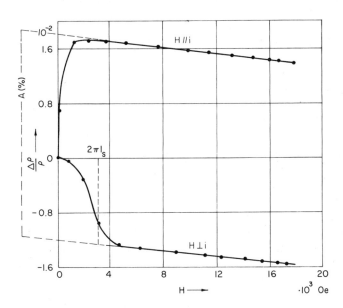

Fig. 9.5. Linear variation in the magnetoresistance of polycrystalline Ni in relation to the field, i = measurement current (based on Englert [9.24]).

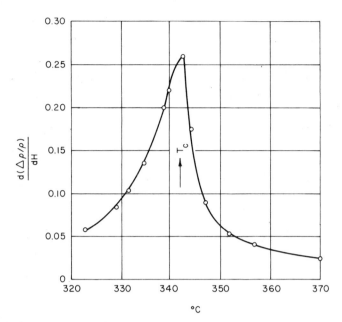

Fig. 9.6. Linear variation in the magnetoresistance per field unit of Ni in relation to temperature around T_C (based on Gerlach and Schneiderhahn [9.25]).

negative slope is more marked here, however (Ni -40×10^{-6}). The difference in the values of the two magnetoresistances is particularly high in ferromagnetics. When extrapolated for $H = 0$ from linear laws in high fields, it constitutes the *ferromagnetic anisotropy*, which is independent of the original conditions (Fig. 9.5.)

The longitudinal magnetoresistance is often positive at low temperatures (Shirakawa [9.27] has found $+0\,16$ for a Co–Ni alloy at 77°K), but its algebraic value drops when the temperature rises, so that the effect is always negative when one approaches T_C, if this temperature is high enough. An (algebraic) minimum is then observed at T_C, corresponding to that for I. It is negative, since the role of the field is to restore the magnetic order, partly destroyed by thermal agitation. Then, at high temperatures, the low values characteristic of paramagnetics are found again. Figure 9.7 shows the shape of one of these curves for a monocrystal of ferrite $MnFe_2O_4$. It will be noted that the effect is negative at low temperatures. This is generally the case for semiconductors, as well as for alloys with a high coercitive field. The existence of a small effect at a temperature slightly higher than T_C is explained by the fact that this is defined by extrapolation for $H = 0$. A local order frequently survives for $H \neq 0$, as is shown by the "tails" of magnetization or specific heat curves above T_C.

9.2. Magnetoresistance

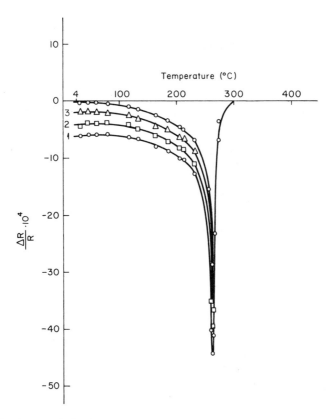

Fig. 9.7. Linear variation in the longitudinal magnetoresistance of manganese ferrite in relation to temperature for saturating fields of 1960 (1), 1700 (2), 1439 (3) and 1178 Oe (4) (based on Belov and Talalaeva [9.28]).

Surveys of magnetoresistance have been met with several times already. Here, we shall confine ourselves to recalling three typical examples. First there is the phosphide MnP, the metamagnetic phase of which, below 50°K, naturally gives an enormous magnetoresistance. In Section 4.4, the work of Suzuki et al. [9.29] was mentioned. Figure 9.8 reproduces their results at 4.2°K for the transverse magnetoresistance, which is negative as in semiconductors. A similar curve has been obtained by the same authors in connection with the increase in the Seebeck effect. Next, there is the study by Samokhvalov and Fakidov [9.30] of the oxide Fe_3O_4 (cf. Fig. 6.21 in Section 6.3). The longitudinal and transverse magnetoresistances are both negative, but do not exceed a few thousandths. Finally, the studies by Fakidov et al. [9.31, 9.32] and Zotov and Shur [9.33] of the telluride CrTe or $Cr_{1-x}Te$ again show two negative magnetoresistances, almost double those for Fe_3O_4

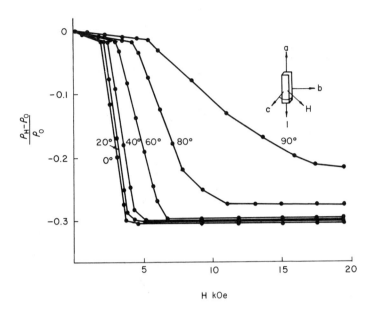

Fig. 9.8. Linear variation in the transverse magnetoresistance of MnP at 4.2°K, in relation to the magnetic field, for different values of the angle (H, c). Metamagnetism appears around 2 kOe (based on Suzuki et al. [9.29]).

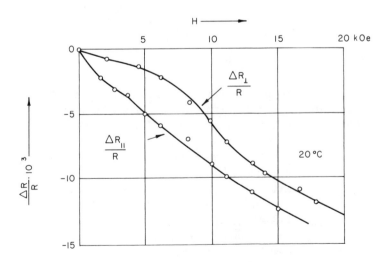

Fig. 9.9. Linear variation in the magnetoresistances of CrTe in relation to the magnetic field (based on Fakidov and Grazhdankina [9.31]).

9.2. Magnetoresistance

(Fig. 9.9). The longitudinal effect reaches a maximum of 22 thousandths at $T_C = 58°C$, for a field of 15 kOe. The authors refer to Vonsovskii's theory [9.34], to explain why the two magnetoresistances have the same sign.

Shortly after World War II, Volger [9.35, 9.36] reported on the existence of mixed ferromagnetic manganites $La_{1-x}Sr_xMnO_3$, which crystallize in the structure of perovskite. Figure 9.10 gives some properties for the composition $x = 0.2$, which at the time drew attention to the behavior of magnetic semiconductors. Maximum resistivity occurs around 305°K (a), and the maximum slope in terms of the field around 295°K. Although the saturation magnetization curve (c) does not on its own enable one to determine T_C accurately, it may still be assumed that the curve for $H = 0$ would cut the T axis slightly below 300°K. This corresponds well with the magnetoresistance peak temperature (b).

Turov and Shavrov [9.37] have recently discussed galvanomagnetic effects in antiferromagnetics and applied their theory to calculating the magnetoresistance of CrSb and α-Fe_2O_3.

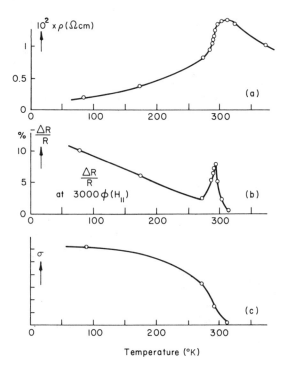

Fig. 9.10. Linear variations in relation to temperature of the resistivity (a), magnetoresistance (b) and saturation magnetization (c) of manganite $La_{0.8}Sr_{0.2}MnO_3$ (based on Volger [9.36]).

9.3. Astrov Effect

The idea of a magnetoelectric coupling involving the appearance of a magnetic moment induced by an electrical field had been vaguely put forward by some writers, but the first serious theoretical analysis of this effect, and the prediction that the induced magnetic field is proportional to the applied electrical field, is found in the work by Landau and Lifshitz [9.38], in a chapter devoted to magnetic symmetry. Shortly afterwards, Dzyaloshinskii [9.39], working on the basis of the symmetry of the crystallographic structure $D5_1$ (corundum) and the magnetic structure recently obtained by neutron diffraction [9.40] (cf. Fig. 6.29, page 187) and the study of magnetic susceptibility [9.41], pointed out that the oxide Cr_2O_3 would be particularly suitable for observation of this effect. In the symmetry transformations which are peculiar to this magnetic structure, two expressions remain invariable in the thermodynamic potential, and this invariability corresponds to a linear equation between the induction and field intensities in the substance:

$$\mathbf{D} = \varepsilon \mathbf{E} + \alpha \mathbf{H} \tag{9.7}$$

$$\mathbf{B} = \mu \mathbf{H} + \alpha \mathbf{E} \tag{9.8}$$

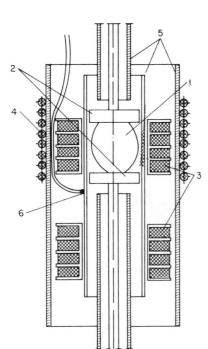

Fig. 9.11. Diagram of the apparatus to measure the Astrov effect (based on Astrov [9.42]).

9.3. Astrov Effect

Astrov [9.42] discovered it in 1959 in a monocrystal of Cr_2O_3 (1), placed in the apparatus illustrated diagrammatically in Fig. 9.11 between two electrodes (2), between which an alternating electromotive force with a frequency of 10^4 Hz is set up. The appearance of a magnetic moment produces a signal in the astatic arrangement of the coils (3), and this is sent into the amplifier with maximum noise level 10^{-7} V. The temperature is kept steady for 10 min by means of the resistance (4), and controlled by the thermocouple (6). Efficient electrostatic shielding (5) protects the whole unit. Figure 9.12 shows, in arbitrary units, the signals corresponding to an effective field of 430 (1) and 230 (2) V cm^{-1} at various temperatures. The addition of an outside solenoid next allowed an alternating magnetic field with an intensity of 6 Oe and a frequency of 10^4 Hz to be set up, and the magnetic susceptibility (3) to be measured. It can be seen that the temperature at which the signal disappears corresponds to the Neel temperature of Cr_2O_3, here found to be 312°K. In a later, more precise experiment, Astrov [9.43] measured the proportionality coefficient α of the Eq. (9.8) parallel to the axis of the crystal ($\alpha\|$) and in the base plane ($\alpha\perp$). Figure 9.13 shows the marked difference shown by the variations in these two parameters in relation to temperature. Astrov also speculates on the significance of these curves. Folen et al. [9.44] have confirmed earlier findings.

Rado and Folen [9.45] have observed the reverse effect (induced by a magnetic field) and shown that the changes that occur in these two effects at T_N provide the first direct proof of the existence of antiferromagnetic "domains." Rado [9.46] has linked the variation in the magnetic susceptibilities of Cr_2O_3 with the antiferromagnetic susceptibilities, by considering an imaginary magnetic field produced in the crystal by the applied electrical field. The physical origin of the effect is attributed to a distortion in the electron cloud round the magnetic atom by the electrical field, and the transmission of this distortion to the spins by the spin–orbit coupling, thus causing the appearance of induced magnetization. A calculation based on this theory gives results that correspond to the experimental findings. Rado and Folen [9.47] have repeated the preceding interpretation based on the role of the spin–orbit coupling and of the crystal field and applied electrical field. Date et al. [9.48], after observing the influence of an electrical field on absorption by magnetic resonance in ruby, have expressed their disagreement with the preceding explanation, and suggest attributing the origin of the effect to an interaction in the same sublattice. Rado and Folen [9.49] have presented a digest on magnetoelectric effects, including the Astrov effect, the inverse effect, and the piezomagnetic, and piezoelectromagnetic effects.

Rado [9.50] has confirmed the occurrence of the spin–orbit coupling, on the basis of a statistical mechanical calculation, involving an approximation of the molecular field and the free energy. This atomic model explains the two

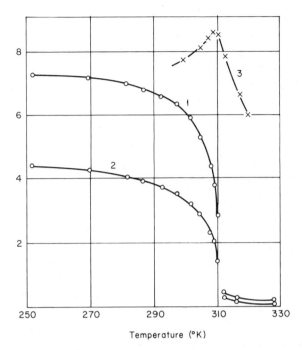

Fig. 9.12. Linear variation in the signal proportional to the magnetic moment (1, 2) and the magnetic susceptibility (3) of Cr_2O_3 in relation to temperature (based on Astrov [9.42]).

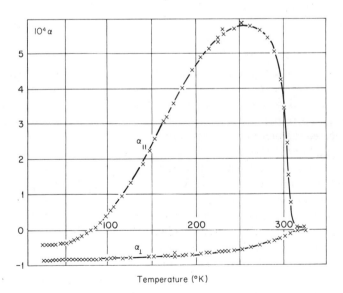

Fig. 9.13. Linear variation in the coefficients $\alpha\|$ and $\alpha\perp$ of Cr_2O_3 in relation to temperature (based on Astrov [9.43]).

9.3. Astrov Effect

electromagnetic effects and leads to the calculation of a magnetoelectric contribution to the magnetic susceptibility. Izuyama and Pratt [9.51] have shown that the Astrov effect depends on the square of the spin–orbit interaction, and discussed the case of Cr_2O_3. O'Dell [9.52, 9.53] has observed both effects on polycrystal discs of Cr_2O_3 cooled slowly through T_N in magnetic and electrical fields. Alexander and Shtrickman [9.54] attribute the Astrov effect to the change in the internal exchange interaction of the sublattice with the electrical field. Lal *et al.* [9.55] have studied dielectric constant variation at T_N, linking it up with Rado's model.

In the same crystallographic structure, $D5_1$, Foner and Hanabusa [9.56] have observed the inverse effect in mixed $Cr_{2-x}Al_xO_3$ crystals, and its reversibility when the magnetic field reaches the coercitive field. Al'shin and Astrov [9.57] have also observed the inverse effect in polycrystal samples of Ti_2O_3, and have drawn conclusions on the most likely magnetic structure (cf. Fig. 6.29, Section 6.4). The compound $Ga_{2-x}Fe_xO_3$ ($x \sim 1$) crystallizes in a noncentrosymmetrical orthorhombic structure. Remeika [9.58], who discovered it, has pointed out its ferromagnetic and piezoelectric nature, and Wood [9.59] has worked out its crystallographic structure. Nowlin and Jones [9.60, 9.61] have measured the height of the forbidden band and the magnetic susceptibility.

Rado [9.62, 9.63] studied a monocrystal of $Ga_{2-x}Fe_xO_3$ with the composition $x = 1.08$, and showed that it presented the two magnetoelectric effects described above. He therefore observed these effects for the first time in a crystal possessing spontaneous magnetization. The application of a static magnetic field along axis c induces a linear effect along axis b, while the application of the field along a induces a quadratic effect. The coefficient α is 4×10^{-4} at 77°K, namely one order higher than in Cr_2O_3. The application of an alternating electrical field with a frequency of 1 kHz along axis b induces a magnetic moment along axis c. The coefficient α in this case is found to be 3×10^{-4} at 77°K, corresponding satisfactorily with the preceding value. For the variation in α in relation to temperature, see Fig. 9.14. These observations involve marked differences with regard to earlier research on Cr_2O_3. First, the induced polarization and the applied field are here perpendicular to each other, whereas they are collinear in Cr_2O_3. Second, Rado for the first time observed a nonlinear magnetoelectric effect. He has proposed a theoretical model to explain these two particularities, but has admitted that ignorance of the magnetic structure and, to some degree, of the exact atomic crystallographic structure, is regrettable.

Schelleng [9.64] has found a high anisotropy in these effects at 77°K, with an effective anisotropic field of 40 KOe. Bertaut *et al.* [9.65] mention that the compound is ferrimagnetic, and define its crystallographic and magnetic structure. Rado [9.66] has extended his measurements to low temperatures. Finally, Mercier [9.67] has recently repeated Rado's measurements for the

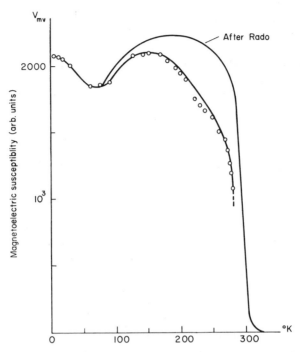

Fig. 9.14. Linear variation in the induced magnetic susceptibility in $Fe_{1.15}Ga_{0.85}O_3$ in relation to temperature in a permanent field of 3 kOe (based on Mercier [9.67]).

composition $Fe_{1.15}Ga_{0.85}O_3$ (Fig. 9.14). He has also given a fairly complete description of the apparatus and technique used to study the electromagnetic effects in Cr_2O_3 and $Ga_{2-x}Fe_xO_3$.

9.4. Possible Applications

Our only aim here is to give a few examples of possible applications of the various magnetoelectric effects described in this chapter. This is a pioneering field, in which applied research has not yet been carried out in any intensive way, and in which, above all, the discovery of new materials may still alter the terms of the problem. We shall consider, in turn, the prospects opened up by changes in magnetic structure, the high positive temperature coefficients of the resistivity, the magnetoresistance and the Astrov effect.

A change in the magnetic structure often brings about a sudden marked variation in resistivity, within a temperature interval of a few degrees. One of the best-known examples is given by the $Mn_{2-x}Cr_xSb$ solid solutions, which crystallize in the quadratic system, with a metallic-type conduction, and

9.4. Possible Applications

present a transition of the first order, from antiferromagnetism to ferrimagnetism, when the temperature rises. Jarrett *et al.* [9.68] have discovered that this transition is accompanied by a resistivity drop, and Biersted [9.69] has shown that the hysteresis decreases as the transition temperature rises. Figure 9.15 shows the phenomenon for various compositions. For $x = 0.1$, a resistivity drop of 27% occurs at 309°K, with 1° hysteresis. There is an immediate application to thermal regulators or detectors. Similar phase changes may be met with in nonmagnetic compounds. It was seen, for instance, in Sections 5.1 and 5.3, that V_2O_3 and VO_2 both present a semiconductor–metal transition, at 140°K and 70°C, respectively, when the temperature rises. The resistivity drop is a large one (10^5 for V_2O_3) and Futaki [9.70] has suggested its use in "critical temperature resistors." The material is prepared by sintering a mixture of V_2O_5 and oxides with stable valency in a reducing atmosphere. In the transition zone, the negative temperature coefficient of the resistivity is 30 times higher than that of usual thermistors.

It was seen in Section 9.1 that a high positive temperature coefficient for the resistivity can sometimes be observed at temperatures immediately below T_C. This situation is similar to the one observed in certain ferroelectric compounds (substitute titanates), in which valency induction creates carriers which are scattered by the increasing disorder of the crystal lattice when approaching the ferroelectric Curie temperature. Suchet [9.71] has suggested that, as an approximation, the following two equations might be applied in turn for the variation in resistivity in relation to temperature:

$$\rho = A \exp[B/(T_C - T)] \qquad (T < T_C) \qquad (9.9)$$

$$\rho = A \exp[B/T] \qquad (T > T_C) \qquad (9.10)$$

If a sample with a very small volume is used as a nonlinear element, voltage and current will correspond to the equations:

$$V = KI^n \qquad (9.11)$$

$$n = d(\ln V)/d(\ln I) = f(T) \qquad (9.12)$$

where ln refers to the Naperian logarithm. A simple calculation allows this function to be included in the graph $\ln V$–$\ln I$ in Fig. 9.16; $n = 1$ at the room temperature T_0, becomes infinite for T_1' and would reach -1 for T_C if Eq. (9.9) still applied. The operation around T_1' is used for current regulation (baretters). Above T_C, Eq. (9.10) shows that $n = -1$ around $2T_0$ and becomes zero at a temperature T_2, where voltage regulation is possible.

The phenomenon of magnetoresistance is already being used in the electronics industry, and there are components with two connections in which the resistance, of around 1000 Ω, increases by about 4% per kilo-Oersted applied. Depending on the temperature range being sought, the materials used are

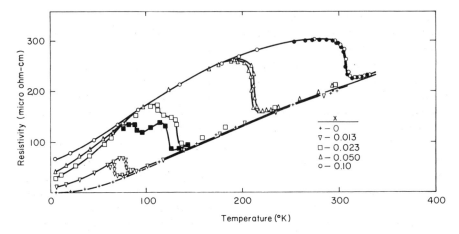

Fig. 9.15. Linear variation in the resistivity of $Mn_{2-x}Cr_xSb$ solid solutions in relation to temperature. The black dots refer to results recorded during cooling (based on Bierstedt [9.69]).

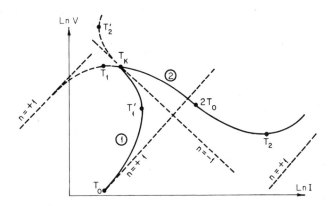

Fig. 9.16. Logarithmic variation in the tension at the terminals of a nonlinear element around T) in relation to the logarithm of current passing through it. [Based on Suchet, presented before the 8th session of the Société Francaise des Electriciens on 10 February 1955. *Bull. Soc. Elect.* (*Paris*) **5**, 274 (1955).]

InSb, Bi, or other metals, but transition element compounds suggest the possibility of more marked effects, particularly when they are ferro- or ferrimagnetic. Recent studies of thiochromites are particularly promising, and Haas *et al.* [9.72] have mentioned maximum magnetoresistances of a few tenths at 122°K, in fields of less than 1 kOe, for a magnetic semiconductor with composition $Cd_{0.98}Ga_{0.02}Cr_2S_4$. Apart from measuring magnetic field and producing variable resistances, such high values suggest the possibility of

9.4. Possible Applications

proper *commutation* without moving mechanical contacts. The importance of this application comes from the efforts made in the space field and the concept of reliability that has been introduced. The existence of moving contacts means a considerable reduction in the durability of a device.

Commutation by magnetoresistance can be used in converting alternating into direct current, low-frequency modulation and oscillation, etc. We feel that the most interesting application, however, lies in the improvements that it will make possible in low-powered motors; dc current motors offer very important advantages, such as the high starting torque, and the wide range of speeds within which they can operate. However, in their conventional form at any rate, they also present two major drawbacks, the frailty of the complex parts providing the rotating contact (collector and brushes) and the marked variation in speed in relation to the charge, in the absence of regulation. For low-powered motors, a routine method has been to replace the fixed inductor (stator) by a permanent magnet, in a field of which a wound armature rotates. The recent discovery of permanent magnets with very high coercitive fields, such as "Ferroxdure," has suggested the possibility of using them in a rotating inductor (rotor), as is shown in Fig. 9.17. This method has already been adopted for low-powered generators (bicycle or moped dynamos), but commutation of the different coils of the fixed armature (stator) depending on the position of the magnet naturally becomes necessary when motors are involved.

Dieulesaint [9.73] has described a motor, called "type II," in which the commutation is provided by means of a rotating segment driven by the rotor and coming periodically between a light source and a germanium photodiode with a very low time constant. The signal for this is sent to the anode of a power valve, the grid of which is connected to a master generator with an angular frequency close to that of the angular speed of "natural" rotation of the motor, and the plate to the winding of the fixed armature. In this way one obtains a dc motor without rotating contacts, which behaves like a self-starting synchronous motor. The use of components with a high magnetoresistance, altering the intensity in the armature circuit by means of silicon thyristors or transistors would provide a more practical solution than the rotating segment proposed by Dieulesaint.

So far as we know, there are no projects for the application of the Astrov effect. Materials that can produce electroinductive effects have always aroused great interest, because of the major applications they make possible, but here the effect is much too slight to be exploitable. The research done by Ascher *et al.* [9.74] in another direction for the moment seems more promising. At low temperatures, the boracite $Ni_3B_7O_{13}I$ presents ferroelectricity and slight ferromagnetism. The coupling between the spontaneous polarization along [001] and spontaneous magnetization along [110] is sufficiently strong for the latter to rotate by 90° when the former is reversed.

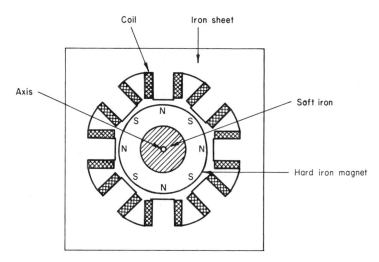

Fig. 9.17. Rough diagram of a dc motor without rotating contact (based on Dieulesaint [9.73]).

REFERENCES

[9.1] R. Becker and W. Döring, "Ferromagnetismus." Springer, Berlin, 1939.
[9.2] N. F. Mott, *Proc. Phys. Soc.* **47**, 571 (1935).
[9.3] N. F. Mott, *Proc. Roy. Soc.* **153**, 699 (1936).
[9.4] T. Kasuya, *Progr. Theor. Phys.* **16**, 58 (1956); **22**, 227 (1959).
[9.5] J. Friedel and P. G. de Gennes, *J. Phys. Chem. Solids* **4**, 71 (1958).
[9.6] J. Kondo, *Progr. Theor. Phys.* **27**, 772 (1962).
[9.7] T. van Peski-Tinbergen, and A. J. Dekker, *Physica*, **29**, 917 (1963).
[9.8] B. R. Coles, *Advan. Phys.* **7**, 40 (1958).
[9.9] R. J. Weiss and A. S. Marotta, *J. Phys. Chem. Solids*, **9**, 302 (1959).
[9.10] J. P. Suchet, *Phys. Status Solidi*, **2**, 167 (1962).
[9.11] Yu. P. Irkhin and E. A. Turov, *Fiz. Metal. Metalloved.* **4**, 9 (1957).
[9.12] J. P. Suchet, "Séminaires de Chimie de l'Etat Solide," Vol. 1, p. 37, S.E.D.E.S., Paris 1969.
[9.13] C. Haas, *Phys. Rev.* **168**, 531 (1968).
[9.14] S. V. Vonsovskii and Ya. I. Shur. "Ferromagnetism." Gostekhizdat, Moscow and Leningrad, 1948.
[9.15] S. V. Vonsovskii, "Moderne Lehre von Magnetismus." D.V.W., Berlin 1956.
[9.16] L. F. Bates, "Modern Magnetism," Cambridge Univ. Press, London and New York, 1951.
[9.17] K. P. Belov, "Ferromagnetism." Gostekhizdat, Moscow and Leningrad, 1951.
[9.18] R. M. Bozorth, "Ferromagnetism." Van Nostrand, Princeton, New Jersey, 1951.
[9.19] J. P. Jan, *Solid State Phys.* **5**, 66 (1957).
[9.20] K. P. Belov, "Magnitnye Prevachcheniya." Gosudarst. Izdat. (Fiz. Mat.), Moscow, 1959; *Engl. transl.* Consultants Bureau, New York, 1961.
[9.21] Y. Tanabe, *Sci. Rep. Res. Inst. Tôhoku Univ.* **A3**, 91 (1951).

[9.22] Y. Toyozawa, *Proc. Int. Conf. Phys. Semicond.* (*Exeter* 1962), p. 104. Inst. Phys. and Phys. Soc., London, 1962.
[9.23] L. W. McKeehan, *Phys. Rev.* **36**, 948 (1930).
[9.24] E. Englert, *Ann. Phys. Leipzig* **14**, 589 (1932).
[9.25] W. Gerlach and K. Schneiderhahn, *Ann. Phys. Leipzig* **6**, 772 (1930).
[9.26] H. H. Potter, *Phil. Mag.* **13**, 233 (1932).
[9.27] Y. Shirakawa, *Sci. Rep. Tôhoku Imp. Univ.* **27**, 485 (1939).
[9.28] K. P. Belov and E. V. Talalaeva, *Zh. Eksp. Teor. Fiz.* **33**, 1517 (1957).
[9.29] T. Suzuki, Y. Matsumura, and E. Hirahara, *J. Phys. Soc. Japan* **21**, 1446 (1966).
[9.30] A. A. Samokhvalov and I. G. Fakidov, *Fiz. Metal. Metalloved.* **4**, 249 (1957).
[9.31] I. G. Fakidov, N. P. Grazhdankina, and I. K. Kikoin, *Dokl. Akad. Nauk SSSR* **68**, 491 (1949).
[9.32] I. G. Fakidov and N. P. Grazhdankina, *Dokl. Akad. Nauk SSSR* **66**, 847 (1949).
[9.33] T. D. Zotov and Ya. S. Shur, *Dokl. Akad. Nauk SSSR* **86**, 267 (1952).
[9.34] S. V. Vonsovskii, *Zh. Tech. Fiz.* **18**, 145 (1948).
[9.35] J. Volger, "Semiconducting Materials" (H. K. Henisch, ed.), p. 162. Butterworths, London and Washington, D.C., 1951.
[9.36] J. Volger, *Physica* **20**, 49 (1954).
[9.37] E. A. Turov and V. G. Shavrov, Ferro i Antiferromagn. Simp. (Krasnoyarsk 1962), *Izv. Akad. Nauk SSSR, Ser. Fiz.* **27**, 1458 (1963).
[9.38] L. D. Landau and E. M. Lifshitz, "Elektrodinamika sploshnyx sred." Gostekhizdat, Moscow and Leningrad, 1957; *Engl. transl.* Addison-Wesley, Reading, Massachusetts, 1960.
[9.39] I. E. Dzyaloshinskii, *Zh. Eksp. Teor. Fiz.* **37**, 881 (1959).
[9.40] B. N. Brockhouse, *J. Chem. Phys.* **21**, 961 (1953).
[9.41] T. R. McGuire, E. J. Scott, and F. H. Grannis, *Phys. Rev.* **98**, 1562 (1955).
[9.42] D. N. Astrov, *Zh. Eksp. Teor. Fiz.* **38**, 984 (1960).
[9.43] D. N. Astrov, *Zh. Eksp. Teor. Fiz.* **40**, 1035 (1961).
[9.44] V. J. Folen, G. T. Rado, and E. W. Stalder, *Phys. Rev. Lett.* **6**, 607 (1961).
[9.45] G. T. Rado and V. J. Folen, *Phys. Rev. Lett.* **7**, 310 (1961).
[9.46] G. T. Rado, *Phys. Rev. Lett.* **6**, 609 (1961).
[9.47] G. T. Rado and V. J. Folen, Magn. Cryst. Conf., I Magnetism (Kyoto 1961), *J. Phys. Soc. Japan suppl. B-I* **17**, 244 (1962).
[9.48] M. Date, J. Kanamori, and M. Tachiki, *J. Phys. Soc. Japan* **16**, 2589 (1961).
[9.49] G. T. Rado and V. J. Folen, *J. Appl. Phys. suppl.* **33**, 1126 (1962).
[9.50] G. T. Rado, *Phys. Rev.* **128**, 2546 (1962).
[9.51] T. Izuyama and G. W. Pratt Jr., Magnetism Conf. (Pittsburgh 1962), *J. Appl. Phys.* **34** (Pt. 2), 1226 (1963).
[9.52] T. H. O'dell, *Phil. Mag.* **10**, 899 (1964).
[9.53] T. H. O'dell, *Phil. Mag.* **13**, 921 (1966).
[9.54] S. Alexander and S. Shtrickman, *Solid State Commun.* **4**, 115 (1966).
[9.55] H. B. Lal, R. Srivastava, and K. G. Srivastava, *Phys. Rev.* **154**, 505 (1967).
[9.56] S. Foner and M. Hanabusa, Magnetism Conf. (Pittsburgh 1962), *J. Appl. Phys.* **34** (Pt. 2), 1246 (1963).
[9.57] B. I. Al'shin and D. N. Astrov, *Zh. Eksp. Teor. Fiz.* **44**, 1195 (1963).
[9.58] J. P. Remeika, *J. Appl. Phys. suppl.* **31**, 263 (1960).
[9.59] E. A. Wood, *Acta Crystallogr.* **13**, 682 (1960).
[9.60] C. H. Nowlin and R. V. Jones, *J. Appl. Phys.* **34**, 1262 (1963).
[9.61] C. H. Nowlin, Thesis, Harvard Univ., Cambridge, Massachusetts, 1963.
[9.62] G. T. Rado, *Phys. Rev. Lett.* **13**, 335 (1964).

[9.63] G. T. Rado, *Proc. Int. Conf. Magn.*, *Nottingham 1964* p. 361. Inst. Phys. Phys. Soc., London, 1964.
[9.64] J. H. Schelleng, *J. Appl. Phys.* **36**, 1024 (1965).
[9.65] E. F. Bertaut, G. Bassi, G. Buisson, J. Chappert, A. Delapalme, R. Pauthenet, H. P. Rebouillat, and R. Aleonard, *J. Phys.* **27**, 433 (1966).
[9.66] G. T. Rado, *J. Appl. Phys.* **37**, 1403 (1966).
[9.67] M. Mercier, *Rev. Phys. Appl.* [*Paris*] **2**, 109 (1967).
[9.68] H. S. Jarrett, P. E. Bierstedt, F. J. Darnell, and M. Sparks, *J. Appl. Phys. suppl.* **32**, 57S (1964).
[9.69] P. E. Bierstedt, *Phys. Rev.* **132**, 669 (1963).
[9.70] H. Futaki, *Jap. J. Appl. Phys.* **4**, 28 (1965).
[9.71] J. P. Suchet, *Bull. Soc. Electr.* (*Paris*) **5**, 274 (1955).
[9.72] C. Haas, A. M. van Run, P. F. Bongers, and W. Albers, *Solid State Commun.* **5**, 657 (1967).
[9.73] E. Dieulesaint, Engr-Dr. thesis, Paris, 1960.
[9.74] E. Ascher, H. Rieder, H. Schmid, and H. Stössel, *Conf. Magn. Magn. Mater.* San Francisco, 1965.

Chapter 10 | Hall Magnetoelectric Effects

10.1. ORDINARY HALL EFFECT

The Hall effect was described briefly in Part One (Section 2.4). We shall deal with it in greater detail, before considering the effect in the magnetic materials. Most of the inspiration for this section comes from Lindberg's digest [10.1]. Readers may also refer to the digest by Pistoulet [10.2], and the work by Putley [10.3].

The Hall effect occurs when a substance carrying an electrical current is the site of a magnetic induction in a perpendicular direction. Let us consider a rectangular parallelepiped with the dimensions a, b, and c in the directions x, y, and z. Let us assume the existence of an induction B along the first, and a current I along the second. An electrical field E_H then appears along the third:

$$E_H = -\operatorname{grad} V_H = RBI/ac \qquad (10.1)$$

This is the Hall effect, and grad V_H refers to the potential gradient along the z direction; R is a coefficient independent of B or I, called the "Hall coefficient." In a homogeneous material, the field E_H produces a voltage V_H, known as the "Hall voltage," and in a diamagnetic material the induction B is equal to the external magnetic field H, so that

$$V_H = -RHI/a \qquad (10.2)$$

where a is the thickness of the sample in the direction of the applied magnetic field.

The origin of the Hall effect lies in the force to which a moving charged particle is subject in a magnetic field (Lorentz force). A curve is produced in its rectilinear trajectory by the field H, so that the charges gather on the sides of the sample, until the effect of the lateral electrical field thus set up counterbalances the deviation caused by the field H (Fig. 10.1). The trace of the electrical equipotential surfaces on the plane of Fig. 10.2 is thus no longer perpendicular to the direction of the current, but revolves by a certain angle, generally small:

$$\theta = \tan \theta = E_H/E_y \tag{10.3}$$

where E_y is the electrical field responsible for the passage of the current I. Taking (10.1) into account, there occurs, in a diamagnetic material

$$\theta \sim RH\sigma \tag{10.4}$$

The force to which a moving particle is subject when equilibrium is attained can therefore be expressed in two different ways. First, Eq. (10.1) gives:

$$eE_H = eRHI/ac \tag{10.5}$$

Second, the force eE_z caused by the lateral electrical field exactly counterbalances the effect of the Lorentz forces, which can be expressed here as:

$$eE_z = eHI/Neac \tag{10.6}$$

where N is the carrier density. By making (10.5) and (10.6) equal, one obtains the value of the Hall coefficient:

$$R = 1/Ne \tag{10.7}$$

Measurement of this coefficient thus makes it possible to attain the density of particles N directly, when the calculations given above are valid, in other words when a large number of free particles (metal or semiconductor) are involved. In fact, a more exact theory produces the formula

$$R = (3\pi/8)/Ne \tag{10.8}$$

In Section 2.4, it was seen that simultaneous knowledge of the coefficient R and the conductivity σ allowed one to obtain a very important component, the mobility of the majority carriers

$$\mu = R\sigma \tag{10.9}$$

so that the Hall angle given in (10.4) may also be written as

$$\theta \sim \mu H \tag{10.10}$$

This equation reveals the link existing between the Hall effect and the carrier mobility.

The equations given above all assume that one of the types of carriers is clearly predominant. If this is not so, the conductivity depends on two terms

10.1. Ordinary Hall Effect

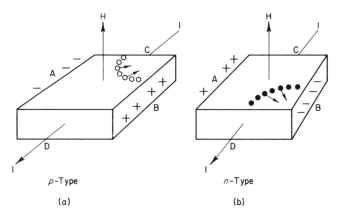

Fig. 10.1. Hall effect caused by a magnetic field H on a current of holes (a) or electrons (b) in a parallelepiped ABCD.

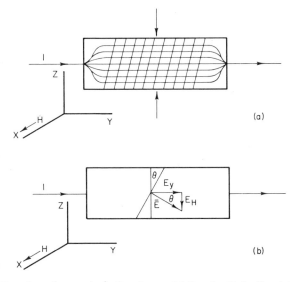

Fig. 10.2. Rotation of an angle θ of equipotential lines by Hall effect (a) and decomposition of the total electric field E resulting from the initial field E_y and Hall field E_H (b) (based on Lindberg [10.1]).

$$\sigma = Ne\mu_n + Pe\mu_p \qquad (10.11)$$

with

$$R = -(3\pi/8)(N\beta^2 - P)/(N\beta - P)^2 e \qquad (10.12)$$

if $\beta = \mu_n/\mu_p$ [10.2]. It usually becomes impossible to obtain N, P, μ_n and μ_p in any simple way. The situation is naturally even more complex if, in certain temperature ranges, the excitation and transfer mechanisms create comparable numbers of carriers with different mobility. Only approximate results can then be obtained, and the greatest caution is necessary.

Lindberg [10.1] has reviewed the various physical effects that occur simultaneously with the Hall effect, making it difficult to measure it. The first is the impossibility of aligning perfectly the two electrodes measuring the Hall voltage (Fig. 10.3a). Even for a zero field, one is measuring a voltage V_0 of ohmic origin, often enormous in relation to V_H. This voltage theoretically depends only on the current passing through the sample and its resistivity. However, as has been pointed out by Okamura and Torizuka [10.4], fluctuations in relation to the magnetic field may be observed if the sample has a high magnetoresistance. But this does not change its sign when the field is reversed. Figure 10.3b shows one way of reducing the parasitic ohmic effect by means of an additional electrode and a potentiometer. The sensibility of the measurement is also reduced.

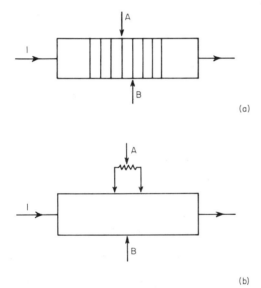

Fig. 10.3. Parasitic ohmic effect upsetting measurement of Hall effect: (a) in the absence of a magnetic field, the electrodes A and B are not on the same equipotential; (b) remedy provided by a third electrode with a potentiometric setup.

Other associated effects are thermal in origin. The Ettingshausen effect consists of the appearance of a temperature gradient along $0z$, caused by the effect of the field H on the perpendicular current I. The Righi–Leduc effect consists of the appearance of the same temperature gradient along $0z$, produced by the effect of the field H on a possible temperature gradient along $0y$ (this gradient always exists, either because of thermal heterogeneity in the measurement area or, more simply, through a Peltier effect produced by the passage of the current I in the connections with the sample). The temperature gradient appearing along z is always accompanied by a parasitic voltage. The Seebeck effect contributes to this, since the electrodes A and B (Fig. 10.3) are no longer at the same temperature, and also the Nernst effect involving the voltage resulting from the effect of the field H on the temperature gradient along y.

By means of four successive measurements, with direct current and field, one can obtain between the electrodes A and B the values V_1 $(+H, +I)$, V_2 $(+H, -I)$, V_3 $(-H, -I)$ and V_4 $(-H, +I)$. One can then write [10.1]:

$$V_H + V_E = (V_1 - V_2 + V_3 - V_4)/4 \qquad (10.13)$$

In the case of diamagnetic semiconductors, the tension V_E resulting from the Ettingshausen effect is low compared with V_H (5% for Ge) and may therefore be ignored. One can also operate with a dc field and an alternating current, with the consequent advantage of eliminating the Ettingshausen effect if the frequency is sufficient. The time for establishment of the temperature gradient is estimated to be 1 min. The voltage V_H is finally given by a complex function involving the sum and difference of the angular velocities corresponding to H and I [10.1].

10.2. EXTRAORDINARY HALL EFFECT

The effect of ferromagnetic materials was discovered by Hall [10.5] in 1881. It was shortly afterwards found to depend on the intensity of magnetization [10.6]. Figure 10.4a and b illustrate the work of Smith [10.7] on Ni, showing the break in the curves at saturation point in the ferromagnetic region, and the marked variation of R in relation to temperature. Smith and Sears [10.8] next showed that there were two terms, proportional to the intensity of magnetization and to the applied field, respectively. In permalloys, V_H increases in relation to the field, reaches a maximum, and then becomes negative. Pugh [10.9, 10.10] has pointed out that the induction should replace the field in the smaller term, so that the Hall voltage is finally expressed, by adopting positive directions so that V_H and BI have the same sign

$$V_H = (R_0 B + R_1 M) I/a \qquad (10.14)$$

where a is always the thickness of the sample in the direction of magnetic induction and R_0 and R_1 are constant in relation to B, M, and I. Webster [10.11] has studied the effect in the different directions of an iron monocrystal, and shown that the second term is isotropic. Finally, Vonsovskii et al. [10.12], Patrakhin [10.13], and other authors have demonstrated theoretically that R_1 varies in the same way as the square of the spontaneous magnetization

$$R_1 = (\sigma_0^2 - \sigma_s^2) \times C^{te} \qquad (10.15)$$

where σ_0 and σ_s refer to this magnetization at 0°K and at the temperature T.

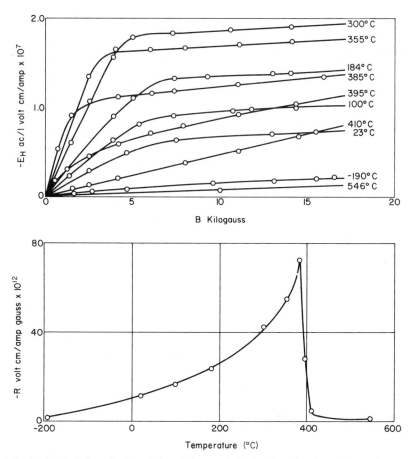

Fig. 10.4. Variations in $E_H ac/I$ in relation to B (a) (top) and R (worked out from the initial slopes of the preceding curves) in relation to temperature (b) (bottom) (based on Smith [10.7], recalculated by Pugh and Rostoker [10.14]).

In a comparative study of various transition metals and their alloys, Pugh and Rostoker [10.14] have discussed the nature of the ordinary ($R_0 BI/a$) and extraordinary ($R_1 MI/a$) effects. Reinterpreting Smith's results in accordance with Eq. (10.14), they have found the coefficient R_0 for Ni of -0.61×10^{-12} V cm A^{-1}Oe^{-1} (-0.61×10^{-4} cm^3 C^{-1}) at room temperature, varying little in relation to temperature, and have pointed out that it comes between those for the nonmagnetic metals next to it in the periodic table: Mn -0.93×10^{-12} and Cu -0.55×10^{-12}. They therefore estimate that the coefficient R_0 of magnetic materials has the same physical significance as the coefficient R for non magnetic materials. This opinion has been generally accepted. For the

10.2. Extraordinary Hall Effect

coefficient R_1, they give the value -74.9×10^{-12} V cm $A^{-1}Oe^{-1}$ (-74.9×10^{-4} cm^3 C^{-1}) for Ni at room temperature (5 times higher towards the Curie point) and point out the impossibility of distinguishing between the ordinary and extraordinary effects in paramagnetic materials, where magnetization is proportional to the field.

It would be possible to continue using Eq. (10.2) in the case of ferromagnetic materials, by bringing in an effective field:

$$H_{eff} = H + 4\pi\alpha M_s \tag{10.16}$$

where M_s refers to the intensity of magnetization at saturation and α to a parameter usually higher than one. The measurements carried out by Jan and Gijsman [10.15] on Ni at low temperatures seem to indicate that α there moves towards the value 2, whereas it reaches nearly 100 at the Curie point. The same authors [10.14] speculate on the possible distinction between two separate coefficients for s and d electrons. Another interpretation is based on the interaction of the spin of a moving electron with the orbits of other electrons [10.16] or with its own [10.17]. Jan and Gijsman [10.15] estimate that the effect thus produced would be more marked when the electron involved is not entirely free, which is generally the case for d electrons.

Akulov and Cheremushkina [10.18] and Karplus and Luttinger [10.19] have extended this point of view and shown that the term R_1, defined as above, contains the square of the resistivity of the sample, to a great extent reflecting its variation in relation to temperature. The extraordinary Hall effect would therefore result from the spin–orbit interaction of the polarized conduction electrons. Smit [10.20, 10.21] discuss the work of Karplus et al., taking into account the asymmetrical dispersal of the conduction electrons by impurities. A theoretical digest by Jan [10.22] concentrates on resiting the Hall effect within the general framework of galvanomagnetic effects. Finally, Maranzana [10.23] has recently put forward a complete mathematical theory which seem to correspond better with experimental results.

Volkov [10.24] has also suggested distinguishing between the Hall coefficients involving magnetization (corresponding to a constant term when saturation is reached) and the paramagnetic processes (corresponding to a term proportional to the field above saturation). Belov and Svirina [10.25] and Samokhvalov and Fakidov [10.26] have used this type of formula to study ferrites and magnetite around the Curie point, where the saturation magnetization M_s is low compared with the magnetization M_p resulting from paramagnetic processes:

$$V_H = (R'_0 H + R'_1 M_s + R_2 M_p) I/a \tag{10.17}$$

The derivation of this equation thus shows a Hall "susceptibility" χ_H and a paramagnetic process "susceptibility" χ_p [10.25]:

$$\delta V_H/\delta H = (R'_0 + R_2 \delta M_p/\delta H) I/a \qquad (10.18)$$

$$\chi_H = (R'_0 + R_2 \chi_p) I/a \qquad (10.19)$$

Figure 10.5 shows how the investigation of a Ni–Zn ferrite allows R'_0 and R_2 to be worked out. One obtains $N = 10^{19}$ cm^{-3} and $\mu_n = 0.08$ cm^2 V^{-1} sec^{-1}.

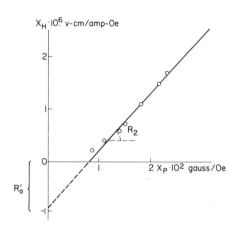

Fig. 10.5. Variation in the Hall "susceptibility" in relation to the "susceptibility" of the paramagnetic processes for a nickel–zinc ferrite (based on Belov and Svirina [10.25]).

It was seen above that, according to the first theories of the extraordinary effect, R_1 was proportional to ρ^2. For many ferromagnetic crystals, experiments produce a proportionality equation in ρ^n, with n slightly less than 2 [10.27]. Samokhvalov and Fakidov [10.26], however, have found an equation $n \sim 4.4$ for Fe$_3$O$_4$ between 0 and 100°C. Lavine [10.28] has defined a parameter $r = R_1/\rho^2$, which he finds to be around -1 to -2 for Ni at room temperature (-2 to -7 at -200°C) and around 1 to 3, more or less independently of temperature, for alloys with 80% nickel. For magnetite and nickel ferrite, Fig. 10.6 shows the complexity of the variation obtained, far removed from the uniformity predicted by Karplus and Luttinger. In addition, there is no explanation of why the sign of R_1 is positive in Fe and Co and negative in Ni, nor why it changes around 400°C in magnetite and nickel ferrite. Neither does there seem to be any correspondence between the signs of R_0 and R_1, which are both positive in Fe and both negative in Ni, but, according to Foner [10.29], $R_0 < 0$ and $R_1 > 1$ in Co at room temperature.

The measurement of the Hall effect in magnetic crystals raises the same problems as those mentioned in the previous section (parasitic ohmic and thermal effects), but the existence of a new term linked to magnetization and thus involving a hysteresis in relation to the applied field requires demagnetization of the sample before each measurement is taken. This should be done in the conventional way, by succeeding reversals and reductions of the applied

10.2. Extraordinary Hall Effect

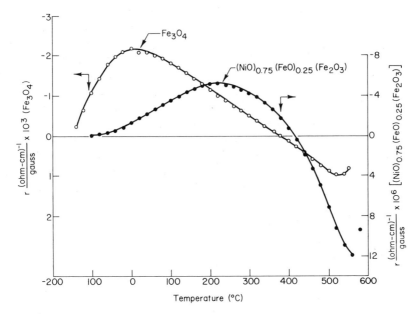

Fig. 10.6. Variation in the parameter $r = R_1/\rho^2$ in relation to temperature, for magnetite and a nickel ferrite (based on Lavine [10.28]).

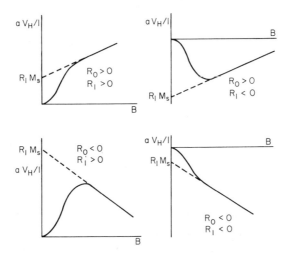

Fig. 10.7. Different possible shapes of curve for aV_H/I in relation to induction (if a is very small compared with the other dimensions, the demagnetizing factor is around 4π and the induction is more or less equal to the applied field).

field, down to zero. Demagnetization by heating to above the Curie point is not generally recommended, because of the thermal gradients which it produces. Figure 10.7 shows the different cases that can arise and the value of the term $R_1 M_s$. The slope of the straight line gives R_0. When saturation can be reached easily, Serre and Suchet [10.30] have proposed a simplified method, in which two measurements are made of each temperature, one at a field H_1, slightly above saturation, and the other at a field $H_2 = H_1 + 4000$ Oe. Each time, the sum of the residual thermoelectric effects in the measurement circuit, the different parasitic effects and the effect resulting from the extraordinary term are compensated for by a countervoltage. The successive inversion of I and H, as in the previous section, gives voltage differences ΔV_1, ΔV_2, ΔV_3, and ΔV_4, from which one obtains, by assuming the Ettingshausen effect to be negligible and R_2 low:

$$R_0 \sim a(\Delta V_1 - \Delta V_2 + \Delta V_3 - \Delta V_4)a/16000I \qquad (10.20)$$

10.3. Experimental Research

The first difficulty in any measurement of the Hall effect is providing the four or five connections. Pressure contacts generally lead to instability, and are really used only at high temperatures. Alloy or electric spark welds have to be adapted to each material, and they involve a fresh investigation each time. In the case of measurements covering a wide temperature range, the expansion coefficients of the connection wire and the crystal obviously have to be roughly the same.

Van der Pauw [10.31] has shown that small connections can be fixed arbitrarily on the perimeter of a thin plate with parallel sides of any shape and of thickness a. Let A, B, C, and D be the four contacts; the resistivity and Hall constant are given by simple equations and, in particular, in terms of the magnetic induction B, one has

$$R_H = [(R_{BD,AC})_{H=0} - (R_{BD,AC})_H]a/B \qquad (10.21)$$

the resistance $R_{BD,AC}$ being defined as the ratio of the potential difference between C and A and the current injected between B and D

$$R_{BD,AC} = (V_C - V_A)/I_{BD} \qquad (10.22)$$

If the surface area of the connections is not negligible, the disturbance involved can be reduced by four slots partly isolating the regions where they occur.

The apparatus needed to measure the Hall effect in magnetic semiconductors is fairly complex, because of the generally low voltages of the ordinary effect, and the presence of a much more marked extraordinary effect. Lavine [10.32, 10.33] has described the apparatus he has used to study magnetite and a ferrite. This has to be capable of detecting approximately 10^{-18} W, with a noise level

10.3. Experimental Research

of around 10^{-9} V. He also considers that the stability of the current in the sample should be 10^{-5}, and that of its temperature 10^{-3}°C, which more or less limited measurements at room temperature. He chose to use an alternating current with a frequency of 1 kHz, and a periodically reversed dc field. Figure 10.8 gives a diagrammatic representation of the electronic appliance used: the first amplifier gives a calibrated signal and a reference phase voltage, the second supplies the sample current and the countervoltages, and the two oscilloscopes are used, respectively, to compare the phases (in order to determine the sign of the output voltage) and to observe the shape of the wave for signals of medium amplitude. Similar apparatus were used later by Fujime et al. [10.34] and other scientists.

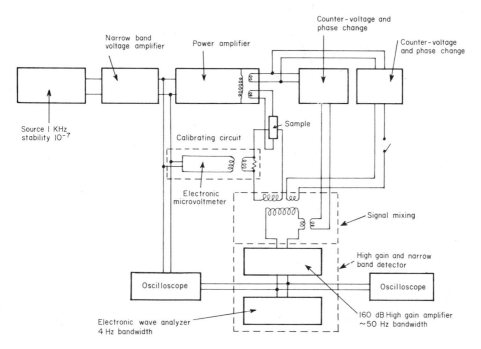

Fig. 10.8. Diagrammatic representation of an apparatus to measure the ordinary Hall effect, using an alternating current in the sample. The field is perpendicular to the plane of the figure (based on Lavine [10.33]).

The use of a direct current in the sample simplifies the apparatus, but introduces other difficulties, because of the parasitic voltages caused by the Seebeck effects in the different measurement circuit welds, and by the impulses induced by slight variations of level in the stabilized supply of the electromagnet. Figure 10.9 gives a diagrammatic representation of the electronic

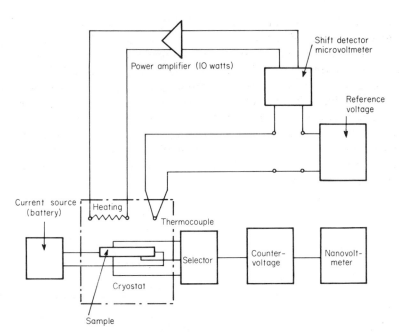

Fig. 10.9. Diagrammatic representation of an apparatus to measure the ordinary Hall effect, using a direct current in the sample. The field is perpendicular to the plane of the figure.

appliance used in the author's laboratory: a selector enables the Hall effect to be studied separately from the magnetoresistance or the resistivity; one counter voltage is supplied by mercury cells, and another, finer, is incorporated in the nanovoltmeter.

The earliest—and most important—experimental research on the Hall effect in magnetic semiconductors involved magnetite and ferrites. Okamura and Torizuka [10.4] used sintered samples of Fe_3O_4 with pressure contacts, and a weak applied field (2400 Oe), and their results should therefore be viewed with caution. In addition, the method of demagnetization is not stated. Foner [10.35] and then Samokhvalov and Fakidov [10.36], provided the first serious results by obtaining $R_H(H)$ curves for a nickel ferrite and magnetite, corresponding, in the diagrams in Fig. 10.7, to $R_0 < 0$ and $R_1 > 0$ (the curve for Fe_3O_4 is reproduced in Fig. 6.20 in Section 6.3). The authors use Eq. (10.16) and quote, for the ferrite, $R_0 = -5 \times 10^{-10}$ V cm $A^{-1}Oe^{-1}$ (-5×10^{-2} cm^3 C^{-1}) and $\alpha \sim -70$, and, for a specimen of magnetite, $R_0 = -2 \times 10^{-2}$ cm^3 C^{-1} and $\alpha \sim +20$. Lavine [10.33] has given these results with greater precision, with, for magnetite, $R_0 \leqslant -1.8 \times 10^{-11}$ V cm $A^{-1}Oe^{-1}$ (-1.8×10^{-3} cm^3 C^{-1}), $\mu \leqslant 0.45$ cm^2 V^{-1}sec^{-1} and $R_1 = -3.28 \times 10^{-11}$ V cm

$A^{-1}Oe^{-1}$ (-3.28×10^{-3} cm^3 C^{-1}) and, for the ferrite Ni$_{0.75}$Fe$_{2.25}$O$_4$, $R_0 = -1.4 \times 10^{-10}$ V cm $A^{-1}Oe^{-1}$, $\mu = 0.0476$ cm^2 V^{-1}sec^{-1} and $R_1 = -16.4 \times 10^{-8}$ V cm $A^{-1}Oe^{-1}$. Samokhvalov and Fakidov [10.26], however, have confirmed for magnetite the positive sign of R_1, which decreases when the temperature rises, matching the negative temperature coefficient of the resistivity. It can therefore be concluded from these results that the conduction is due to approximately 10^{21} electrons cm^{-3}, but the value and sign of R_1 seem to depend very much on the samples.

Experimental scientists also took a fairly early interest in chromium telluride. Kikoin et al. [10.37] have mentioned an exceptional value for the coefficient R_1 in the composition CrTe: -600 to -650×10^{-10} V cm A^{-1} Oe^{-1} (-6.5 cm^3 C^{-1}), 100 times higher in absolute value than those of Fe or Ni for the same standardized temperature values T/T_C. They present curves (probably inverted) of the type $R_0 < 0$ and $R_1 > 0$ in Fig. 10.7, and confirm Eq. (10.15). In Eq. (10.17), the coefficient R_2 would be 32.5×10^{-8} V cm $A^{-1}Oe^{-1}$. The research was pursued later by Nogami [10.38], using polycrystal samples containing 52 and 53% Te atoms. He finds $R_0 \sim 10^{-9}$ m^3 C^{-1} (10^{-3} cm^3 C^{-1}) and $R_1 \sim -10^{-8}$ to -10^{-7} m^3 C^{-1} (-0.01 to -0.1 cm^3 C^{-1}), in other words, for the latter, a much lower value than Kikoin et al. In the paramagnetic region, Nogami has defined the coefficient

$$R_0^* = R_0 + [C/(T-T_C)](R_1 - R_0) \qquad (10.23)$$

where C is the Curie–Weiss law constant $\chi = C/(T-T_C)$. Its value, around -10^{-10} to -10^{-9} m^3 C^{-1}, corresponds to $R_0 > 0$. At high temperatures, however, R_0^* becomes independent of the temperature, while remaining negative. Finally, Serre and Suchet [10.30] have measured R_0 in terms of x in different compositions $\square_x$Cr$_{1-x}$Te and found that the sign is negative for $x < \frac{1}{4}$ and positive for $x > \frac{1}{4}$ (Fig. 10.10). This corresponds to the sign variation for the Seebeck effect [10.39]. A possible explanation was given in Section 7.4.

Several manganese compounds are magnetic. Phosphides containing between 33 and 53% P atoms have been studied by Fakidov and Krasovskii [10.40], who have obtained $R_0 = 3 \times 10^{-4}$ cm^3 C^{-1} for MnP (Fig. 10.11), which corresponds, with the usual hypotheses, to 2×10^{22} holes cm^{-3}, in other words, a transfer density close to the density of the Mn atoms. R_1 is an increasing function of the temperature, corresponding to the metallic character of the resistivity. The antimonide MnSb has been studied by Kikoin et al. [10.41], who have confirmed Eq. (10.15) as well as the resulting equation between R_1 and the magnetoresistance, and then by Nogami et al. [10.42], who have measured, at 15°C, $R \sim 14 \times 10^{-9}$ m^3 C^{-1} (14×10^{-3} cm^3 C^{-1}) and $R_1 \sim 2 \times 10^{-9}$ m^3 C^{-1} (2×10^{-3} cm^3 C^{-1}), and also the variations in these coefficients between -211 and 15°C. As for MnP, a transfer density of 10^{22} cm^{-3} can be worked out, 100 times lower than for ferromagnetic metals.

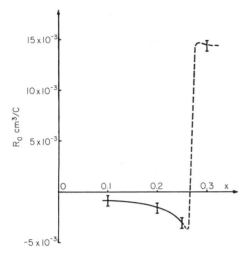

Fig. 10.10. Linear variation in the ordinary Hall coefficient in relation to the composition $\square_x Cr_{1-x}Te$ (based on Serre and Suchet [10.30]).

Fig. 10.11. Linear variation in aV_H/I in relation to H (close to B) for MnP compound (based on Fakidov and Krasovskii [10.40]).

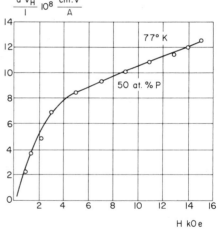

In addition, R_0 increases by a factor of 4 between -100 and $15°C$ (appearance of negative carriers?), while the mobility remains roughly constant, at around 0.02 cm^2 V^{-1}sec^{-1}. Wasscher [10.43] has shown that the Hall effect in the telluride MnTe presents a slight hysteresis below $260°K$, reflecting slight ferrimagnetism. Turov [10.44] had predicted this particularity in antiferromagnetic compounds with the $B8_1$ structure, in which the magnetization of the two sublattices is perpendicular to axis c.

Iron sulfide, finally, has been studied in its almost stoichiometric form FeS by Fujime et al. [10.34]; $R_0 = 7 \times 10^{-5}$ cm^3 C^{-1}, whatever the temperature and crystal direction, corresponding to a transfer density of 8.5×10^{22} cm^{-3}. Figure 10.12 shows the variation in conductivity and mobility of the holes in relation to temperature for the two crystallographic directions, and in particular close to the transformation point $T_\alpha = 138°$C. Theodossiou [10.45], on the other hand, has concentrated on natural pyrrhotite crystals Fe$_7$S$_8$, using a direct current in the sample (Ettingshausen effect assumed to be negligible). The curious sign change found at 165°K in small plates cut perpendicular to axis c (magnetic field parallel to the axis) is reproduced in Fig. 10.13, and the slight variations in the coefficients R_0 and R_1 in plates cut parallel to the axis are shown in Fig. 10.14. No conductivity anisotropy has been observed in these crystals (10^3 mhos cm^{-1}). The hole mobility is 0.7 cm^2 V^{-1}sec^{-1} at 77°K, corresponding to a transfer mechanism, but R_1 is found to be proportional to ρ^7, casting some doubt on the value of the measurements.

In conclusion, the use of Hall effect measurements in studying magnetic semiconductor compounds is beginning to overcome experimental difficulties, and seems likely to provide valuable information on transport phenomena.

10.4. Possible Applications

Although the Hall effect has been known for a long time, its applications are recent. Pearson [10.46] suggested using the effect in a plate of Ge to measure magnetic fields, but the expression "Hall generator" really dates only from the discovery by Welker [10.47] of the properties of the compound InSb and the very high mobility of its carriers. Let us first consider the applications of the effect in diamagnetic compounds such as InAs and InSb, making particular use of the excellent digest by Weiss [10.48]; we shall then say a few words on possible applications of ferro- or ferrimagnetic compounds, which still belong to the field of speculation.

It was seen in Section 10.1 that the sensitivity of V_H to B and I was measured by the ratio R_H/a or $1/Nea$. The material used in a Hall generator will therefore be a very thin plate of a crystal with a low carrier density, i.e., semiconducting. It is also necessary, from a practical viewpoint, to bring in the voltage V which, when applied to the sample, causes the passage of a current I. One obtains [10.49]

$$V_H = R_H \sigma V Bc/b = \mu V Bc/b \tag{10.24}$$

where μ is the mobility of the majority carriers. The available power, proportional to V_H^2, therefore depends on the product $(\mu B)^2$ [10.47]. Magnetic induction does not exceed the applied field, and the capital importance of the mobility is therefore revealed.

In most of the known applications of the Hall effect in diamagnetic crystals, the variable is the induction B (or the applied field). The simplest of them is the

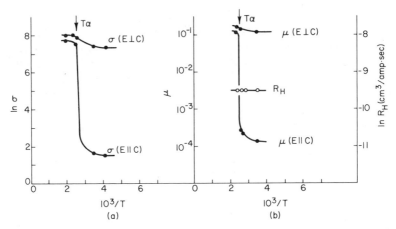

Fig. 10.12. Variations in the conductivity, ordinary Hall coefficient, and hole mobility in the sulfide FeS. The electrical field E is applied perpendicular or parallel to axis c (based on Fujime *et al.* [10.34]).

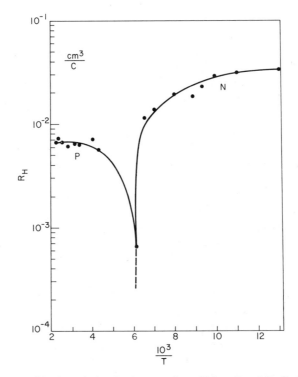

Fig. 10.13. Logarithmic variation in the overall coefficient R_H of Fe_7S_8 in relation to the absolute temperature reciprocal for $H \| c$ and $I \perp c$ with $H = 6850$ Oe (based on Theodossiou [10.45]).

10.4. Possible Applications

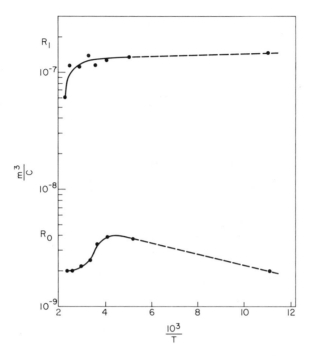

Fig. 10.14. Logarithmic variations in the coefficients R_0 and R_1 of Fe_7Se_8 in relation to the absolute temperature reciprocal for $H \perp c$ (based on Theodossiou [10.45]).

measurement of magnetic fields, but the background noise from the semiconductor means that fields weaker than one millioersted cannot be measured. This is why, in measuring weak fields such as the Earth's field (magnetometers), the lines of force of the field are concentrated by a long μm-metal rod in the middle of which the generator is housed in a gap of 20–30 μm. Using the same principle, and with a generator placed in an open magnetic circuit of soft iron, the changes of position of a small permanent magnet moving parallel to the circuit can be detected. The axis NS is perpendicular to the movement, the flux in the circuit is cancelled out and changes its sign when passing at right angles to the generator [10.50]. Figure 10.15 shows the shape of the signal received. An impulse generator may be designed on the same principle, if a wheel carrying small permanent magnets rotates at high speed in front of the device [10.51].

These examples bring us to the reading of data stored in a magnetic form. It is a case that occurs for bank checks (magnetic figures), mail (coded magnetic addresses), etc. But the most familiar application in this field is the magnetic tape-reading head found, for instance, in a tape recorder. Figure 10.16 illustrates the functioning of a head operated by the longitudinal magnetization of

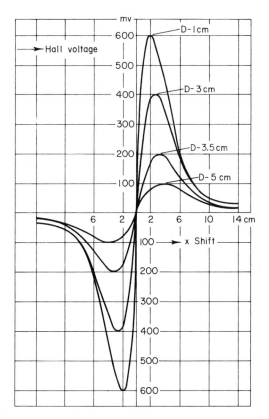

Fig. 10.15. Variation in the voltage supplied by a Hall generator in an open magnetic circuit in relation to the displacement x, parallel to the circuit, of a small permanent magnet. The curves relate to different distances between the system and the trajectory of the magnet (based on Engel et al. [10.50]).

the tape. Heads are produced in which the effective gap does not exceed 10 μm [10.52].

The Hall generator can also be housed in a partly open magnetic circuit containing a permanent magnet. If the design of the circuit allows, the flux passing through the generator can vary considerably when an object made of iron passes nearby. Detectors of this sort have various uses (on production lines, remote-controlled checks, etc.).

The measurements of high-intensity currents by means of the magnetic field they create is familiar. Figure 10.17 shows the principle for these measurements, in which Hall generators are used to measure the flux passing through the magnetic circuit. The dimensions of the circuit and the width of the gaps are calculated so that the soft iron in the circuit is saturated, corresponding to

10.4. Possible Applications

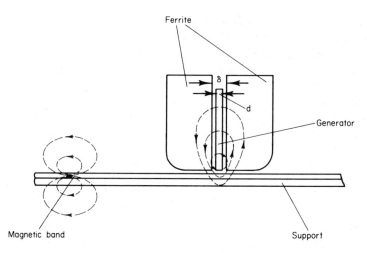

Fig. 10.16. Operation of a magnetic tape-reading head using a Hall generator (based on Weiss [10.48]).

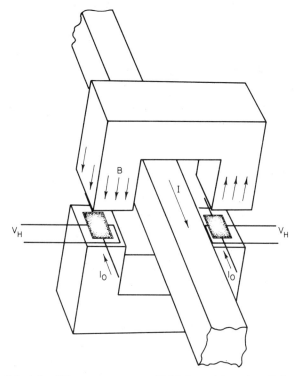

Fig. 10.17. Principle of the measurement of high intensities using Hall generators fed by an auxiliary current I_0 (based on Weiss [10.48]).

a field of around 10^4 Oe in the gap. Saturation means that the measurement is unaffected by the presence of magnetic fields induced by other electrical cables or iron objects nearby. Accuracy is around 0.2% [10.53, 10.54].

Another application in which the variable is the induction consists of the modulation of weak dc currents by an ac field.

Applications in which the variable is the current I are generally based on the nonreciprocal quadrupolar quality of the Hall generator, and involve devices known in electronics under the names of gyrators, isolators, and circulators. The four electrodes may be arranged so that a voltage applied between A and B produces a Hall voltage between C and D, without the reverse applying. One can even imagine a hexagonal plate with six electrodes ABCDEF, in which a current AD would be transferred to BE, a current BE to CF and a current CF to AD.

Finally, the Hall generator has been suggested as a calculating device to multiply two components, B and I, both considered as variables.

Let us now consider the prospects opened up by ferro- or ferrimagnetic semiconductors. If one tries, as in Eq. (10.24), to show up the tension V responsible for the passage of the current I, leaving aside the terms which intervene close to the Curie point and in the paramagnetic region, one finds [10.49]

$$V_H = [R_0 B + R_1 M]V\sigma c/b = R_0 \sigma[B + MR_1/R_0]Vc/b$$
$$V_H = \mu[\mu_0 H + (4\pi - F_d + R_1/R_0)\chi H]Vc/b \qquad (10.25)$$

where μ is the mobility of the majority carriers, μ_0 the permeability of the vacuum and F_d the demagnetizing factor. Let us assume the sample to be very thin ($F_d = 4\pi$) and square ($b = c$):

$$V_H = \mu\mu_0 HV(1 + \chi R_1/R_0) \qquad (10.26)$$

The Hall tension then assumes the particularly simple form:

$$V_H = \mu\mu^* HV \qquad (10.27)$$

assuming an "effective magnetic permeability"

$$\mu^* = \mu_0(1 + \chi R_1/R_0) \qquad (10.28)$$

If the susceptibility is very low (diamagnetism), $\mu^* \sim \mu_0$, and Eq. (10.24) again applies. If it is very high (ferromagnetism),

$$\mu^* \sim \mu_0 \chi R_1/R_0 \sim \mu_0 \mu' R_1/4\pi R_0 \qquad (10.29)$$

where μ' is the relative magnetic permeability of the crystal for the field H. In fact, in known magnetic semiconductors μ^* does not differ from $\mu_0\mu'$ by more than a factor of 10. This argument has left out the phenomenon of magnetoresistance, which, strictly speaking, rules out the proposition that $\sigma = Ne\mu$, because of the intervention of a magnetic scattering term (cf. equation (9.4) in chapter 9). But the temperature range close to the Curie point

10.4. Possible Applications

has been deliberately left out, so that the magnetoresistance is low. In addition, the applications of the Hall effect require a small number of carriers and, in this case, magnetoresistance always remains low.

The author [10.49] has suggested, as a way of assessing the possibilities of each magnetic semiconductor, allotting them a figurative point on the logarithmic graph in Fig. 10.18. If the mobility is measured in $m^2 V^{-1} sec^{-1}$ and the magnetic permeability in $H m^{-1}$, the product $\mu\mu^*$ would constitute a "factor of merit" characteristic of the material and can be measured in $H m V^{-1} sec^{-1}$. The figurative points corresponding to compounds with the same factor of merit will be on the same straight line parallel to the second diagonal. Solid solutions such as GeTe–MnTe [10.55] have a very low μ^*. Magnetic metals (Fe, Co, Ni) and pseudometals (MnAs, MnSb, CrTe) have a low μ and ferrites a very low μ. R_1/R_0 is 100 for Ni [10.14] and approximately 120 for a Ni–Zn ferrite [10.25]. In this last case, one finds $\log \mu\mu^* \sim -7.4$. In order to compete with InSb, a magnetic semiconductor would have to have a factor $\mu\mu^*$ close to 10^{-6} or even 10^{-5} $H m V^{-1} sec^{-1}$, which in fact more or less requires an excitation mechanism. Thiospinel-type materials are in this respect more promising.

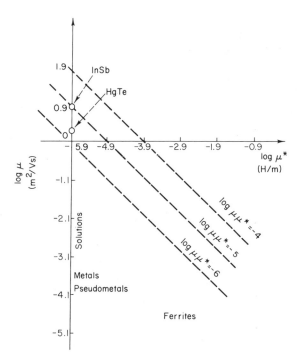

Fig. 10.18. Doubly logarithmic graph showing the mobility of the majority carriers in relation to the effective magnetic permeability (based on Suchet [10.49]).

REFERENCES

[10.1] O. Lindberg, *Proc. IRE* **40**, 1414 (1952).
[10.2] B. Pistoulet, *Onde Elec.* **35**, 71 (1955).
[10.3] E. H. Putley, "The Hall Effect and Related Phenomena." Butterworths, London and Washington, D.C., 1960.
[10.4] T. Okamura and Y. Torizuka, *Sci. Rep. Res. Inst. Tôhoku Univ.*, Ser. A. **2**, 352 (1950).
[10.5] E. H. Hall, *Phil. Mag.* **12**, 157 (1881).
[10.6] A. Kundt, *Wied. Ann.* **49**, 257 (1893).
[10.7] A. W. Smith, *Phys. Rev.* **30**, 1 (1910).
[10.8] A. W. Smith and R. W. Sears, *Phys. Rev.* **34**, 1466 (1929).
[10.9] E. M. Pugh, *Phys. Rev.* **32**, 824 (1928).
[10.10] E. M. Pugh, *Phys. Rev.* **36**, 1503 (1930).
[10.11] W. L. Webster, *Proc. Cambridge Phil. Soc.* **23**, 800 (1925).
[10.12] S. V. Vonsovskii, L. Ya. Kovelev, and K. P. Rodionov, *Izv. Akad. Nauk SSSR, Ser. Fiz.* **16**, 560 (1952).
[10.13] N. P. Patrakhin, *Izv. Akad. Nauk SSSR, Ser Fiz.* **16**, 584 (1952).
[10.14] E. M. Pugh and N. Rostoker, *Rev. Mod. Phys.* **25**, 151 (1953).
[10.15] J. P. Jan and H. M. Gijsman, *Physica* **5**, 277 (1952).
[10.16] V. Rudnitsky, *Zh. Eksp. Teor. Fiz.* **9**, 262 (1939).
[10.17] A. G. Samoilovich and B. L. Konkov, *Zh. Eksper. Teor. Fiz.* **20**, 783 (1950).
[10.18] N. S. Akulov and A. B. Cheremushkina, *Dokl. Akad. Nauk SSSR* **98**, 1, 35 (1954).
[10.19] R. Karplus and J. M. Luttinger, *Phys. Rev.* **95**, 1154 (1954).
[10.20] J. Smit, *Physica* **21**, 877 (1955).
[10.21] J. Smit, *Physica* **24**, 39 (1958).
[10.22] J. P. Jan, *Solid State Phys.* **5**, 66 (1957).
[10.23] F. E. Maranzana, *Phys. Rev.* **160**, 421 (1967).
[10.24] D. I. Volkov, *Vestn. Mosk. Gos. Univ.* **3**, 567 (1954).
[10.25] K. P. Belov and E. P. Svirina, *Zh. Eksp. Teor. Fiz.* **37**, 1212 (1959).
[10.26] A. A. Samokhvalov and I. G. Fakidov, *Fiz. Tverd. Tela* **2**, 414 (1960).
[10.27] E. M. Pugh *Phys. Rev.* **97**, 647 (1955).
[10.28] J. M. Lavine, *Phys. Rev.* **123**, 1273 (1961).
[10.29] S. Foner, *Phys. Rev.* **91**, 20 (1953).
[10.30] J. Serre and J. P. Suchet, *C.R. Acad. Sci. Paris* **264B**, 1412 (1967).
[10.31] L. J. van Der Pauw, *Philips Res. Rep.* **13**, 1 (1958).
[10.32] J. M. Lavine, *Rev. Sci. Instr.* **29**, 970 (1958).
[10.33] J. M. Lavine, *Phys. Rev.* **114**, 482 (1959).
[10.34] S. Fujime, M. Murakami, and E. Hirahara, *J. Phys. Soc. Japan* **16**, 183 (1961).
[10.35] S. Foner, *Phys. Rev.* **88**, 955 (1953).
[10.36] A. A. Samokhvalov and I. G. Fakidov, *Fiz. Metal. Metalloved.* **4**, 249 (1957).
[10.37] I. K. Kikoin, E. M. Buryak and Yu. A. Muromkin, *Dokl. Akad. Nauk SSSR* **125**, 1011 (1959).
[10.38] M. Nogami, *Jap. J. Appl. Phys.* **5**, 134 (1966).
[10.39] W. Albers and C. Haas, *Congr. Int. Phys. Semicond., Paris, 1964* p. 1261. Dunod, Paris, 1964.
[10.40] I. G. Fakidov and V. P. Krasovskii. *Fiz. Metal. Metalloved.* **7**, 302 (1959).
[10.41] I. K. Kikoin, N. A. Babushkina, and T. N. Igosheva, *Fiz. Metal. Metalloved.* **10**, 488 (1960).
[10.42] M. Nogami, M. Sekinobu, and H. Doi, *Jap. J. Appl. Phys.* **3**, 572 (1964).
[10.43] J. D. Wasscher, *Solid State Commun.* **3**, 169 (1965).

[10.44] E. A. Turov, *Zh. Eksp. Teor. Fiz.* **42**, 1582 (1962).
[10.45] A. Theodossiou, *Phys. Rev.* **137A**, 1321 (1965).
[10.46] G. L. Pearson, *Rev. Sci. Instr.* **19**, 263 (1948).
[10.47] H. Welker, *Elektrotechn. Z.* **76A**, 513 (1955).
[10.48] H. Weiss, *Solid State Electron.* **7**, 279 (1964).
[10.49] J. P. Suchet, *C.R. Acad. Sci. Paris* **262B**, 127 (1966).
[10.50] W. Engel, F. Kuhrt, and H. J. Lippmann, *Elektrotechn. Z.* **81A**, 323 (1960).
[10.51] J. Brunner, *Siemens Z. (All.)* **36**, 521 (1962).
[10.52] F. Kuhrt, G. Stark, and F. Wolf, *Elektron. Rundsch.* **13**, 407 (1959).
[10.53] F. Kuhrt and K. Maaz, *Elektrotechn. Z.* **77**, 487 (1956).
[10.54] K. Maaz and R. Schmid, *Elektrotechn. Z.* **78**, 734 (1957).
[10.55] M. Chomentowski, H. Rodot, G. Villers, and M. Rodot, *C.R. Acad. Sci. Paris* **261**, 2198 (1965).

Chapter 11 **Electro- and Magnetooptical Effects**

11.1. Forced Birefringencies

An isotropic transparent medium can become anisotropic under an external influence. For instance, a block of glass or a cubic crystal, compressed along the x-direction assumes the characteristics of a uniaxial crystal of which the ellipsoid of the indices would have a symmetry of revolution around $0x$. The properties of the medium under an external influence are thus defined entirely by knowledge of the ordinary and extraordinary indices n_0 and n_e, relating to a given direction of propagation.

For theories of crystal optics, readers may refer to educational books such as those by Bruhat [11.1] or Fleury and Mathieu [11.2]. Light vibrations contained in the wave plane (perpendicular to the direction of propagation) correspond, in a given medium, to perfectly determined indices n_1, n_2, and n_3 along the three coordinate axes, defining the ellipsoid of the indices by the equation

$$x^2/n_1^2 + y^2/n_2^2 + z^2/n_3^2 = 1 \tag{11.1}$$

The directions of the light rays coming from a given incident ray are calculated from the ellipsis of intersection between the wave plane and the ellipsoid.

The basic property of an anisotropic medium is that any incident ray generally corresponds to two transmitted rays, known as ordinary and extraordinary. This is the *birefringency* phenomenon. The indices n_0 and n_e, referring to their direction of propagation, correspond, respectively, to the indices n_2 and $n_3 < n < n_1$, which refer to the vibrations of the wave plane. Since these indices are different, the result is that during the passage through the

11.1. Forced Birefringencies

anisotropic medium, the corresponding waves are propagated at different velocities, and on leaving the medium present a certain path difference which depends on the length l of the medium:

$$\delta = (n_e - n_0)l \qquad (11.2)$$

This corresponds to a phase difference in the waves

$$2\varphi = 2\pi\delta/\lambda \qquad (11.3)$$

where λ is the wavelength of the monochromatic light used. If a polarizer has been placed on the light path, in front of the uniaxial medium, at 45° to the optical axis E, the vibration at the outlet of the medium will be elliptical (Fig. 11.1) with one of the axes in the direction OP and the ratio of the axes:

$$b/a = \tan \varphi \qquad (11.4)$$

Measurement of the lag δ therefore simply involves measuring the ratio of the axes of an elliptical vibration of known direction. This vibration is received on a quarter-wave plate, a neutral line of which coincides with direction OP of the original vibration. A rectilinear vibration OR is thus reestablished, and the angle φ with OP is measured by means of a penumbra analyzer [11.1].

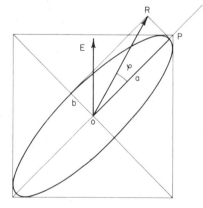

Fig. 11.1. Elliptical vibration obtained from a rectilinearly polarized vibration OP, after passing through an anisotropic medium, the optical axis E of which forms an angle of 45° with OP. [Based on Bruhat, "Cours de Physique Générale: Optique." Masson, Paris, 1959.]

Electrooptical effects have recently been covered in the digest by Billard [11.3], which also deals with their application to modulation. In 1875, Kerr discovered that most liquids, placed in an electrical field E, acquire the properties of a uniaxial crystal whose optical axis coincides with the direction of the field. The effect is generally observed on a light ray perpendicular to the field. Experiments show that the difference of the indices is proportional to the square of the field

$$n_e - n_0 = K\lambda E^2 \qquad (11.5)$$

where K is the Kerr constant, often positive and around 10^{-7} esu. From this, one obtains

$$\varphi = KlE^2 \tag{11.6}$$

It is accepted that this effect results from a directional influence exerted by the electrical field on molecules of the liquid, so that the value of the Kerr constant corresponds to the anisotropy of the molecule. The theory of molecular orientation provides that the sum of the three main indices $(n_e + 2n_0)$ is unaltered by the field, meaning that

$$(n_e - n)/(n_0 - n) = -2 \tag{11.7}$$

This estimate is fairly well confirmed by experiment. The time of establishment of the birefringency is around 10^{-8} sec.

A few years later, Pockels [11.4] showed that the accidental birefringency of piezoelectric crystals such as $NaClO_3$, α-quartz, and Seignette salt, when submitted to an electrical field, is not limited to that involved in the simultaneous mechanical distortions. Some crystals like RbH_2PO_4 [11.5] even possess a high direct electrooptical effect, corresponding to Eq. (11.5). If the frequency is very high, mechanical distortions of the crystal cannot follow the rapid variation in the field, and the birefringency resulting from all the piezoelectric and photoelastic effects is negligible compared with that caused by the direct electrooptical effect, due mainly to the displacement of the ions [11.6]. Recent work, which we shall return to in Section 11.4, concern ferroelectric compounds, potassium diphosphate (KDP) and ammonium diphosphate (ADP) [11.7].

Magnetooptical birefringency effects were discovered by Cotton and Mouton in liquids such as nitrobenzene, and by Voight in gases, shortly after the Kerr electrooptical effect. The magnetic birefringency complies with the law, similar to (11.5)

$$n_e - n_0 = C\lambda B^2 \tag{11.8}$$

where C is the Cotton–Mouton constant, around 10^{-12} emu in nitrobenzene, and B the magnetic induction in the medium in question. Here too, the effect is observed on a ray perpendicular to the field. From it, one obtains an expression, similar to (11.6), for the phase difference between the two vibrations. This phenomenon is explained, like electrical birefringency, by the molecular orientation theory. It has also been studied in paramagnetic liquids such as solutions of rare earth salts or in oxygen at $90°K$. Recently, Suits *et al.* [11.8] have observed an intense effect in monocrystals of the paramagnetic compound EuF_2. Some mesomorphous substances such as azoxyanisol have a marked magnetic anisotropy, which confers an exceptionally high magnetic birefringency on them.

11.2. Faraday Effect

Ferromagnetic transparent media can naturally give an even higher birefringency. The effect was observed by Majorana in colloidal suspensions of $Fe(OH)_3$, even before Cotton and Mouton carried out their work. It has been observed in particular by Dillon [11.9] in the garnet $Y_3Fe_5O_{12}$, and then more recently by Suits *et al.* [11.8, 11.10] on the metamagnetic compound EuSe at 4.2°K. Figure 11.2 shows that the law prevailing for high fields, in other words in the ferromagnetic domain, is one in B^2. We shall return to these points in Section 11.3.

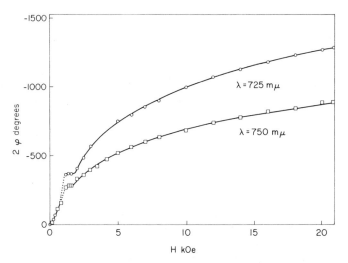

Fig. 11.2. Linear variation in the phase difference between ordinary and extraordinary rays at the outlet of a EuSe crystal of 157 μm in relation to the applied field (based on Suits and Argyle [11.10]).

11.2. FARADAY EFFECT

In 1846, Faraday discovered that an anisotropic transparent medium, placed in a magnetic field, produced a rotation of the polarization plane of a light ray passing through parallel to the field direction. The rotatory power of a medium has been interpreted by Fresnel as a difference in absorption for the right-hand and left-hand circular vibrations (circular dichroism) [11.1]. The propagation velocities v and v' correspond to indices $n = c/v$ and $n' = c/v'$, introducing a path difference, after passing through a length l of the medium

$$\delta = (n' - n)l \tag{11.9}$$

and a phase difference in the waves 2φ like in (11.3). The angle of rotation is given by

$$\varphi = V_R l B \tag{11.10}$$

where V_R is the Verdet constant, often around 0.01 to 0.1 minutes of arc per centimeter-gauss, and B the magnetic induction in the medium in question. The rotation changes direction at the same time as the field, so that a light ray passing back through the medium after reflection presents a rotation 2φ. This is an important difference from natural rotatory polarization, for which the same experiment would produce zero total rotation. In the latter case, a light ray, whatever its direction of propagation, meets with the same influences bound up with the special molecular structure. The direction of rotation is therefore related to the direction of propagation, and changes for a fixed observer, when passing through the medium and back.

The Verdet constant is positive in diamagnetic solids and liquids such as glass, water or the carbon sulfide CS_2 (0.042 min cm^{-1} G^{-1}). Its variation with the wavelength is symmetrical in relation to the absorption bands. In contrast, it is negative and fairly high in absolute value in most paramagnetic media, notably solutions of rare earth salts. The monocrystal europium selenide EuSe at room temperature (paramagnetic) reaches -9.7 min cm^{-1} G^{-1} for $\lambda = 0.67$ μm, for instance. Its variation with the wavelength is thus asymmetrical in relation to the absorption bands. Finally, it reaches enormous levels in ferromagnetic substances such as Fe, Co, or Ni in very thin layers [10.2]. The same is true in the semiconductor crystals which we are particularly interested in, such as the garnet $Y_3Fe_5O_{12}$.

The relationship existing between the wavelengths in which the Faraday rotation is high and those producing resonance absorption has been pointed out by Kastler [11.11] and other authors. Dillon [11.9] was the first to mention the resemblance between the optical absorption and rotatory power curves for the garnet $Y_3Fe_5O_{12}$ (Fig. 11.3). It will be seen later that the relation between absorption and rotatory power has also been observed in other magnetic compounds. This resemblance seems to suggest that the electronic transitions that lie at the origin of the absorption peaks concern the electrons responsible for the magnetism, in other words those in the levels 3d (Fe(III), Cr(III)) and 4f (Eu(II)). It is also known that in general the influence of the electronic transitions causes a variation in the constant V_R in relation to the wavelength in $1/\lambda^2$ in diamagnetic semiconductors [11.12]. It also seems to apply to paramagnetic compounds. Suits et al. [11.8] have found for EuF_2 the equation

$$V_R = C^{te}/(\lambda_t^2 - \lambda^2) \tag{11.11}$$

where λ_t is a characteristic wavelength of the ion Eu^{2+}.

The magnetic rotatory power is not linked to any special molecular structure, like the natural rotatory power. It is a very widespread phenomenon, the theory for which is connected to that of diamagnetism. Consideration of the Laplace forces exerted on each electron describing an orbit (assumed to be circular and perpendicular to the propagation direction $0x$ of the wave)

11.2. Faraday Effect

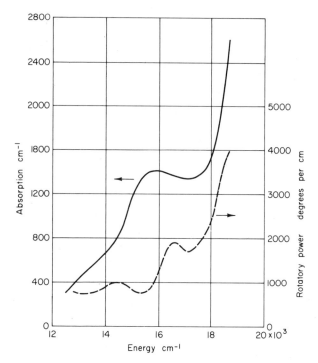

Fig. 11.3. Linear variations in the absorption coefficient (Beer–Lambert law) and specific rotatory power (Verdet law) of yttrium and iron garnet (based on Dillon [11.9]).

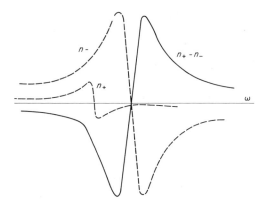

Fig. 11.4. Diagrammatic variation in the indices n_+, n_-, and their difference in relation to the angular velocity ω of the vector E of the circular vibration in a paramagnetic medium ($n_+ < n_-$) (based on Fleury and Mathieu [11.2]).

shows that the originally isotropic medium acquires two indices n_+ and n_- if it is placed in the magnetic induction field parallel to Ox. These refer, respectively, to circular vibrations moving in the same direction as the magnetizing current (right-hand) and those moving in the opposite direction (left-hand). It is shown that the rotation occurs in the same direction as the magnetizing current, and that the angle may be expressed as follows:

$$\varphi = \pi(n_+ - n_-)l/\lambda \qquad (11.12)$$

The existence of an atomic magnetic moment in paramagnetic media increases the number of electrons rotating in the same direction as the magnetizing current and, consequently, the ratio n_-/n_+. The difference $n_+ - n_-$ may therefore turn out to be negative at low frequencies. By this theory, one finds symmetric variations (diamagnetic bodies) and asymmetrical variations (paramagnetic bodies) in terms of the wavelength near an absorption band. Figure 11.4 gives a diagrammatic representation of the variation of $n_+ - n_-$ in the second case ($n_+ < n_-$); ω is the angular velocity at which the field E turns in the circularly polarized wave [11.2]. For a rigorous mathematical theory of the effect, readers may refer to Ballhausen's [11.13] book, and in particular, Smith's digest [11.14].

In recent years, research has concentrated on the compounds of europium (EuO: $T_C = 69°K$, EuSe and Eu$_2$SiO$_4$: $T_C = 7°K$) and chromium trihalogenides (CrI$_3$: $T_C = 68°K$, CrBr$_3$: $T_C = 32.5°K$, and CrCl$_3$: $T_C = 16.8°K$). The results will be given in Section 11.3. Frequently 4f–5d transitions have been suggested in the case of europium.

Until now, little attention has been given to the problem of transparency, which is fundamental to the Cotton–Mouton and Faraday effects. It is known that a semiconductor is transparent only for wavelengths above the absorption threshold λ_s, corresponding to a forbidden band height of $1.25/\lambda_s$ eV. In fact, this transparency is maximum only in the wavelength band immediately following the threshold (Fig. 11.5). At higher wavelengths, the absorption begins to increase again, more or less rapidly depending on the purity (and perfection) of the crystal, and absorption peaks are then found, corresponding to the crystal's own vibrations. There therefore exists a relation between the height of the forbidden band of a magnetic semiconductor and the approximate wavelength at which its magnetooptical effects by transparency can be easily observed.

Kerr had observed another effect involving the rotation of the polarization plane of the incident wave. This is the magnetooptical effect which bears his name, caused by the reflection on the polished surface of a magnetized piece of iron. It is known that reflection on an insulating surface such as glass introduces a phase difference equal to π. Reflection on a metallic surface, on the other hand, introduces a complex phase difference, the value of which differs for the

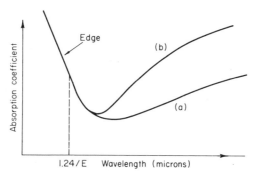

Fig. 11.5. Variation in the optical absorption in relation to the wavelength of a pure (a) or impure (b) semiconductor crystal.

two inverse circular vibrations carried. This effect is sensitive to the magnetization of the reflecting surface and the rotation attains approximately 20 minutes for saturated Fe [11.1]. The phenomenon has been studied, for instance, by Krinchik and Nuralieva [11.15] in Ni, and by Greiner and Fan [11.16] in EuO and EuS compounds.

11.3. Experimental Research

This section will be confined to magnetic birefringency (Cotton–Mouton effect) and the magnetic rotatory power (Faraday and Kerr effects), since the Pockels effect has not yet been observed in transition metals or rare earth compounds.

Dillon [11.9] has studied the magnetic birefringency of a plate of garnet $Y_3Fe_5O_{12}$, 2.5 μm thick parallel to the (110) plane, and has mentioned a possible observation of the magnetic domains. Suits et al. [11.10] have studied a plate of EuSe 157 μm thick submerged in liquid helium, and lit by a monochromator with a chopper at 150 Hz and a photomultiplier. The effect of the field was shown in Fig. 11.2. Suits et al. [11.8] have also studied EuF_2 which, although paramagnetic, presents a fairly high effect.

Shen and Bloembergen [11.17], in connection with the study of the magnetic rotatory power of monocrystals of CaF_2 doped with various rare earth ions, describe the apparatus used for their measurements (Fig. 11.6). At the lower end of the Dewar flask, the light ray passes through four windows and two pieces of glass, submerged in nitrogen and liquid helium, respectively, before reaching the sample. It is polarized by a Polaroid sheet placed immediately beneath the sample. The top of the inside tube of the Dewar, filled with He, is closed by a window that is strainless at room temperature. This arrangement reduces to a minimum the depolarization resulting from the multiple passages

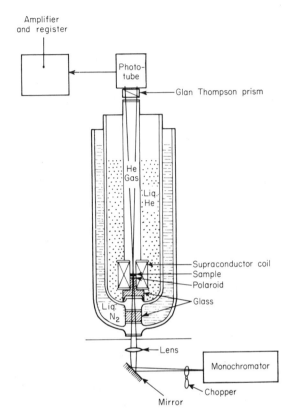

Fig. 11.6. Apparatus used to study the Faraday effect at low temperatures (based on Shen and Bloembergen [11.17]).

through windows and liquefied gases. An intense magnetic field (up to 45 kOe), parallel to the light rays, is obtained by means of a supraconductor winding. Source consists of a tungsten-filament lamp and a monochromator centered on 0.5 μm. The beam emitted is interrupted and the current detected by the photomultiplier amplified before observation. The accuracy of the measurement is estimated at a few minutes of arc.

More recently, in connection with similar work, Boccara [11.18] has described a slightly different apparatus, designed to optimize the signal/noise ratio. He places a fairly powerful field generator in a compact optical assembly (Fig. 11.7). A small permanent magnet produces 3–4 kOe in a 22-mm gap. The polar parts are pierced, since the return of the optical beam at right angles by two mirrors would have involved a Kerr magnetooptical effect of the same order as the rotation to be measured. The introduction of this magnet in the actual tail of the cryostat allows one to reduce, by an important factor, the

11.3. Experimental Research

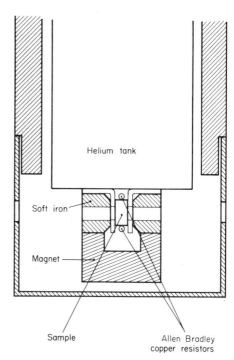

Fig. 11.7. Apparatus used to study the Faraday effect at low temperatures (based on Boccara [11.18]).

geometrical extent of the spectrometer, and the signal/noise ratio is finally comparable to what would be obtained with an ordinary electromagnet. The magnet measures $26 \times 24 \times 22$ cm, and is hard-soldered onto the tail of the cryostat with an alloy that melts at 100°C. The sample is cooled by two plates of copper soldered to the central tank. Two Allen–Bradley resistors glued to it measure its temperature, which is uniform to within 0.2°K.

Dillon [11.9] has measured the Faraday rotation of the garnet $Y_3Fe_5O_{12}$, and has used the effect to observe the magnetic domains. Kooy [11.19] has done the same for the magnetobaryte $BaFe_{12}O_{19}$, and Sherwood et al. [11.20] for the ferrite $Li_{0.5}Fe_{2.5}O_4$, the magnetoplombite $PbFe_{12}O_{19}$, and compounds with a perovskite structure $LFeO_3$. Shafer et al. [11.21] have studied the orthosilicate Eu_2SiO_4, orthorhombic in structure, and found $V_R = 2.5$ min $Oe^{-1}cm^{-1}$ at 20°C for a wavelength of around 2.55 μm. Much higher values are expected at 4.2°K, in the ferromagnetic domain. Le Craw et al. [11.22] have studied optical absorption in specially pure crystals of ferrite $MgFe_2O_4$ and garnet $Y_3Fe_5O_{12}$ (Fig. 11.8), and defined their factor of merit as the ratio between the rotatory power and the weakening of the light in decibels (Fig. 11.9). Suits et al. [11.8] have studied the paramagnetic EuF_2, the Verdet constant of which decreases with the energy of the photons, very much like the

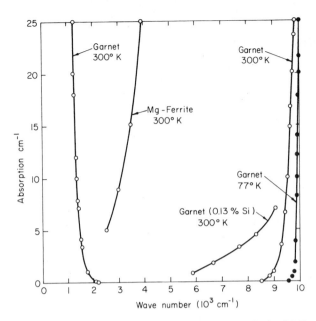

Fig. 11.8. Linear variation in the absorption coefficient of the ferrite MgFe$_2$O$_4$ and the garnet Y$_3$Fe$_5$O$_{12}$ in relation to the energy of photons in the infrared range. The coefficient is $\leqslant 0.03$ between 2500 and 9000 cm^{-1} (based on Le Craw *et al.* [11.22]).

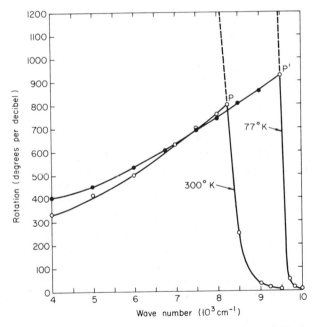

Fig. 11.9. Linear variation in the factor of merit of the garnet Y$_3$Fe$_5$O$_{12}$ in relation to the energy of the photons in the infrared range. The dots to the left of P and P' indicate only a lower limit. The probable curve is shown as a dotted line (based on Le Craw *et al.* [11.22]).

11.3. Experimental Research

absorption coefficient, a glass containing 30.5% EuO molecules, also paramagnetic at atmospheric temperature, and the compound EuSe, the factor of merit of which at 4.2°K is of the same order as that of garnet at room temperature.

Dillon et al. [11.23] have studied a completely different class of compounds, chromium trihalogenides. Since transition metal halogenides were not included in Part Two of this book, because of lack of information on the semiconductibility, Fig. 11.10 shows the structure in which $CrBr_3$, $CrCl_3$ at low

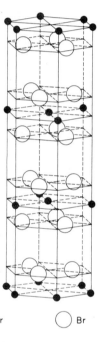

Fig. 11.10. Representation of the structure of $CrBr_3$ (based on Tsubokawa [11.24]).

● Cr   ◯ Br

temperature, and probably CrI_3 crystallize. It is a layer structure in which the Cr—X bonds are markedly covalent [11.25]. CrI_3 and $CrBr_3$ are ferromagnetic and $CrCl_3$ metamagnetic at low temperatures. Figure 11.11 gives the variation found by Dillon et al. in the absorption coefficient in relation to the energy of the photons near the main thresholds, as well as the variation in the "specific magnetic rotation" in degrees per centimeter (without indicating the corresponding induction value). They also interpret the various absorption peaks of $CrCl_3$ and $CrBr_3$ in terms of the assumed MO energy diagram in Fig. 11.12, where the transitions (1), (2), and (3) correspond in turn to "internal" processes (3d–4p in the Cr atom), and "σ" and "π" processes (shifts from Br to Cr from a B or AB t_{1u} orbital). This diagram assumes that the Cr atom is in the field of six Br neighbors, forming a perfect octahedron.

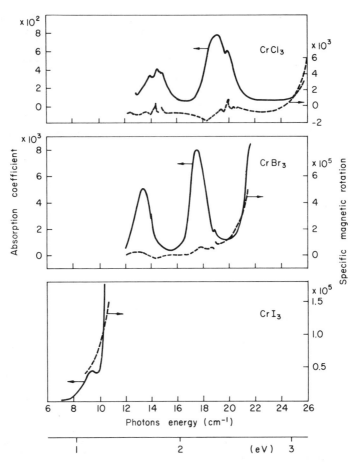

Fig. 11.11. Linear variations in the absorption coefficient (solid line) and specific rotation (broken line) of chromium trihalogenides (based on Dillon et al. [11.23]). The article does not indicate the value of the induction in the sample.

Table 11.1 recapitulates the factors of merit defined by Le Craw et al. [11.22], as mentioned by different authors for some ferro- or ferrimagnetic materials. The value of iron is given for comparison. Scientific literature often mentions "specific" rotations of several thousands of degrees per centimeter, but the magnetic induction (probably saturation induction) is not stated.

Greiner and Fan [11.16] have studied the Kerr magnetooptical effect on cleavage surfaces of EuO and EuS monocrystals at the temperature of liquid He, in other words insulating, but ferromagnetic surfaces. The polarization plane was more or less the same as the incidence plane, the angle of incidence of 30° and the field in the plane of the sample 2500 Oe (longitudinal effect).

11.4 Possible Applications

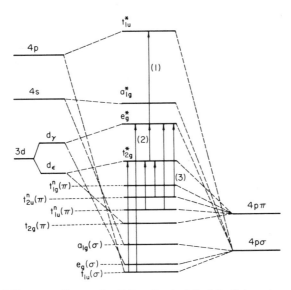

Fig. 11.12. MO energy diagram for CrBr$_3$. On the left, AO of chromium; in the middle, MO of CrBr$_3$; on the right, MOs of the 6 bromine neighbors (based on Dillon *et al.* [11.23]).

TABLE 11.1

Compound	Factor of merit (degrees/dB)	Temperature (°K)	Wavelength (μm)	Wave number (cm^{-1})	Author
Y$_3$Fe$_5$O$_{12}$	925	77	1.04	9600	[11.22]
EuSe	850	4.2	0.755	13,200	[11.8]
Y$_3$Fe$_5$O$_{12}$	800	300	1.2	8300	[11.22]
CrBr$_3$	30	1.5	0.493	20,300	[11.22]
Fe	0.7	300	1	10,000	[11.26]

The Kerr rotation, shown in Fig. 11.13, is of a higher scale than that for ferromagnetic metals. From these values, the authors have calculated the diagonal dielectric tensor factor, and have found that this reaches almost 10^{-2}, compared with 10^{-3} for Fe. They consider that the Kerr resonance observed corresponds to the 4f–5d transition in the Eu atom.

11.4. POSSIBLE APPLICATIONS

The main, in fact almost the only application of electro- and magnetooptical effects is the modulation of visible or infrared light, and of laser beams in particular, with a view to remote data communication.

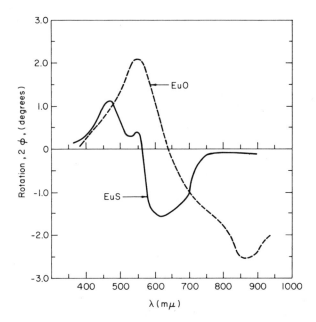

Fig. 11.13. Linear variation in the Kerr longitudinal rotation of EuO (12°K) and EuS (8°K) in relation to the wavelength (based on Greiner and Fan [11.16]).

Electrooptical modulators have recently been dealt with in an excellent digest by Billard [11.3], from which the summary below has been taken. The first references to Pockels-effect modulators date back to 1948 [11.27, 11.28]. Study of their characteristics shows the weakness of the usable opening [11.29], and research turned to cubic crystals. Frequency modulation reached 3.15 GHz in 1957 [11.30], 9.25 GHz in 1961 [11.31], and could go up to 36 GHz [11.32]. The wave band can reach 1 GHz [11.33]. Suggestions were made in turn for the use of a large number of crystals presenting high electrooptical coefficients. One might mention, for instance, Seignette salt, which would allow modulation in the infrared range [11.34].

Different types of modulators may be imagined. In the one shown in Fig. 11.14, described by Buhrer [11.35], the difference in the transit times of the two privileged vibrations of a fairly thick plate of a birefringent crystal is used to convert a polarization state modulation into an amplitude modulation without loss of intensity. In contrast, Kulcke *et al.* [11.36] have converted polarization-state modulation into spatial modulation by the Pockels effect in plates with no phase-lag in the absence of strain, and in half-wave plates under strain. They have thus obtained intermittent lateral displacements of the beam.

In other cases, the phenomenon recalls the modulation of a radio-electric wave on a musical frequency, and one finds the decomposition of the carrier

11.4. Possible Applications

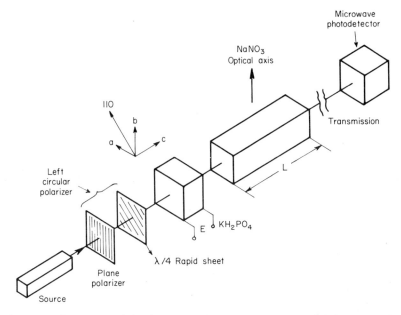

Fig. 11.14. Conversion of the electrooptical polarization state modulation to amplitude modulation by passing through a thick birefractive blade (based on Buhrer [11.35]).

wave with the appearance of lateral bands. Connes *et al.* [11.37] have used the Kerr effect to modulate the spectrum of a multiple-wave interferometer at high frequency. From Eqs. (11.5) and (11.7), one obtains:

$$n_0 - n = -\tfrac{1}{3}K\lambda E^2 \\ n_e - n = +\tfrac{2}{3}K\lambda E^2 \tag{11.13}$$

If E is the sum of a direct polarization field E_p and an alternating field $E_m \cos 2\pi Nt$ ($E_m < E_p$), the path differences of the two vibrations are

$$\delta_y = (n_0 - n)l \sim -\tfrac{1}{3}Kl\lambda(E_p^2 + 2E_p E_m \cos 2\pi Nt) \\ \delta_x = (n_e - n)l \sim +\tfrac{2}{3}Kl\lambda(E_p^2 + 2E_p E_m \cos 2\pi Nt) \tag{11.14}$$

The variable parts of the phase-lags are thus

$$\varphi_y = -m \cos 2\pi Nt \\ \varphi_x = +2m \cos 2\pi Nt \tag{11.15}$$

with

$$m = \tfrac{4}{3}KlE_p E_m$$

Let us assume that the initial vibration has the form $a \cos 2\pi vt$ and that m is very much less than one: the spectrum of the *phase-modulated* wave is then reduced to the carrier wave and the first lateral bands (Fig. 11.15). Analysis by

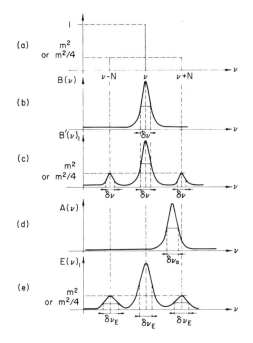

Fig. 11.15. Decomposition of an optical line by modulation (a, b, c) and spectrometrical analysis (d, e) (based on Connes et al. [11.37]).

means of a spectroscope with a finite power of resolution naturally widens the lines. The same authors refer to the cases of *carrier-suppression modulation* if the Kerr cell, between a crossed polarizer and analyser, is placed at 45° to the field E, and of *single lateral band modulation* if a special cell can produce a rotating field E providing a frequency shift.

Variable-transmission modulators (Kerr or Pockels cells placed between rectilinear polarizers) are used for the photographic recording of sound on films [11.28, 11.38]. Their use has been suggested to transmit television images [11.39]. As for frequency modulators, they mainly concern lasers. An optical heterodyne transmitting–receiving set with a laser source is described by Peters [11.40]. A linear optical modulator has been described by Ohm [11.7]: a crystal of potassium diphosphate (KDP) is mounted in a laser cavity, between the terminal mirror and a plate of quartz with antireflecting coating. The longitudinal modes of an optical cavity are $c/2L$ apart, where L is the length of the optical path in the cavity. If the linear phase-difference introduced by the crystal is 2φ, one finds that the frequency difference is

$$\Delta f = c\varphi/L\pi \qquad (11.16)$$

in other words, substituting numerical values, ± 44 MHz. The sensitivity of the frequency emitted to the applied voltage is 0.4 MHz V^{-1}, and the power of the laser is reduced from 1.2 to 0.4 mW by the presence of the crystal. The

11.4. Possible Applications

maximum modulation frequency is limited by distortion to a small fraction of $c/2L$ (which equals 0.7 GHz).

Much less attention has been paid up to now to magnetooptical modulators than to the preceding ones. Porter et al. [11.41] have described one type using the Faraday effect in a disc of garnet $Y_3Fe_5O_{12}$, placed in the center of an axial hole drilled in a high permeability ferrite pot (Fig. 11.16). The winding of this pot creates a sinusoidal magnetic field, parallel to the path of the light ray polarized at P, the polarization plane of which therefore revolves by an angle proportional to the induction in the garnet. Amplitude modulation is obtained after passing through the analyzer A. Because of the high demagnetizing factor of the garnet plate one has to concentrate the light beam into a very small volume, in order to be able to set up an intense field, and the heating of the sample creates difficulties. However, 20% of the light of a tungsten-filament lamp transmitted by a 100-μm plate has been successfully modulated.

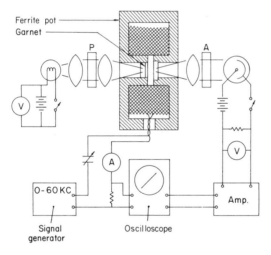

Fig. 11.16. Diagrammatic representation of a first type of Faraday effect modulator (based on Porter et al. [11.41]).

Dillon [11.42] and Anderson [11.43] have described another type, in which the sample is placed in a waveguide fed by a klystron operating in the X band (approximately 3 cm) (Fig. 11.17). A static magnetic field H saturating the magnetic semiconductor sample is applied perpendicular to the path of the light ray, so that there is no Faraday effect under normal conditions. On the other hand, if the frequency of the signal sent to the wave guide reaches that of the ferro- or ferrimagnetic resonance of the sample, the saturation magnetization intensity vector M_s precesses round the direction of H and the polarization plane revolves proportionally to the component m_s. Amplitude

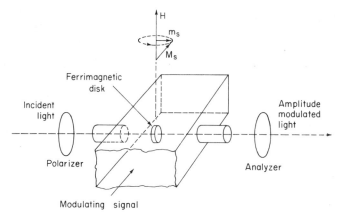

Fig. 11.17. Diagrammatic representation of a second type of Faraday effect modulator operating at the ferro- or ferrimagnetic resonance frequency (based on Dillon [11.42] and Anderson [11.43]).

modulation is again obtained. The wavelengths at which the factor of merit (ratio between the rotatory power and weakening) is maximum are shown in Table 11.2 for $Y_3Fe_5O_{12}$ and $CrBr_3$, as well as the optimum thickness and the level of modulation obtained [11.43]. The resonance frequency is around 9 to 10 GHz for the saturating field. It then increases in proportion to the field.

TABLE 11.2

Compound	Wavelength (μm)	Thickness (μm)	Modulation level (%) at 10 GHz	Modulated power (mW)
$Y_3Fe_5O_{12}$	0.85	92	0.1	65
$CrBr_3$	0.48	25	2.3	17

Zitter and Spencer [11.44] have recently described a third type, in which a bar of garnet is placed inside the optical cavity of a helium–neon laser, the emission of which on a wavelength of 3.39 μm is polarized by plates of calcite. The bar is surrounded by a solenoid resonating at 6.6 MHz, so that modulation by the Faraday effect occurs at the frequency of 13.2 MHz. For a field-saturating sample, rotations of 184 and 87° cm^{-1} have been obtained on wavelengths of 1.15 and 3.39 μm, respectively. Absorption is practically nil at the latter. Rotation, and consequently modulation, are increased by the presence of the garnet in the optical cavity of the laser [11.45].

The most powerful lasers at present produced are gas lasers, particularly carbon dioxide molecular lasers, some of which supply 300 W with an efficiency of 10%, whereas the efficiency of atomic lasers is not more than 0.1% [11.46]. The CO_2 laser emits in the infrared range at a wavelength of approximately 10 μm, and this is particularly useful for data communication, since it corresponds to an "atmospheric window" in which the steam and other atmospheric components do not present any absorption bands. In accordance with the remark made on transparency in Section 11.2, it can be seen that the semiconductors most suitable for its modulation will have to have a narrow forbidden band gap, of around 0.2 or 0.3 eV. The garnet $Y_3Fe_5O_{12}$ is not suitable, since its first absorption peaks, resulting from the lattice, occur at around 10 μm.

To complete this review of the possible applications of the magnetooptical effects of magnetic semiconductors, one might mention the amplification of spin waves. Spector [11.47] has recently suggested this possibility, by the application of an electrical field with a frequency close to that of the resonance frequency. This application, based on an interaction with free electrons, therefore concerns hyperfrequency waves in the 10- to 50-GHz range.

REFERENCES

[11.1] G. Bruhat, "Cours de Physique Générale: Optique." Masson, Paris, 1959.
[11.2] P. Fleury and J.-P. Mathieu, "Physique Générale et Expérimentale: Lumière." Eyrolles, Paris, 1961.
[11.3] J. Billard, *Rev. Phys. Appl.* **1**, 311 (1966).
[11.4] F. Pockels, *Abh. Gött. Ges. Wiss. Math. Phys. Kl.* **39**, 1 (1894).
[11.5] M. Beck and H. Gränicher, *Helv. Phys. Acta* **23**, 522 (1950).
[11.6] D. A. Kleinman, *Phys. Rev.* **126** 1977 (1962).
[11.7] E. A. Ohm, *Appl. Opt.* **6**, 1233 (1967).
[11.8] J. C. Suits, B. E. Argyle, and M. J. Freiser, *J. Appl. Phys.* **37**, 1391 (1966).
[11.9] J. F. Dillon, *J. Appl. Phys.* **29**, 539 (1958).
[11.10] J. C. Suits and B. E. Argyle, *Phys. Rev. Lett.* **14**, 687 (1965).
[11.11] A. Kastler, *C.R. Acad. Sci. Paris* **232**, 953 (1951).
[11.12] G. S. Krinchik and M. V. Chetkin, *Zh. Eksp. Teor. Fiz.* **38**, 1643 (1960).
[11.13] C. J. Ballhausen, "Introduction to Ligand Field Theory." McGraw Hill, New York, 1962.
[11.14] S. D. Smith, "Handbuch der Physik" (S. Flügge, ed.), Vol. XXV/2a, p. 234. Springer, Berlin, 1967.
[11.15] G. S. Krinchik and R. D. Nuralieva, *Zh. Eksp. Teor. Fiz.* **36**, 1022 (1959).
[11.16] J. H. Greiner and G. J. Fan, *Appl. Phys. Lett.* **9**, 27 (1966).
[11.17] Y. R. Shen and N. Bloembergen, *Phys. Rev.* **133A**, 515 (1964).
[11.18] A. C. Boccara, *Rev. Phys. Appl.* **2**, 142 (1967).
[11.19] C. Kooy, *Philips Tech. Rev.* **19**, 286 (1958).
[11.20] R. C. Sherwood, J. P. Remeika, and H. J. Williams, *J. Appl. Phys.* **30**, 217 (1959).
[11.21] M. W. Shafer, T. R. McGuire, and J. C. Suits, *Phys. Rev. Lett.* **11**, 251 (1963).
[11.22] R. C. le Craw, D. L. Wood, J. F. Dillon, Jr., and J. P. Remeika, *Appl. Phys. Lett.* **7**, 27 (1965).

[11.23] J. F. Dillon, Jr., H. Kamimura, and J. P. Remeika, *J. Phys. Chem. Solids* **27**, 1531 (1966).
[11.24] I. Tsubokawa, *J. Phys. Soc. Japan* **15**, 1664 (1960).
[11.25] A. F. Wells, "Structural Inorganic Chemistry." Oxford Univ. Press, London and New York, 1950.
[11.26] D. O. Smith, "Optical Information Processing." Technol. Press, Cambridge, Massachusetts, 1965.
[11.27] H. Jaffe, *Phys. Rev.* **73**, 95 (1948).
[11.28] G. D. Gotschall, *J. Soc. Motion Picture Eng.* **51**, 13 (1948).
[11.29] B. H. Billings, *J. Opt. Soc. Amer.* **39**, 797, 802 (1949).
[11.30] Y. Fujisawa, *Electrotechn. J. Japan* **3**, 23 (1957).
[11.31] I. P. Kaminow, *Phys. Rev. Lett.* **6**, 528 (1961).
[11.32] O. C. Jones, *J. Sci. Instr.* **41**, 653 (1964).
[11.33] C. J. Peters, *Proc. IEEE* **51**, 147 (1963).
[11.34] M. P. Bernard, *C.R. Acad. Sci. Paris* **250**, 1235 (1960).
[11.35] C. F. Buhrer, *Proc. IEEE* **51**, 1151 (1963).
 11.36] W. Kulcke, T. J. Harris, K. Kosanke, and E. Max, *IBM J. Res. Develop.* **8**, 64 (1964).
[11.37] P. Connes, Duong Hong Tuan, and J. Pinard, *J. Phys. Radium* **23**, 173 (1962).
[11.38] R. O'B. Carpenter, *J. Acoust. Soc. Amer.* **25**, 1144 (1953).
[11.39] W. J. Hannan, J. Bordogna, and T. E. Penn, *Proc. IEEE* **53**, 171 (1965).
[11.40] C. J. Peters, *Appl. Opt.* **4**, 857 (1965).
[11.41] C. S. Porter, E. G. Spencer, and R. C. le Craw, *J. Appl. Phys.* **29**, 495 (1958).
[11.42] J. F. Dillon, U.S. Patent 2.974.568 (February 15, 1957).
[11.43] L. K. Anderson, *J. Appl. Phys.* **34** (Part 2), 1230 (1963).
[11.44] R. N. Zitter and E. G. Spencer, *J. Appl. Phys.* **37**, 1089 (1966).
[11.45] R. Rosenberg, C. B. Rubinstein, and D. R. Herriott, *Appl. Opt.* **3**, 1079 (1964).
[11.46] X. Ziegler and C. Frapard, *Onde Elec.* **46**, 423 (1966).
[11.47] H. N. Spector, *Solid State Commun.* **6**, 811 (1968).

APPENDIX

The thousand or so references given in Part Two concern all the articles summarized in the *Physics Abstracts*, on 30 June 1967, referring to the electrical properties of binary compounds of transition metals, rare earths, and actinides. A number of articles concerning the crystallographic structure, magnetic properties or physical theories of these compounds have also been mentioned, to the extent that they might help to explain the electrical properties. In what follows we have tried to show how work on these compounds has progressed in time, in the three continents where significant scientific activity at present takes place. Only the nationality of the scientific review publishing them is mentioned, however, and this inevitably involves errors. *Acta Cryst.*, *J. Phys. Chem. Solids* and *Solid State Comm.* are counted as being in Europe. We are, of course, aware of the arbitrary nature of this classification.

The main experimental results that have increased our knowledge of the electrical properties of binary compounds which do not have typically metallic properties are summarized below, in a very approximate and by no means exhaustive list:

Year	Europe (including the Soviet Union)	America	Asia (not including the Soviet Union)
Before 1950	57	9	4
1950–1954	83	57	13
1955–1959	107	50	23
1960–1964	209	136	50
1965–1969 (first half only)	132	55	28

Year	Compound	Result	Authors
1936	Fe_3O_4	Inverted structure	E. J. W. Verwey and F. de Boer
1944	L_2O_3	L(II) influence on ρ	M. Foëx
1946	V_2O_3	Semicond.–metal at $-100°C$	M. Foëx
1949	WO_3	Ferroelectricity (?) $< 20°C$	T. Okada, K. Hirakawa, F. Irie, and B. T. Matthias
1950	CoO(Li), NiO(Li)	Valency induction	E. J. W. Verwey, P. W. Haayman, J. Romeyn, and G. W. Van Oosterhout
1951	MnAs	Maximum ρ at T_C	C. Guillaud
1953	$Fe_{1-x}S$	Order of vacancies	E. F. Bertaut
1954	VO_2	ρ variation at $70°C$	J. Jaffray and D. Dumas
1956	$Fe_{1-x}Se$	Order of vacancies	A. Okazaki and K. Hirakawa
1957	CoO(Li), NiO(Li)	Exp. law $\mu(T)$	R. R. Heikes and W. D. Johnston
1957	CrSb	ρ maximum at T_N	T. Suzuoka
1957	Fe_2Si_5, $CrSi_2$, $MnSi_2$	High values for α	P. V. Gel'd
1960	βMnO_2	Ferroelectricity (??) $< 50°C$	V. G. Bhide and R. V. Damle
1960	$Fe_{1-x}S$	Magnetic order	J. T. Sparks, W. Mead, A. J. Kirschbaum, and W. Marshall
1961	Cr_2O_3	Astrov effect	D. N. Astrov
1961	$Fe_{1-x}S$	μ variation at $138°C$	S. Fujime, M. Murakami, and E. Hirahara
1961	Fe_3O_4	Mössbauer effect at $120°K$	R. Bauminger, S. G. Cohen, A. Marinov, S. Ofer, and E. Segal
1962	$Fe_{1-x}O$	Valency induction	D. S. Tannhauser
1963	NiO(Li)	Exp. law $N(T)$	Ya. M. Ksendzov, L. N. Ansel'm, L. L. Vasil'eva, and V. M. Latisheva
1963	Ti_2O_3	Astrov effect	B. I. Al'shin and D. N. Astrov
1964	EuO, EuS, EuSe	E_G variation at T_C	G. Busch, P. Junod, and P. Wachter
1964	MnTe	α irregularity $< T_N$	J. Wasscher and C. Haas
1964	Gd_2S_3, Gd_2Se_3	N influence on T_C	F. Holtzberg, T. R. MacGuire, S. Methfessel, and J. C. Suits
1965	Cr—X	Nature of bond	A. Z. Men'shikov and S. A. Nemnonov
1966	MnP	ρ irregularity at $50°K$	T. Suzuki, Y. Matsumura, and E. Hirahara
1966	V_2O_3	Mössbauer effect at $-100°C$	T. Shinjo and K. Kosuge
1966	CoO	Exp. law $\mu(T)$	V. P. Zhuze and A. I. Shelykh
1966	EuO	Magnetooptical effects	J. C. Suits, B. E. Argyle, and M. J. Freiser
1967	VO_2	Mössbauer effect at $70°C$	K. Kosuge

Index of Formulas

ADP 324

Ac 238
AcAs 244
AcC 241
AcC$_2$ 241
AcN 241
AcP 242
AcS 260
Ac$_2$S$_3$ 36
AcSb 245
Ac$_2$Se$_3$ 36

Ag$_x$V$_2$O$_5$ 154
Ag$_x$WO$_3$ 154

AlB$_2$ 82, 89, 235
Al$_x$Cr$_{2-x}$O$_3$ 128
α-Al$_2$O$_3$ 124, 139, 184, 190
γ-Al$_2$O$_3$ 190

Am 238

As 27
AsH$_3$ 15, 22

AuSb$_2$ 86

BaFe$_{12}$O$_{19}$ 331
BaS 260
BaTiO$_3$ 145

Bi 27, 294
Bi$_2$OF$_4$ 27

Bk 238

C 8
CH$_4$ 8, 15, 22
CS$_2$ 326

CaB$_6$ 235, 237, 240
CaC$_2$ 235, 237
CaF$_2$ 26, 27, 102, 236, 329
CaTiO$_3$ 151
Ca$_{1-x}$Y$_x$F$_{2+x}$ 27
Ca$_x$Zr$_{1-x}$O$_{2-x}$ 147

Cb, see Nb

Cd$_{0.98}$Ga$_{0.02}$Cr$_2$S$_4$ 294
CdI$_2$ 196
Cd(OH)$_2$ 196

Ce 238, 282
CeAs 244, 245
CeB$_6$ 239
CeC 241
CeC$_2$ 241
Ce$_2$C$_3$ 241
CeN 241, 242
CeO$_2$ 27, 249, 251
Ce$_2$O$_3$ 236, 246, 249, 344

CeP 242, 243
Ce$_{1-x}$Pr$_x$O$_y$ 249
CeS 257, 258
Ce$_2$S$_3$ 36, 260
Ce$_{3-x}$S$_4$ 260
Ce$_3$S$_4$ 236, 260
CeSb 245, 246
CeSe 262, 263, 264
CeSe$_2$ 264
Ce$_2$Se$_3$ 36, 237, 264
CeSi$_2$ 241
CeTe 265, 266
Ce$_2$Te$_3$ 267
Ce$_{1-x}$Th$_x$O$_2$ 251
Ce$_{1-x}$Y$_x$O$_{2-0.5x}$ 27

Cf 238

Cm 238

Co 58, 306, 319, 326
^{57}Co 161, 162, 164
CoAs 34, 83, 106
CoAs$_2$ 35, 86, 106, 109
CoAs$_3$ 103, 106, 109
Co$_2$As 106
Co$_5$As$_2$ 106
CoB 89
Co$_2$B 89
CoBr$_2$ 35
Co$_3$C 90

Index of Formulas

CoCl$_2$ 35
Co$_{1-x}$Cr$_x$Si 98
CoF$_2$ 35
Co$_x$Fe$_{3-x}$O$_4$ 185
Co$_x$Fe$_{1-x}$S 212
Co$_{1-x}$Fe$_x$Si 97
CoGe 102
CoGe$_2$ 102
Co$_{1+x}$Ge 102
CoI$_2$ 35
Co$_{1-x}$Li$_x$O 174, 344
Co$_{1-x}$Mn$_x$Si 97
Co–Ni 284
Co$_{1-x}$Ni$_x$Si 98
CoO 29, 34, 167, 170, 171, 174, 344
Co$_2$O$_3$ 190
Co$_3$O$_4$ 184
CoP 103
CoP$_2$ 35, 103
CoP$_3$ 103, 106
Co$_2$P 103
CoS 196, 212
CoS$_2$ 201, 203, 212
Co$_{1-x}$S 196
Co$_3$S$_4$ 204
CoSb 34, 109
CoSb$_2$ 35, 86, 105, 109, 111
CoSb$_3$ 109, 112
CoSe 196
CoSe$_2$ 202, 203, 218
Co$_{1-x}$Se 196, 218
Co$_7$Se$_8$ 196
CoSi 95, 97, 99
CoSi$_2$ 95
CoSn 102
CoSn$_2$ 102
Co$_{1+x}$ 102
CoTe 34, 196
CoTe$_2$ 202, 203, 228, 229
Co$_{1-x}$Te 196, 228
Co$_2$Te$_3$ 34

Cr$_{2-x}$Al$_x$O$_3$ 128, 291
CrAs 106, 114
Cr$_2$As 106
Cr$_3$As$_2$ 106
CrB 89
CrB$_2$ 89
Cr$_2$B 89

CrBr$_3$ 328, 333, 335, 340
CrCl$_3$ 328, 333
Cr$_x$Co$_{1-x}$Si 98
CrGe 102
CrI$_3$ 328, 333
Cr$_x$Mn$_{1-x}$Sb 110
Cr$_x$Mn$_{1-x}$Te 225
Cr$_x$Mn$_{2-x}$Sb 109, 292
Cr$_{1-x}$Mn$_x$O$_2$ 145
CrN 91, 92
CrO$_2$ 35, 139, 142, 145, 146, 148, 149, 155, 225, 280
CrO$_3$ 148, 149
Cr$_2$O$_3$ 124, 125, 128, 129, 155, 184, 186, 190, 288, 289, 291, 292, 344
CrP 103
CrS 34, 196, 203, 205, 344
Cr$_{1-x}$S 204
Cr$_2$S$_3$ 34, 196, 204
Cr$_3$S$_4$ 217
Cr$_5$S$_6$ 205
Cr$_7$S$_8$ 196, 205
CrSb 34, 73, 83, 109, 110, 112, 287, 344
CrSb$_x$Te$_{1-x}$ 223
CrSb$_2$ 35, 85, 86, 109, 111
CrSe 34, 196, 198, 203, 214, 344
CrSe$_x$Te$_{1-x}$ 214, 223
Cr$_{1-x}$Se 214
Cr$_2$Se$_3$ 34, 196, 214
Cr$_3$Se$_4$ 214
Cr$_5$Se$_8$ 214
Cr$_7$Se$_8$ 196, 214
CrSi 95, 97
CrSi$_2$ 79, 87, 95, 98, 99, 102, 103, 344
Cr$_5$Si$_3$ 95
CrSn 102
CrTe 34, 73, 196, 203, 220, 223, 285, 311, 319, 344
Cr$_{1-x}$Te 196, 221, 285
Cr$_2$Te$_3$ 34, 73, 196, 220, 221, 225
Cr$_3$Te$_4$ 221

Cr$_5$Te$_6$ 221
Cr$_7$Te$_8$ 196

CsCl 236

Cu 304
CuAl$_2$ 89, 102
CuBr 29
CuFeS$_2$ 29
Cu$_{1-x}$Li$_x$O 174
CuO 173, 174
CuS$_2$ 203, 213
Cu$_2$Sb 106, 235, 238
Cu$_x$V$_2$O$_5$ 154
Cu$_x$WO$_3$ 154

Dy 238
DyAs 244
DyB$_6$ 239
DyC 241
DyC$_2$ 241
DyN 241
Dy$_2$O$_3$ 236, 247, 249
DyP 242, 243, 244
DyS 257
Dy$_{3-x}$S$_4$ 260
DySe 262
DySb 245
DySi$_2$ 241
DyTe 265

Er 238
ErAs 244
ErB$_6$ 239
ErC 241
ErC$_2$ 241
ErN 241
Er$_2$O$_3$ 35, 236, 249
ErP 242, 243
ErS 257
Er$_{3-x}$S$_4$ 260
ErSb 245, 246
ErSe 262, 263
ErSe$_2$ 265
Er$_2$Se$_3$ 265
ErSi$_2$ 241
ErTe 265, 266, 267
Er$_2$Te$_3$ 267

Eu 238
EuAs 244
EuB$_6$ 239, 240, 241

Index of Formulas

EuC 241
EuC$_2$ 241
EuF$_2$ 324, 326, 329, 331
Eu$_x$Gd$_{2-x}$Se$_{3-0.5x}$ 265
Eu$_{1-x}$Gd$_x$S 260
Eu$_{1-x}$Gd$_x$Se 264
Eu$_{1-x}$La$_x$S 260
Eu$_{1-x}$La$_x$Se 264
EuN 241, 242
EuO 29, 34, 246, 258, 260, 264, 328, 329, 333, 334, 344
Eu$_2$O$_3$ 236, 247, 249
Eu$_3$O$_4$ 246
EuP 242
EuS 29, 34, 257, 258, 263, 264, 329, 334, 344
EuS$_x$Se$_{1-x}$ 260, 264
Eu$_{3-x}$S$_4$ 260
EuSb 245
EuSe 29, 34, 262, 263, 264, 325, 326, 328, 333, 335, 344
EuSe$_x$Te$_{1-x}$ 264, 267
EuSi$_2$ 241
Eu$_2$SiO$_4$ 328, 331
EuTe 29, 34, 265, 266, 267

FH 15

α-Fe 58, 163, 278, 306, 319, 326, 329, 335
^{57}Fe 161, 162, 164
FeAs 34, 106, 114
FeAsS 85, 103
FeAs$_2$ 29, 35, 86, 106
Fe$_2$As 106, 238
FeB 89, 95
Fe$_2$B 89
FeBr$_2$ 35
Fe$_3$C 90, 103
FeCl$_2$ 29, 35
Fe$_x$Co$_{1-x}$Si 97
Fe$_{1-x}$Co$_x$S 212
Fe$_{3-x}$Co$_x$O$_4$ 184
FeF$_2$ 35
Fe$_x$Ga$_{2-x}$O$_3$ 291, 292
Fe$_{1.15}$Ga$_{0.85}$O$_3$ 292
FeGe 102
FeGe$_2$ 102

Fe$_{1+x}$Ge 102
FeI$_2$ 35
Fe$_x$Mn$_{2-x}$O$_3$ 124
Fe$_x$Mn$_{3-x}$O$_4$ 124, 183
Fe–Ni 306
FeO 29, 34, 61, 64, 67, 167, 170, 181
Fe$_{1-x}$O 63, 167, 190, 199, 344
α-Fe$_2$O$_3$ 128, 166, 184, 186, 188, 190, 287
γ-Fe$_2$O$_3$ 182, 190
Fe$_3$O$_4$ 124, 177, 179, 181, 183, 190, 285, 306, 310, 344
Fe$_{21.33}$O$_{32}$ 190
Fe(OH)$_3$ 325
FeP 103, 105, 114
FeP$_2$ 35, 85, 86, 103, 112
Fe$_2$P 102, 103
FeS 34, 67, 196, 198, 203, 204, 208, 312
FeS$_2$ 29, 35, 85, 203, 210
Fe$_{1-x}$S 208, 344
Fe$_2$S$_3$ 34
Fe$_3$S$_4$ 204
Fe$_7$S$_8$ 196, 208, 210, 217, 313
FeSb 34, 112
FeSb$_2$ 35, 86, 109, 111
Fe$_3$Sb$_2$ 109
FeSe 34, 67, 196, 203, 204
FeSe$_2$ 35, 203, 218
Fe$_{1-x}$Se 217, 218, 344
Fe$_2$Se$_3$ 34
Fe$_3$Se$_4$ 217
Fe$_7$Se$_8$ 196, 217, 218
FeSi 95, 97, 102
FeSi$_2$ 95
Fe$_2$Si$_5$ 95, 97, 103, 344
Fe$_5$Si$_3$ 95
FeSn 102
FeSn$_2$ 102
Fe$_{1+x}$Sn 102
FeTe 34, 196, 203
FeTe$_2$ 35, 202, 203, 226
Fe$_{1-x}$Te 72, 196
Fe$_2$Te$_3$ 34, 64, 67, 196, 226, 229
Fe$_3$Te$_4$ 226

FeTiO$_3$ 188
Fe$_{2-x}$Ti$_x$O$_3$ 188, 190, 225

Ga$_{0.02}$Cd$_{0.98}$Cr$_2$S$_4$ 294
Ga$_{0.85}$Fe$_{1.15}$O$_3$ 292
Ga$_{2-x}$Fe$_x$O$_3$ 291, 292

Gd 58, 238
GdAs 34, 244
GdB$_6$ 239, 241
GdC 241
GdC$_2$ 241
Gd$_x$Eu$_{1-x}$S 260
Gd$_x$Eu$_{1-x}$Se 264
Gd$_{2-x}$Eu$_x$Se$_{3-0.5x}$ 265
Gd$_{2-x}$Y$_x$Se$_{3-0.5x}$ 265
GdN 34, 241, 242
Gd$_2$O$_3$ 35, 236, 247, 249
GdP 34, 242, 243, 244
GdS 257
Gd$_2$S$_3$ 36, 260, 344
Gd$_{3-x}$S$_4$ 260
GdSb 34, 245, 246
Gd$_4$Sb$_3$ 246
GdSe 262, 263
Gd$_2$Se$_3$ 36, 264, 344
GdSi$_2$ 241
GdTe 265, 266
Gd$_2$Te$_3$ 267

Ge 8, 29, 39, 303
Ge$_x$Mn$_{1-x}$Te 226, 319

H$_2$ 4, 9, 12
H$_2^+$ 9, 11, 12
HBr 21
HCl 21
HF 21
HI 21
H$_2$O 15, 16, 22
H$_2$S 15, 16, 22
H$_2$Se 15, 16, 22
H$_2$Te 15, 22

He$_2^+$ 12

HfB$_2$ 89
HfC 90, 241
HfC$_x$ 90
HfGe$_2$ 102
Hf$_5$Ge$_3$ 102
HfN 91, 92
HfO$_2$ 139, 147

Index of Formulas

HfP_2 105
HfS_2 202, 204
HfS_3 204
$HfSe_2$ 202, 213
$HfSe_3$ 213
$HfSi$ 95
$HfSi_2$ 95
Hf_5Si_3 95
$HfSn_2$ 102
Hf_5Sn_3 102
$HfTe$ 219

$HgTe$ 31

Ho 238
$HoAs$ 244
HoB_6 239
HoC 241
HoC_2 241
HoN 241
Ho_2O_3 236, 249
HoP 242, 243
HoS 257
$Ho_{3-x}S_4$ 260
$HoSb$ 245, 246
$HoSb_{1-x}Te_x$ 246, 267
$HoSe$ 262
$HoSi_2$ 241
$HoTe$ 265

$InAs$ 313
$InSb$ 29, 31, 39, 294, 313, 319

$IrAs_2$ 86
$IrAs_3$ 106, 109
Ir_2As 106
$IrGe_2$ 102
$Ir_{1+x}Ge$ 102
IrP_2 86, 103, 105
IrP_3 103, 106
Ir_2P 103
$IrS_{2.8}$ 203
$IrSb$ 109
$IrSb_2$ 86, 109
$IrSb_3$ 109
$IrSe_{2.8}$ 203
$IrSi$ 95
$IrSi_3$ 95
Ir_5Si_3 95
$IrSn_2$ 102

$Ir_{1+x}Sn$ 102
$IrTe_2$ 202
$IrTe_{2.8}$ 203

KDP 324, 338

K_xReO_3 154
K_xTiO_2 154

La 238
$LaAs$ 34, 244
LaB_6 239, 240, 241
LaC 241
LaC_2 241
$La_xEu_{1-x}S$ 260
$La_xEu_{1-x}Se$ 264
LaN 34, 241
La_2O_3 35, 235, 236, 246, 247, 249, 344
LaP 34, 242
LaS 257, 258
La_2S_3 36, 260
$La_{3-x}S_4$ 260
$LaSb$ 34, 245
$LaSe$ 262, 263
La_2Se_3 36
$LaSi_2$ 241
$La_{1-x}Sr_xMnO_3$ 287
$LaTe$ 265, 266, 267
$LaTe_2$ 267
La_2Te_3 267
$La_xZr_{1-x}O_{2-0.5x}$ 147

Li_2 12
$Li_xCo_{1-x}O$ 174, 344
$Li_xCu_{1-x}O$ 174
$Li_{0.5}Fe_{2.5}O_4$ 331
LiH 18
$Li_xMn_{1-x}O$ 123, 174
$Li_xMn_{1-x}Se$ 216
$Li_xNi_{1-x}O$ 61, 62, 63, 174, 344
$Li_xNi_{1-x}S$ 213
Li_2O 174
$Li_xV_2O_5$ 154
Li_xWO_3 154

Lu 238
$LuAs$ 34, 244
LuB_6 239
LuC 241
LuC_2 241

LuN 34, 241
Lu_2O_3 35, 236, 247, 249
LuP 34, 242
LuS 257
Lu_2S_3 36
$Lu_{3-x}S_4$ 260
$LuSb$ 34, 245
$LuSe$ 262
Lu_2Se_3 36
$LuSi_2$ 241
$LuTe$ 265

$MgAl_2O_4$ 177
$MgFe_2O_4$ 331
MgO 190

Mn 58, 304
$MnAs$ 34, 73, 82, 106, 110, 114, 319, 344
$MnAs_{1-x}P_x$ 106
$MnAs_{1-x}Sb_x$ 106
Mn_2As 106
Mn_3As_2 106
MnB 89
Mn_2B 89
$MnBi$ 34, 58, 82
$MnBr_2$ 35, 214
$MnCl_2$ 35
$Mn_xCo_{1-x}Si$ 97
$Mn_xCr_{1-x}O_2$ 145, 155
$Mn_{1-x}Cr_xSb$ 110
$Mn_{2-x}Cr_xSb$ 109, 292
$Mn_{1-x}Cr_xTe$ 225
MnF_2 35
$MnFe_2O_4$ 284
$Mn_{2-x}Fe_xO_3$ 124
$Mn_{3-x}Fe_xO_4$ 124, 183
$MnGe_2$ 102
$Mn_{1-x}Ge_xTe$ 226, 319
$Mn_{1+x}Ge$ 102
Mn_5Ge_3 102
MnI_2 35
$Mn_{1-x}Li_xO$ 123, 174
$Mn_{1-x}Li_xSe$ 216
MnN 91
MnO 29, 34, 47, 68, 119, 123, 167, 174
β-MnO_2 35, 143, 145, 146, 148, 151, 154, 155, 344
γ-MnO_2 145

Index of Formulas

α-Mn$_2$O$_3$ 124, 190, 235, 236
γ-Mn$_2$O$_3$ 124, 190
Mn$_3$O$_4$ 123, 183, 184, 190
MnP 29, 73, 79, 82, 95, 102, 103, 104, 105, 106, 114, 285, 311, 344
Mn$_2$P 104
α-MnS 29, 34, 68, 203, 204, 207
β-MnS 207
γ-MnS 207
MnS$_2$ 35, 203
MnSb 34, 58, 73, 82, 83, 109, 110, 111, 114, 311, 319
Mn$_2$Sb 109
MnSe 29, 34, 68, 203, 213, 216
MnSe$_2$ 35, 203, 217
MnSi 95, 97
MnSi$_2$ 98, 99, 344
Mn$_4$Si$_7$ 99, 103
Mn$_5$Si$_3$ 95, 102
MnSn$_2$ 102
Mn$_{1+x}$Sn 102
Mn$_5$Sn$_3$ 102
MnTe 34, 68, 142, 196, 198, 203, 225, 229, 280, 312, 344
MnTe$_2$ 35, 203, 226
Mn$_{1-x}$Te 196

MoAs$_2$ 106
MoB 89
MoB$_2$ 89
Mo$_2$B 89
MoC 90
MoN 91
MoO$_2$ 139, 148, 149
MoO$_3$ 148, 149, 152, 154
Mo$_n$O$_{3n-1}$ 148
γ-Mo$_4$O$_{11}$ 149, 152
β-Mo$_8$O$_{23}$ 149, 152
β'-Mo$_9$O$_{26}$ 149, 152
MoP 103
MoP$_2$ 103, 105
MoS$_2$ 201, 202, 207
MoS$_3$ 204
MoSe$_2$ 202, 214, 216

MoSe$_3$ 213
MoSi$_2$ 87, 95
Mo$_5$Si$_3$ 95
Mo$_{1-x}$Ta$_x$Se$_2$ 216
MoTe$_2$ 202, 223
Mo$_x$W$_{1-x}$Se$_2$ 216
Mo$_x$W$_{1-x}$Te$_2$ 224

NH$_3$ 15, 16, 22

NaClO$_3$ 324
Na$_x$ReO$_3$ 154
Na$_x$TiO$_2$ 154
Na$_x$V$_2$O$_5$ 154
Na$_x$WO$_3$ 154

NbAs$_2$ 105, 106
NbB 89
NbB$_2$ 89
NbC 90
NbGe$_2$ 102
Nb$_5$Ge$_3$ 102
NbN 91, 92
NbO 119
NbO$_2$ 147
α-Nb$_2$O$_5$ 139, 150, 151
α-Nb$_2$O$_{5-x}$ 150
NbP$_2$ 103, 105
NbS 204
NbS$_2$ 202
NbSi$_2$ 95
Nb$_5$Si$_3$ 95
NbSn$_2$ 102
Nb$_5$Sn$_3$ 102
NbTe 219
NbTe$_2$ 220
Nb$_{2-x}$W$_x$O$_5$ 150

Nd 238
NdAs 244
NdB$_6$ 239
NdC 241
NdC$_2$ 241
NdN 241
Nd$_2$O$_3$ 35, 236, 246, 249, 344
NdP 242
NdS 257, 258
Nd$_2$S$_3$ 260
Nd$_{3-x}$S$_4$ 260
NdSb 245, 246

NdSe 262, 263
Nd$_2$Se$_3$ 265
NdSi$_2$ 241
Nd$_x$Sm$_{1-x}$Se 264
NdTe 265, 266
NdTe$_2$ 267
Nd$_2$Te$_3$ 267
Ni 58, 277, 282, 284, 304, 305, 306, 319, 326
NiAs 29, 79, 82, 102, 106
NiAs$_2$ 35, 85, 86, 106, 109
NiAs$_3$ 109
NiB 89
Ni$_2$B 89
Ni$_3$B$_7$O$_{13}$I 295
NiBr$_2$ 35
Ni$_3$C 90
NiCl$_2$ 35
Ni–Co 284
Ni$_x$Co$_{1-x}$Si 98
NiF$_2$ 35
Ni–Fe 306
NiFe$_2$O$_4$ 182, 306, 310
Ni$_{0.75}$Fe$_{2.25}$O$_4$ 311
NiGe 102
Ni$_{1+x}$Ge 102
NiI$_2$ 35
Ni$_2$In 79, 82, 102
Ni$_{1-x}$Li$_x$O 61, 62, 63, 174, 344
Ni$_{1-x}$Li$_x$S 213
NiO 29, 34, 46, 47, 61, 62, 64, 167, 171, 172, 174
Ni$_2$O$_3$ 190
Ni$_3$O$_4$ 184
NiP$_2$ 35, 103
NiP$_3$ 103, 106
NiS 212
NiS$_2$ 203, 213
Ni$_3$S$_4$ 204
NiSb 34, 109, 111
NiSb$_2$ 35, 86, 109
NiSb$_{2.2}$ 112
NiSe 196, 219
NiSe$_2$ 202, 203, 219
Ni$_7$Se$_8$ 196
NiSi 95
NiSi$_2$ 95
NiSn 102

Index of Formulas

$Ni_{1+x}Sn$ 102
NiTe 34, 197, 199, 228
$NiTe_2$ 197, 199, 202, 228
Ni_2Te_3 34

No 238

Np 238

$OsAs_2$ 86, 106
OsC 90
OsP_2 86
OsS_2 203
$OsSb_2$ 86, 109
$OsSe_2$ 203
OsSi 95
Os_5Si_3 95
$OsTe_2$ 203

PH_3 15, 22

Pa 238
PaN_2 241

$PbCl_2$ 103
$PbFe_{12}O_{19}$ 331
PbS 27, 28

Pd 277
$PdAs_2$ 86, 106
PdGe 102
$Pd_{1+x}Ge$ 102
PdP_2 103
PdP_3 103, 106
PdS 204
PdS_2 29, 203
$Pd_{2.2}S$ 213
Pd_4S 213
PdSb 109, 245
$PdSb_2$ 86, 109, 111
Pd_5Sb_3 109
$PdSe_2$ 202, 203
PdSi 95
PdSn 102
$Pd_{1+x}Sn$ 102
PdTe 219, 229
$PdTe_2$ 202, 228, 229

Pm 238
PmAs 244
PmB_6 239
PmC 241
PmC_2 241
PmN 241
Pm_2O_3 236, 249
PmP 242
PmS 257
$Pm_{3-x}S_4$ 260
PmSb 245
PmSe 262
$PmSi_2$ 241
PmTe 265

Pr 238
PrAs 244, 245
PrB_6 239
PrC 241
PrC_2 241
$Pr_xCe_{1-x}Oy$ 249
PrN 241, 242
PrO_x 246
$PrO_{1.5+x}$ 249
PrO_2 251
Pr_2O_3 236, 246, 249, 344
Pr_6O_{11} 251
PrP 242, 243
PrS 257, 258
Pr_2S_3 260
$Pr_{3-x}S_4$ 260
PrSb 245, 246
PrSe 262, 263
Pr_2Se_3 265
$PrSi_2$ 241
PrTe 265, 266
Pr_2Te_3 267

$PtAs_2$ 86, 106
$PtBi_2$ 86
PtGe 102
$PtGe_2$ 102
$Pt_{1+x}Ge$ 102
PtP_2 86, 103
PtS 29, 204
PtS_2 202
PtSb 109
$PtSb_2$ 86, 109, 112
$PtSe_2$ 202
PtSi 95
PtSn 102
$PtSn_2$ 102
$Pt_{1+x}Sn$ 102
PtTe 219
$PtTe_2$ 202
PtTl 102

Pu 238
PuAs 244
PuB_6 239
PuC 241, 244
PuC_2 241
PuN 241, 244
PuO_{2-x} 251
PuP 242, 244
PuS 260
PuSb 245
$PuSi_2$ 241

RbH_2PO_4 324

$ReAs_2$ 106
ReO_2 139, 148, 155
ReO_3 148, 151, 154, 155
Re_2O_7 148
ReP 103
Re_2P 103
ReS_2 202, 208
Re_2S_7 204
$ReSe_2$ 202
Re_2Se_7 213
ReSi 95
$ReSi_2$ 95
Re_5Si_3 95
$Re_xW_{1-x}O_3$ 154

RhAs 106
$RhAs_2$ 86, 106
$RhAs_3$ 106, 109
Rh_2As 106
$RhBi_2$ 86
RhGe 102
$RhGe_2$ 102
$Rh_{1+x}Ge$ 102
RhP_2 86, 103, 105
RhP_3 103, 106
Rh_2P 103
RhS_2 203
RhSb 109
$RhSb_2$ 86, 109
$RhSb_3$ 109
$RhSe_2$ 203
Rh_2Se_5 219
RhSi 95
Rh_5Si_3 95
RhSn 102
$RhSn_2$ 102
$Rh_{1+x}Sn$ 102

Index of Formulas

RhTe 219
α-RhTe$_2$ 203, 228
β-RhTe$_2$ 202

RuAs 106
RuAs$_2$ 86, 106
RuC 90
RuP$_2$ 86, 103
Ru$_2$P 103
RuS$_2$ 203
RuSb$_2$ 86
RuSe$_2$ 203
RuSi 95
RuTe$_2$ 203

SO$_2$ 150
SO$_3$ 150

Sb 27
SbH$_3$ 15, 22
Sb$_2$S$_3$ 235, 265

Sc 238
ScAs 244
ScB$_6$ 239
ScC 241
ScC$_2$ 241
ScGe$_2$ 241
ScN 241, 242
Sc$_2$O$_3$ 236, 247
ScP 242
ScS 257
Sc$_{3-x}$S$_4$ 260
ScSb 245
ScSe 262
ScSi$_2$ 241
ScTe 265

Si 8, 39

Sm 238
SmAs 244
SmB$_6$ 239, 240
SmC 241
SmC$_2$ 241
SmN 241
Sm$_{1-x}$Nd$_x$Se 264
Sm$_2$O$_3$ 236, 246, 247, 249, 344
SmP 242
SmS 257, 258
Sm$_2$S$_3$ 260

Sm$_{3-x}$S$_4$ 260
SmSb 245
SmSe 262, 263
SmSi$_2$ 241
SmTe 265, 266

Sn 8
SnO$_2$ 190
SnS 27

Sr$_x$La$_{1-x}$MnO$_3$ 287
SrS 27

TaAs$_2$ 106
TaB 89
TaB$_2$ 89
Ta$_2$B 89
TaC 90
TaC$_x$ 91
TaGe$_2$ 102
Ta$_5$Ge$_3$ 102
Ta$_x$Mo$_{1-x}$Se$_2$ 216
TaN 91, 92
Ta$_2$O$_5$ 139, 150, 151, 154, 251
TaP$_2$ 103, 105
TaS$_2$ 202
TaSe$_2$ 214
TaSi$_2$ 95
Ta$_5$Si$_3$ 95
TaSn$_2$ 102
Ta$_5$Sn$_3$ 102
TaTe 219
TaTe$_2$ 202, 220
TaTe$_3$ 220
Ta$_x$W$_{1-x}$Se$_2$ 214

Tb 238
TbAs 244
TbB$_6$ 239
TbC 241
TbC$_2$ 241
TbN 241
Tb$_2$O$_3$ 236, 249
TbO$_2$ 251
TbP 242, 243, 244
TbS 257
Tb$_{3-x}$S$_4$ 260
TbSb 245, 246
TbSe 262
TbSi$_2$ 241
TbTe 265

TcO$_2$ 139
Tc$_2$O$_7$ 148
Tc$_2$S$_7$ 204

Th 238
ThAs 244
ThAs$_2$ 244
Th$_3$As$_4$ 244
ThB$_6$ 239, 240
ThC 241
ThC$_2$ 241
Th$_x$Ce$_{1-x}$O$_2$ 251
ThN 241
Th$_2$N$_3$ 241
ThO$_2$ 251
ThP 242
Th$_3$P$_4$ 29, 235, 236, 242
ThS 260
Th$_2$S$_3$ 36
ThSb 245
ThSb$_2$ 245
Th$_3$Sb$_4$ 245
Th$_2$Se$_3$ 36
β-ThSi$_2$ 241
Th$_{1-x}$Y$_x$O$_{2-0.5x}$ 249, 251

TiAs 106
TiB$_2$ 89
TiC 90, 93, 241
TiC$_x$ 90
TiC$_x$N$_{1-x}$ 92
Ti$_x$Fe$_{2-x}$O$_3$ 188, 190, 225
TiGe$_2$ 102
Ti$_{1+x}$Ge 102
Ti$_5$Ge$_3$ 102
TiN 91, 93
TiN$_x$ 93
TiN$_x$O$_y$ 92, 120
TiO 93, 119, 123
TiO$_x$ 93, 132
TiO$_2$ 29, 35, 69, 130, 132, 139, 148, 154, 171, 188
Ti$_2$O$_3$ 73, 119, 124, 125, 129, 154, 184, 291, 344
Ti$_n$O$_{2n-1}$ 133, 152
TiP 103
TiP$_2$ 105
TiS 203, 204
TiS$_2$ 202, 204

TiS_3 204
Ti_3S_4 204
$TiSb$ 109, 114
$TiSb_2$ 109, 112
Ti_5Sb_2 109
$TiSe$ 196, 197, 203, 204, 213
$TiSe_2$ 197, 202, 213
Ti_3Se_4 213
Ti_5Se_8 213
Ti_7Se_8 196
$TiSi$ 95, 97
$TiSi_2$ 87, 95
Ti_5Si_3 95
$TiSn_2$ 102
$Ti_{1+x}Sn$ 102
Ti_5Sn_3 102
$TiTe$ 197, 203, 220
$TiTe_2$ 197, 202, 220
Ti_3Te_4 220
$Ti_xV_{1-x}C$ 92
$Ti_{1.8}V_{0.2}O_3$ 125

Tl_xWO_3 154

Tm 238
$TmAs$ 244
TmB_6 239
TmC 241
TmC_2 241
TmN 241, 242
Tm_2O_3 236, 249
TmP 242
TmS 257
$Tm_{3-x}S_4$ 260
$TmSb$ 245
$TmSe$ 262
$TmSi_2$ 241
$TmTe$ 265

U 238
UAs 244
UAs_2 244
U_3As_4 244
UC 241
UC_2 241
UN 241
UN_2 241
UO_2 251, 252, 255, 256
UO_{2-x} 255
UO_{2+x} 251, 254, 255
UO_3 252, 255, 256
U_3O_7 252
U_3O_8 235, 252, 255, 256
U_4O_9 252, 254
U_nO_{2n+2} 252
UP 242
UP_2 243
U_3P_4 242
US 260
US_2 235, 262
U_2S_3 36, 262
USb 245
USb_2 245
U_3Sb_4 245
USe_2 265
USe_3 265
U_2Se_3 36, 265
U_3Se_4 265
USi_2 235, 241

VAs 83, 106, 114
V_2As 106
VB 89
VB_2 89
VC 90
VC_x 90, 93
V_5Ge_3 102
VN 91, 92
VO 73, 119, 120, 123, 129
VO_2 35, 73, 139, 140, 145, 148, 154, 155, 166, 293, 344
V_2O_3 120, 124, 126, 128, 129, 130, 139, 140, 154, 155, 166, 184, 190, 293, 344
V_2O_5 139, 148, 150, 154, 293
V_6O_{13} 139
V_nO_{2n-1} 139
VP 103, 112, 114
VP_2 103, 105
VS 196, 204
VS_4 204
$V_{1-x}S$ 196
V_3S_4 204
V_5S_8 204
VSb 109
VSb_2 109, 112
VSe 204, 214
VSe_2 202, 214

V_2Se_3 214
V_3Se_4 214
V_5Se_8 214
VSi_2 95, 98
V_5Si_3 95
V_5Sn_3 102
VTe 203
$V_{0.2}Ti_{0.8}O_3$ 125
$V_{1-x}Ti_xC$ 92

WAs_2 106
WB 89
W_2B 89
WC 82, 90, 103
$W_{1-x}Mo_xSe_2$ 216
$W_{1-x}Mo_xTe_2$ 224
Wn 91
$W_xNb_{2-x}O_5$ 150
WO_2 139, 148, 149
WO_3 149, 151, 154, 344
γ-W_4O_{11} 149, 152, 154
γ-$W_{18}O_{49}$ 149, 152, 154
β-$W_{20}O_{58}$ 149, 152
WP 103
α-WP_2 103, 105
β-WP_2 103, 105
$W_{1-x}Re_xO_3$ 154
WS_2 202, 207
WS_3 204
WSe_2 202, 214, 224
WSi_2 95
W_5Si_3 95
$W_{1-x}Ta_xSe_2$ 214
WTe_2 202

Y 238
YAs 244
YB_6 239, 240
YC 241
YC_2 241
$Y_xCa_{1-x}F_{2+x}$ 27
$Y_xCe_{1-x}O_{2-0.5x}$ 27
YF_3 27
$Y_3Fe_5O_{12}$ 325, 326, 329, 331, 335, 339, 340, 341
$Y_xGd_{2-x}Se_{3-0.5x}$ 265
YGe_2 241
YN 241, 242
Y_2O_3 26, 27, 35, 236, 247
YP 242

Index of Formulas

YS 257
Y_2S_3 36
$Y_{3-x}S_4$ 260
Y_5S_7 260
YSb 245
YSe 262
Y_2Se_3 36, 264
YSi_2 241
YTe 265
$Y_xTh_{1-x}O_{2-0.5x}$ 249
$Y_xZr_{1-x}O_{2-0.5x}$ 147

Yb 238
YbAs 244
YbB_6 239, 240, 241
YbC 241
YbC_2 241
YbN 241, 242
YbO 34
Yb_2O_3 236, 249

YbP 242
YbS 34, 257, 258
$YbS_{1.05}$ 258
Yb_2S_3 260
$Yb_{3-x}S_4$ 260
YbSb 245
YbSe 34, 262, 263
$YbSi_2$ 241
YbTe 34, 265, 266

ZnS 24, 26, 29

ZrAs 106
ZrB_2 89, 90
ZrC 90, 241
ZrC_x 90
$Zr_{1-x}Ca_xO_{2-x}$ 147
$ZrGe_2$ 102
Zr_5Ge_3 102
$Zr_{1-x}La_xO_{2-0.5x}$ 147

ZrN 91, 92
ZrO_2 139, 146
ZrP 103
$ZrP_{0.9}$ 103
ZrP_2 105, 106
ZrS 204
ZrS_2 202, 204
ZrS_3 204
$ZrSe_2$ 202, 213
$ZrSe_3$ 213
ZrSi 95
$ZrSi_2$ 95
Zr_5Si_3 95
$ZrSn_2$ 102
Zr_5Sn_3 102
ZrTe 219
$ZrTe_2$ 202
$ZrTe_3$ 220
$Zr_{1-x}Y_xO_{2-0.5x}$ 147

Index of Structures (Strukturbericht)

A4 24, 26, 30, 33, 55
A7 24, 26, 27
A15 102, 109

B1 27, 29, 31, 46, 49, 58–60, 63, 65, 68, 79, 82, 85, 90, 91, 93, 103, 119, 123, 167, 197, 199, 203, 204, 213, 226, 235, 237, 242, 244, 245, 268
B2 236, 237
B3 24, 26, 29, 33, 49, 55, 59
B4 26
B8$_1$ 29, 34, 58, 60, 63, 68, 79, 82, 102, 103, 106, 109, 112, 196, 197, 199, 200, 203–205, 208, 212–214, 217, 219, 220, 225, 226, 228, 229, 312
B8$_2$ 79, 82, 102, 109
B10 196, 217, 226
B13 196, 212, 213, 219
B17 29, 196, 204
B20 95, 102
B27 89
B29 27
B31 29, 33, 34, 58, 79, 82, 95, 102–104, 106, 109, 112
B34 196, 204
B35 102

C1 26, 35, 95, 102, 103, 106, 139, 146, 147, 235, 236, 241, 251, 252
C2 29, 35, 79, 85, 86, 106, 109, 112, 196, 201, 203
C4 29, 35, 69, 130, 133, 139, 145, 148, 149
C5 130
C6 29, 196, 197, 200, 202, 214, 224, 228, 229
C7 196, 201, 202, 216, 224
C11 235, 237, 241
C11b 87, 95, 103
C16 89, 102, 103, 109, 112
C18 29, 35, 73, 79, 85, 86, 103, 106, 109, 112, 196, 201, 203, 226, 228
C19 29, 33, 35, 196, 201, 202
C21 130
C22 102, 103, 106
C23 103, 105, 106, 235
C27 95, 196, 201
C32 82, 89, 235, 239
C37 196
C38 106, 109, 235, 238, 243, 244, 245
C40 79, 87, 95, 98, 99, 102, 103
C43 139, 143, 147
C49 95, 102, 103
C54 95, 102, 103
C222 235, 252

DO$_2$ 79, 103, 106, 109, 112
DO$_9$ 148, 149, 151, 152
DO$_{10}$ 149, 151
DO$_{11}$ 90, 103
DO$_{19}$ 102
D2$_1$ 235, 237
D5$_1$ 124, 184, 288
D5$_2$ 235, 236, 241
D5$_3$ 235, 236
D5$_8$ 235
D7$_2$ 196, 204
D7$_3$ 29, 36, 235, 236, 242, 244, 245, 260, 264, 265, 268
D8$_8$ 95, 102

EO$_7$ 85, 86, 103, 105
E2$_1$ 151, 152, 287

H1$_1$ 123, 177, 204

L1$_2$ 102. 235

354

Author Index

Numbers in brackets are reference numbers, and indicate that an author's work is referred to, although his name is not cited in the text. Numbers in *italics* show the page on which the complete reference is listed.

A

Abrahams, S. C., 125, *155*, 179 [6.98], 186, 187, *193*, *194*
Abrikosov, N. Kh., 95, 111 [4.64, 4.120, 4.122], 112 [4.64, 4.122, 4.125], *116*, *117*, *118*, 217, *233*
Ackermann, R. J., 253, 254, *272*
Acket, G. A., 126, 135, 138 [5.97], 139, *155*, *157*, 190, *195*
Adachi, K., 210 [7.90] 218 [7.90], *232*
Adler, D., 130 [5.49], *156*
Adler, S. F., 124, 125 [5.20], *155*
Afans'ev, A. Ya., 110 [4.113], *117*
Aigrain, P., 36, 38, 40, *48*, 207 [7.49], *231*
Akiyama, M., 174, *192*
Akulov, N. S., 305, *320*
Albers, W., 221, 222, *233*, 294 [9.72], *298*, 311, *320*
Aleonard, R., 205 [7.41, 7.43], 207 [7.43], 214 [7.41, 7.43], 221 [7.41, 7.43], *230*, 291 [9.65], *298*
Alexander, C., 252 [8.136], *272*, 291, *297*
Allersma, T., 150, *159*
Alperin, H. A., 266 [8.217], *274*
Al'shin, B. I., 125, *155*, 291, *297*
Amaroni, A., 186, *194*

Amiel, J., 143 [5.143], *158*
Andersen, L. A., 255, *272*
Anderson, J. S., 128 [5.37], *156*, 184 [6.151], 186 [6.151], *194*
Anderson, L. K., 339, 340, *342*
Anderson, S., 133 [5.63, 5.64, 5.65], *156*
Ando, R., 152 [5.215], *160*
Andreeva, L. P., 97, *116*
Andrellos, J. C., 267, *274*
Andresen, A. F., 218, 221, *233*, 256, *272*
Andrews, J. O., 171, 173, *191*
Anselin, F., 241 [8.49], *269*
Ansel'm, L. A., 62 [3.37], *75*
Ansel'm, L. N., 174 [6.82], *192*
Anthony, A.-M., 147, *158*, *159*
Antonova, I. I., 266 [8.221], 267 [8.221], *274*
Anzai, S., 225 [7.193], *234*
Apker, L., 151, *159*
Appel, J., 260 [8.198], *273*
Apurina, M. S., 228 [7.217], *234*
Aramu, F., 226, 227, *234*
Argyle, B. E., 246 [8.80], 257 [8.175], 258 [8.179], 262 [8.175, 8.179], 263, 264 [8.80], 265 [8.175], *270*, *272*, *274*, 324 [11.8], 325 [11.8], 326 [11.8], 329 [11.8], 331 [11.8], 335 [11.8], *341*

Ariya, S. M., 119, 126, 148 [5.182], *155*, *159*, 168, *191*, 228 [7.217], *234*
Arizumi, T., 251 [8.116], *271*
Arnold, G. P., 246 [8.70], *270*
Arnott, R., 150 [5.185], *159*
Aronson, S., 241 [8.35], 254, *269*, *272*
Aronsson, B., 79, 106 [4.3], *115*
Arrott, A., 252, *272*
Artemov, K. S., 171 [6.30], *191*
Arthur, P., Jr., *158*
Arvin, M. J., 95 [4.60, 4.61], *116*
Asanabe, S., 97, 98, 99 [4.85], 101 [4.85], *116*, *117*
Asbrink, S., 133 [5.64], *156*
Ascher, E., 295, *298*
Ashida, S., 140 [5.120], *157*
Astrov, D. N., 125, 128, *155*, *156*, 288, 289, 290, 291, *297*
Atoji, M., 241 [8.38, 8.39], *269*
Aubry, J., 168, *191*
Auskern, A. B., 241, *269*
Austin, B. A., 174 [6.85], *193*
Austin, I. G., 63 [3.40], *75*, 127 [5.33], *155*, 171, 174 [6.85], 176, *191*, *193*
Avgustinnik, A. I., 90 [4.33, 4.34, 4.38], *115*

B

Babushkina, N. A., 111 [4.116], *117*, 221 [7.165], *233*, 311, *320*
Baenziger, N. C., 241 [8.46], 242 [8.46], *269*
Bärnigshausen, H., 246 [8.82], *270*
Bailly, F., 18, 19 [1.15], 20 [1.17], 21 [1.15], 22, *23*, 30, 31 [2.11], 32, 33 [2.11], 42, *48*, 55, 56, 58, *74*, *75*
Bakken, R., 255, *272*
Balashova, A. P., 154 [5.240], *160*
Balkanski, M., 40, *48*
Ballhausen, C. J., 49, *74*, 328, *341*
Balta, P., 173 [6.62], *192*
Bamburov, V. G., 246 [8.78, 8.83], *270*
Banewicz, J. J., 214 [7.115], *232*
Bankina, V. F., 111 [4.120], *117*
Barantseva, I. G., 260 [8.205], *273*
Barbezat, S., 220 [7.15], *233*
Barker, A. S., Jr., 140, *158*
Barone, F. J., 136 [5.88], *157*
Barstad, J., 199, *230*
Barth, T. F. W., 177 [6.91], *193*
Basili, R. R., 247 [8.98], *271*

Bassi, G., 250 [7.41], 214 [7.41], 221 [7.41], *230*, 291 [9.65], *298*
Bates, L. F., 281, *296*
Batsanov, S. S., 207, *231*
Bauerle, J. E., 249 [8.100], 251, *271*
Baukus, J., 136 [5.89], *157*
Baum, B. A., 97 [4.73], 99, *116*
Bauminger, R., 182, *194*
Bean, C. P., 106 [4.105], *117*
Beck, M., 324 [11.5], *341*
Becker, J. H., 135 [5.79], 138, *156*, *157*
Becker, R., 277, *296*
Beernsten, D. J., 201 [7.17], 214 [7.17], 216, 223 [7.129], 224, *230*, *232*, *234*
Belov, K. P., 220 [7.158], *233*, 281, 282 [9.20], 285, *296*, *297*, 305, 306 [10.25], 319 [10.25], *320*
Bénard, J., 140 [5.126], *158*
Benoit, R., 58 [3.20], *75*, 208 [7.62, 7.66], 209, 210, 211, 212 [7.62, 7.66, 7.100], 213 [7.62, 7.66, 7.100], *231*, *232*, 240, *268*
Berdyshev, A. A., 217 [7.135], *232*
Berets, J., 150, *159*
Bernard, M., 179 [6.108], *193*, 336 [11.34], *342*
Bertaut, E. F., 196 [7.3, 7.5], 205 [7.41], 207, 208, 213 [7.111], 214 [7.111, 7.123], 220, 221 [7.169], *229*, *230*, *231*, *232*, *233*, 237 [8.13], 241 [8.43], *268*, *269*, 291, *298*
Bethe, H., 49, 58, 59, *74*, *75*
Bevan, D. J. M., 128, *156*, 184, 186, *194*, 246 [8.87c], *270*
Bhatnagar, S. S., 207 [7.56], *231*
Bhide, V. G., 124 [5.15], 143, 144, 145, *155*, *158*, 183, *194*
Bienert, W. B., 99, *117*
Bierstedt, P. E., 109, *117*, 293 [9.68], 294, *298*
Billard, J. 323, 336, *341*
Billings, B. H., 336 [11.29], *342*
Bilz, H., 88, 89, *115*
Bin, M., 210, *231*, *232*
Binder, I., 240, *269*
Birckenstaedt, M., 251 [8.104], *271*
Birks, J. B., 179 [6.110], *193*
Bittner, H., 91 [4.39], 93 [4.39], *116*
Bizette, H., 143 [5.138], 144, *158*, 168 [6.14], 173 [6.63], *191*, *192*, 207 [7.57], *231*
Bliznakov, G. B., 150, *159*
Bloch, F., 61 [3.27], *75*

Block, J., 128, *156*, 174, *192*
Bloembergen, N., 329, 330, *341*
Blum, P., 237 [8.13], 241 [8.43], *268*, *269*
Blumenthal, R. N., 136, *157*
Bobb, L. C., 154 [5.237], *160*
Bober, M., 128, *156*
Boccara, A. C., 330, 331, *341*
Bodine, J. H., 251, *271*
Boehm, M. J., 190, *195*
Bogdanova, N. I., 119, 126, *155*
Bogomolov, V. N., 138, *157*
Bogoroditskii, G. P., 139, *157*
Bogoroditskii, N. P., 247, *271*
Bohla, F., 247, *271*
Bokii, G. B., 88, *115*
Boltaks, B. I., 133, *156*
Bongers, P. F., 140, *157*, 294 [9.72], *298*
Bonnerot, J., 142, *158*
Bordogna, J., 338 [11.39], *342*
Borgonovi, G. 212 [7.97], *232*
Boros, J., 150, *159*
Bosman, A. J., 172, 174, 176, *192*, *193*
Boyd, G. E., 148 [5.177], *159*
Bozorth, R. M., 246 [8.69], *270*, 281, *296*
Brach, B. Ya., 168, *191*
Brauer, G., 246 [8.82], 247 [8.89], *270*
Breckenridge, R. G., 131, 134, 136, 138, *156*
Brenet, J., 143 [5.137, 5.142, 5.143], 145 [5.146, 5.154], 148 [5.179], *158*, *159*
Brewer, L., 257 [8.174], 260 [8.192, 8.197], *272*, *273*
Bridgman, P. W., 207 [7.48], *231*
Brix, P., 246 [8.77], *270*
Brixner, L. H., 214, 223, 224, *232*
Bro, P., 267, *274*
Brockhouse, B. N., 128, *156*, 173 [6.66], *192*
Brockhouse, I. E., 288 [9.40], *297*
Bromley, L. A., 257 [8.174], 260 [8.192], *272*, *273*
Brown, B. E., 201 [7.17], 214 [7.17], 220 [7.152], 224, *230*, *233*, *234*
Brown, D. E., 135 [5.87], *157*
Brown, R. F., 139 [5.115], *157*
Bruck, A., 171, *191*
Bruhat, G., 322, 323 [11.1], 325 [11.1], 329 [11.1], *341*
Brunie, S., 196 [7.5], 204, 213 [7.112], 214 [7.112], *229*, *230*, *232*
Brunner, J., 315 [10.51], *321*

Bruno, M. 251 [8.102, 8.103], *271*
Buhrer, C. F., 336 [11.35], 337, *342*
Buisson, G. 291 [9.65], *298*
Bungett, I., 124 [5.16], *155*, 183 [6.141], *194*
Burdese, A., 252 [8.134], *272*
Buryak, E. M., 220 [7.163], *233*, 311 [10,37], *320*
Busch, G., 106, *117*, 152, *159*, 242, 243, 244 [8.61], 245 [8.61], 246 [8.61, 8.68], 247, 254, 257, 258 [8.176], 259, 260 [8.191], 262, 263, 264 [8.191, 8.214], 265, 266 [8.176], 267 [8.68, 8.191], *269*, *270*, *272*, *273*, *274*
Butler, J. D., 150, *159*

C

Cabannes, F., 147 [5.163], *158*
Cable, J. W., 241 [8.47, 8.48], 243 [8.48], 244 [8.48], 245 [8.48], *269*
Caglioti, G., 173, *192*, 212 [7.97], *232*
Calhoun, B. A., 179 [6.98], *193*
Calhoun, B. P., 179 [6.99], 181, *193*
Callaway, J., 246 [8.71], 258 [8.71], 262 [8.71], *270*
Cardon, F. 132 [5.61], *156*
Carpenter, R. O'B., 338 [11.38], *342*
Carter, F., 260 [8.187, 8.188, 8.199], *273*
Carter, W. S., 207 [7.58], *231*
Casalot, A., 154 [5.223, 5.224], *160*
Castro, P. L., 135 [5.78], 137, 138 [5.78], *156*, *157*
Cauville, R., 139, *157*
Cejchan, O., 188 [6.174], *194*
Cella A. A., 246 [8.87b], *270*
Cerwenka, E., 87 [4.11], *115*
Champion, J. A., 216, 224, *232*
Chapin, D. S., 142, *158*
Chapman, P. R., 148 [5.173], *159*
Chappert, J., 228, *234*, 291 [9.65], *298*
Chen, C. W., 260 [8.187, 8.188], *273*
Cheremushkina, A. B., 305, *320*
Cherki, C., 154 [5.246], *160*
Chetkin, M. V., 326 [11.12], *341*
Chevallier, R., 184, *194*
Chevillot, J. P., 143, 145 [5.146, 5.154], *158*
Chevreton, M., 196 [7.5], 204, 205 [7.41, 7.43], 207 [7.43], 213 [7.111, 7.112], 214 [7.41, 7.43, 7.111, 7.112], 220, 221 [7.41, 7.43], *229*, *230*, *232*, *233*

Chiba, S., 210 [7.77, 7.90], 217, 218 [7.90], 225 [7.137], 226 [7.203], *231*, *232*, *234*
Child, H. R., 241 [8.47], 243, 244, 245, *269*
Chirkin, L. K., 154 [5.238], *160*
Choain-Maurin, C., 148, 152, 153, *159*
Chomentowski, M., 226 [7.197], *234*, 319 [10.55], *321*
Chopoorian, J. A., 152 [5.207], 154 [5.207], *159*
Chopra, E. L., 154 [5.237], *160*
Chrenko, R. M., 142, *158*, 171, *191*
Chukarev, S. A., 148 [5.182], *159*
Chvatik, J., 188 [6.174], *194*
Cirilli, V. 190 [6.185], *195*
Clark, H., 150, *159*
Cloud, W. H., 140 [5.127], *158*
Coburn, J., 136 [5.89], *157*
Coelho, R., 154 [5.246], *160*
Cohen, J., 226 [7.198], *234*
Cohen, S. G., 182 [6.132], *194*
Cole, T., 240 [8.29], *269*
Coles, B. R., 240, *269*, *278*, *296*
Collen, B., 133 [5.63, 5.64], *156*
Companion, A., 254, *272*
Connes, P., 337, 338, *342*
Conroy, L. E., 154, *160*
Cooper, J. R. ., 91 [4.40], *116*
Corenzwit, E., 219 [7.150], 228 [7.150], *233*, 241 [8.44], *269*
Corliss, L. M., 184, *194*, 207 [7.60], 208 [7.64, 7.65], 214, 215, 217 [7.64, 7.65], 226 [7.64, 7.65], *231*, *232*
Cornwell, J. C., 246 [8.87b], *270*
Costa, P., 241 [8.32, 8.34], *269*
Cotton, F. A., 49, 52 [3.2], *74*
Coulson, C. A., 17 [1.11], *23*
Cox, D. E., 164 [6.7], 168 [6.7], 169 [6.7], *191*, 223 [7.179, 7.180], *233*
Cox, N. L., *158*
Crawford, J. A., 129, *156*
Crevecoeur, C., 174, 176, *193*
Croatto, U., 170, *191*, 249, 251 [8.102], *271*
Croft, W. J., 174 [6.80], *192*
Cromer, D. T., 236 [8.5], *268*
Cronemeyer, D. C., 130 [5.55], 131, 132, 133 [5.54], 134 [5.55], 135, *156*, 171, *191*
Crowder, B. L., 152, 154 [5.219], *160*
Curtiss, C. E., 147, *159*
Cutler, M., 260 [8.203, 8.204], 263, *273*
Czanderna, A. W., 251, *271*

D

Daane, A. H., 241 [8.37], *269*
Damle, R. V., 143 [5.147], 144, *158*
Damon, D. H., 112, 113, *118*
Dancy, E. A., 212 [7.102], 213 [7.102], *232*
Danforth, W. E., 251 [8.115, 8.118, 8.120, 8.121, 8.122], *271*
Dani, R. H., 124 [5.15], 143 [5.147], 144 [5.147], *155*, *158*, 183, *194*
Danielson, G. C. 152 [5.218], 154 [5.229, 5.230], *160*
Danon, J., 268 [8.225], *274*
Darken, L. S., 168, *191*
Darnell, F. J., 140 [5.127], *158*, 293 [9.68], *298*
Das, J. N., 143, *158*
Date, M., 289, *297*
Dawson, J. K., 252, 256, *272*
Day, J., 252, *272*
Deb, S. K., 152 [5.207], 154 [5.207], *159*
de Bergevin, F., 214 [7.122], *232*
de Boer, J. H., 45, *48*, 124 [5.14], *155*, 177, 179 [6.94], 183, 184, *193*
de Coninck, R., 255 [8.163], *272*
de Gennes, P. G., 278, *296*
de Graaf, A. M., 268 [8.225], *274*
Dekker, A. J., 278 [9.7], *296*
de La Banda, J. F. G., 129, *156*
Delapalme, A., 205 [7.41], 214 [7.41, 7.122], 221 [7.41], *230*, *232*, 291 [9.65], *298*
Dempsey, E., 88 [4.19], *115*
Denayer, M., 255 [8.154, 8.155, 8.156], *272*
Denker, S. P., 93, *116*, 120, 123, *155*
Derge, G. J., 212 [7.102], 213 [7.102], *232*
Devreese, J., 255 [8.154, 8.155], *272*
De Vries, R. C., 142 [5.135], *158*
Devyatkova, E. D., 225 [7.195], *234*
Didchenko, R., 242, 243, 257, 258, *269*
Diesel, T. J., 135 [5.77], *156*
Dietrich, H., 129, *156*
Dieulesaint, E., 295, *296*, *298*
Dillon, J. F., Jr., 325, 326, 327, 329, 331 [11.22], 332 [11.22], 333, 334 [11.22], 335 [11.22], 339, *341*, *342*
Dixon, J. M., 147 [5.159], *158*
Doclo, R. J., 186, *194*
Dode, M., 255 [8.151], *272*
Döring, W., 277, *296*
Doi, H., 111 [4.117], *117*, 311 [10.42], *320*
Domange, L., 236 [8.8], 260 [8.201], 264 [8.201], *268*, *273*

Domenicali, C. A., 179 [6.103], *193*
Doney, L. M., 147 [5.169], *159*
Doroshenko, A. V., 225, *234*
Drabkin, I. A., 173, *192*
Drotschmann, C., 145, *158*
Druilhe, R., 142, 143, 145, 146, 147, *158*, 218, 223, 225, 227, 228 [7.175], *233*
Dubrovskaya, L. B., 91, *116*, 218 [7.144], *233*
Dudkin, L. D., 88, 103, 111 [4.121, 4.122], 112 [4.13, 4.122, 4.125, 4.126, 4.127], *115, 117, 118*, 218, 228 [7.218], *233, 234*
Dürrschnabel, W., 92 [4.51], *116*
Dugas, C., 207 [7.49], *231*
Dumas, D., 139, *157*
Duong Hong Tuan, 337 [11.37], 338 [11.37], *342*
Duquesnoy, A., 255 [8.157], 256 [8.166], *272*
Dutta, A. K., 207 [7.44, 7.45], *231*
Dwight, K. 205, *230*
Dyatkina, M. E., 17 [1.10], *23*
Dyuldina, K. A., 228 [7.218], *234*
Dzeganovskii, V. P., 98 [4.81], *117*
Dzyaloshinskii, I. E., 186, *194*, 288, *297*

E

Earle, M. D., 133, *156*
Eastman, E. D., 257 [8.174], 260 [8.192], *272, 273*
Ebel, H., 218 [7.145], *233*
Eick, H. A., 239, 241 [8.17, 8.46], 242, *268, 269*
Eliseev, A. A., 266, 267 [8.221, 8.223], *274*
Ellerbeck, L. D., 154, *160*
Elliott, N., 203, 207 [7.60], 208 [7.64, 7.65], 214 [7.119], 215 [7.119], 217 [7.21, 7.64, 7.65], 226 [7.21, 7.64, 7.65], *230, 231, 232*
Ellis, M., 190 [6.181], *195*
Engel, W., 315 [10.50], *321*
Englert, F., 36, 38, *48*, 283, *297*
Enz, U., 258 [8.178], 262 [8.178], 266 [8.178], *273*
Erickson, R. A., 143, *158*
Ern, V., 88, *115*
Esaki, L., 251 [8.116], *271*
Esin, O. A., 150, *159*
Evans, B. J., 210 [7.94], *232*
Eyraud, C., 214 [7.123], *232*, 256, *272*
Eyring, H., 7 [1.4], *23*
Eyring, L., 241 [8.46], 242 [8.46], 246 [8.87c], 247 [8.87d], 249, *269, 270, 271*

F

Fakidov, I. G., 104, 110 [4.113], 111, *117*, 180, 181, *193*, 204, 220 [7.157], 7.160], 221 [7.164], *230, 233*, 285, 286 [9.31], *297*, 305, 306, 310, 311, *320*
Fallot, M., 140 [5.126], *158*
Fan, G. J., 246, 260, 270, 329 [11.16], 334, 336, *341*
Farag, B. S., 145, *158*
Fast, J. F., 258 [8.178], 262 [8.178], 266 [8.178], *273*
Fatseas, G. A., 64, 66 [3.43], 73, *75*, 102, *117*, 228 [7.212], *234*
Fedorchenko, V. P., 260 [8.207], *273*
Fehr, G. A., 151 [5.201], *159*
Feigel'man, F. L., 90 [4.34], *115*
Feinleib, J., 130 [5.49, 5.50], *156*
Feretti, A., 154, *160*
Ferguson, G. A., 190, *195*
Fert, A., 247 [8.94], *271*
Fine, M. E., 150 [5.196], *159*
Finlayson, D. M., 226, *234*
Fischer, G., 106, 111, *117*, 218, *233*
Fischer, W. A., 129, *156*, 168 [6.16], *191*
Fischmeister, H., 213 [7.109], *232*
Fisher, B., 170, 176, *191, 193*
Fitzpatrick, R. L., 260 [8.199a, 8.200, 8.203], *273*
Flahaut, J., 236 [8.8], 260 [8.193, 8.201], 262 [8.193, 8.209], 264 [8.201], *268, 273*
Flanders, P. J., 188 [6.173], *194*
Flechon, J., 105, *117*
Fleury, P., 322, 326 [11.2], 327, 328 [11.2], *341*
Flindt, H. G., 213, *232*
Flinn, P. A., 161, 163, 165, *191*
Flodmark, S., 240, *269*
Foëx, M., 124, 125 [5.21], 126, 127, 152 [5.208, 5.210], *155, 159*, 167, *191*, 246, 249, 250, 251, *270, 271*
Folen, V. J., 289, *297*
Fomenko, V. S., 240 [8.24], *269*
Foner, S., 128, *156*, 291, *297*, 306, 310, *320*
Forrat, F., 214 [7.122], *232*
Frapard, C., 341 [11.47], *342*
Frederikse, H. P. R., 125, 133, 135, *155, 156*
Frei, E. H., 186 [6.165], *194*
Freier, S., 186, *194*

Freiser, M. J., 246 [8.80], 263 [8.213], 264 [8.80], *270, 274*, 324 [11.8], 325 [11.8], 326 [11.8], 329 [11.8], 331 [11.8], 335 [11.8], *341*
Freymann, M., 252 [8.140], *272*
Freymann, R., 252 [8.140], *272*
Friedel, J., 278, *296*
Frindt, R. F., 207 [7.55], 214, 223, *231, 232*
Fuchs, R., 154, *160*
Fujime, S., 172, *192*, 309, 312, 314, *320*
Fujisawa, Y., 336 [11.30], *342*
Fukuoka, N., 225 [7.185], *234*
Futaki, H., 293, *298*

G

Gaidukov, L. G., 221 [7.167], 223 [7.167], *233*
Gal'perin, F. M., 220 [7.151, 7.159], *233*
Galy, J., 154 [5.223, 5.224], *160*
Gammage, R. B., 251, *271*
Gardner, R. F. G., 186 [6.170], 190, *194, 195*
Garrison, M. C., 172 [6.47], *192*
Geballe, T. H., 44 [2.24], *48*
Geiger, H., 170, 174 [6.22], *191*
Gel'd, P. V., 91 [4.43], 95 [4.58], 97 [4.71, 4.73], 98 [4.82], 99 [4.73, 4.82], 101, 102, *116, 117*, 218 [7.144], *233*
George, A. M., 255, 256, *272*
Gerard, A., 188, *194*
Gerdanian, P., 255 [8.151], *272*
Gerlach, W., 282 [9.25], 284, *297*
Germann, R. W., 205 [7.38], *230*
Gibart, P., 145 [5.155], 148 [5.179], *158, 159*
Gijsman, H. M., 305, *320*
Gilleo, M. A., 130, 133 [5.54], *156*, 179, 183, 186, *193*
Gilles, P. W., 260 [8.192], *273*
Gillson, J. L., 154, *160*
Glaser, F. W., 95 [4.59], *116*
Glasser, L., 92 [4.50], *116*
Glemser, O., 148 [5.183], *159*
Gliessman, J. R., 44, 45, *48*
Glushkova, V. B., 148 [5.182], *159*
Gocan, S., 174 [6.75], *192*
Goering, H. L., 252 [8.133], 254 [8.133], *271*
Golabi, S. M., 236 [8.8], *268*
Gol'danskii, V. I., 161, *191*
Goldman, J. E., 184, *194*, 252, *272*
Goldsztaub, S., 126 [5.28], 127 [5.28], *155*

Goldwater, D. L., 251, *271*
Golikova, O. A., 90, *115*
Golubkov, A. V., 225 [7.195], *234*, 258 [8.184, 8.185, 8.186], 259 [8.186], 263 [8.184], 266 [8.184], *273*
Gomankov, V. I., 225 [7.188], *234*
Goncharova, E. V., 258 [8.184, 8.185, 8.186], 259 [8.186], 263 [8.164], 266 [8.164], *273*
Goodenough, J. B., 53, 58, 65, 68, 70, 71, 73, *74, 75*, 104, 106, 108, *117*, 129, 139 [5.48], 141, 142, 148, 154 [5.220], *156, 159, 160*, 174, 183, 184 [6.137], *192, 194*, 210 [7.86], *231*
Goodman, G., 126, *155*
Gorelik, S. I., 133, 139, *156*
Goretzki, H., 91 [4.39], 93 [4.39], *116*
Gorter, E. W., 110, *117*, 214, 215, 220, 223 [7.118], *232*
Gortsema, F. P., 242, 243, 257, 258, *269*
Goto, Y., 190 [6.190], *195*, 205 [7.39], 210 [7.39], *230*
Gotschall, G. D., 336 [11.28], 338 [11.28], *342*
Grabovskii, M. A., 179 [6.104], *193*
Graham, H. C., 147 [5.166], *158*
Gränicher, H., 251 [8.106], *271*, 324 [11.5], *341*
Grannis, F. H., 288 [9.41], *297*
Grant, F. A., 130, 131, 134, *156*
Grazhdankina, N. P., 109, 111, *117*, 204, 220 [7.157, 7.160], 221, 223 [7.167], 225 [7.186], *230, 233, 234*, 285 [9.31], 286 [9.31], *297*
Greenaway, D. L., 204, 213, 214 [7.26], 220, *230*
Greenberg, I. N., 260 [8.199], *273*
Greener, E. H., 132 [5.60], 134 [5.60], 136, 147 [5.170], 150, 151, *156, 157, 159*, 251, *271*
Greenshpan, M., 186 [6.166], *194*
Greig, D., 226 [7.204], *234*
Greiner, J. H., 246, 260, *270*, 329 [11.16], 334, 336, *341*
Gresh, M., 154 [5.244], *160*
Griffiths, L. B., 241 [8.31], *269*
Griffith, R. H., 148 [5.173], *159*
Grimmeiss, H. G., 204, 213, 220, *230*
Grodshtein, E., 238, 240 [8.21], *268*
Grønvold, F., 199 [7.15], 226 [7.201], 228 [7.215], *230, 234*, 252 [8.132], 253, 256 [8.132], *271*

Grossman, L. N., 90 [4.37], *115*
Gruen, D. M., 251 [8.105], 252, *271, 272*
Grunewald, H., 139 [5.102], *157*, 188, *194*
Gschneider, K., Jr., 241 [8.37, 8.45], *269*
Guentert, O. J., 247 [8.93], *271*
Guggenheim, H. J., 140 [5.124], *158*, 229 [7.223], *234*
Guillaud, C., 58 [3.25], *75*, 106, 107, 109 [4.108], *117*, 140 [5.126], *158*, 184 [6.147], *194*, 220 [7.153, 7.154], *233*
Guillien, R., 36, 39 [2.16], *48*
Guillot, A., 147 [5.168], *159*
Guittard, M., 236 [8.8], 260 [8.201], 264 [8.201], *268, 273*
Gu'kov, Yu. K., 256 [8.167], *272*
Gurney, R. W., 61, *75*
Gurry, R. W., 168, *191*
Guseva, L. N., 98 [4.83], *117*

H

Haaijman, P. W., 174 [6.76], 175 [6.76], 179 [6.95, 6.96], *192, 193*
Haas, C., 221, 222, 225 [7.192], 226, *233, 234*, 281, 294, *296, 298*, 311, *320*
Haase, D. J., 263, 265, *274*
Haayman, P. W., 46, *48*, 61 [3.33], *75*
Haemers, J., 150, *159*
Hafner, S. S., 210, *232*
Hagel, W. C., 129, *156*
Hagenmuller, P., 154, *160*
Hahn, H., 204 [7.23, 7.25], 213 [7.25], 220 [7.25], *230*
Haigh, G., 184, 188, *194*
Hakim, R., 150 [5.195], *159*
Hall, E. H., 303, *320*
Hamaguchi, Y., 225 [7.193], *234*
Hamilton, W. C., 179, *193*
Hanabusa, M., *156*, 291 [9.56], *297*
Hanley, T. E., 251 [8.111], 252, *271*
Hannan, W. J. 338 [11.39], *342*
Hansen, M., 89, 90 [4.20], 91 [4.20], 99 [4.20], 102 [4.20], 103 [4.20], 106 [4.20], 109 [4.20], *115*, 204 [7.22], 208 [7.22], 214 [7.22], 217 [7.22], *230*, 236 [8.6], *268*
Hansler, R. L., 91 [4.40], *116*
Harada, T., 225 [7.189], 226 [7.189], *234*
Haraldsen, H., 197, 226 [7.201], 228 [7.7], *230, 234*, 255, *272*
Haraldsen, J. 228 [7.215], *234*
Harder, B., 204 [7.23], *230*

Hardy, A., 154, *160*
Harris, T. J., 336 [11.36], *342*
Hartmann, H., 30 [2.10], *48*
Harvey, B. G., 161, *191*
Harwood, M. G., 138 [5.101], *157*
Hasegawa, K., 190, *195*
Hasiguti, R. R., 135, *157*
Hass, M., 190, *195*
Hastings, J. M., 184 [6.157], *194*, 207 [7.60], 208 [7.64, 7.65], 214 [7.119], 215 [7.119], 217 [7.64, 7.65], 226 [7.64, 7.65], *231, 232*
Haubenreisser, W., 181, *193*
Hauffe, K., 128, 133, *156*, 172, 174, *192*, 213, *232*
Hauschild, U., 148 [5.183], *159*
Heikes, R. R., 61, 62 [3.32], *75*, 88 [4.12], 95, *115, 116*, 123, *155*, 171 [6.39], 174, *191, 192*, 216, *232*, 254, 260 [8.187, 8.188], *272, 273*
Heilmann, E. L., 177, *193*
Heisenberg, W., 57, *75*
Heitler, W., 9, *23*
Henisch, H. K., 134 [5.73], *156*
Henry, W. E., 190, *195*, 247 [8.91], *271*
Henry la Blanchetais, C., 171 [6.34], *191*, 247 [8.90], 251 [8.107], *271*
Heraud, A., 143 [5.142], *158*
Herriott, D. R., 340 [11.45], *342*
Hicks, W. T., 214, 216, 223 [7.127], 224 [7.127], *232*
Hihara, T., 205 [7.36], 207 [7.36], 210 [7.20, 7.21], 211 [7.80], 214 [7.120], *230, 231, 232*
Hillman, P., 186 [6.166], *194*
Hilsum, C., 40, *48*
Himes, R. C., 236 [8.10a], 244 [8.63], 245 [8.63], 262 [8.210], 263 [8.63, 8.211], 264 [8.63, 8.210, 8.211, 8.215a], 266 [8.63, 8.211], 267 [8.83, 8.210], *268, 270, 273, 274*
Hipp, J. C., 148, *159*
Hirahara, E., 104 [4.93], 105 [4.95, 4.96], *117*, 172 [6.49], 182 [6.133], 186 [6.133], 187 [6.133], 188 [6.133], *192, 194*, 210 [7.80, 7.81, 7.92], 211 [7.80], 225 [7.189, 7.190], 226 [7.84, 7.189, 7.206], 228, *231, 232, 234*, 285, 286 [9.29], *297*, 309 [10.34], 312 [10.34], 314 [10.34], *320*
Hirakawa, K., 152 [5.212], *160*, 217, *232, 233*
Hirone, T., 205 [7.35], 208 [7.68], 209, 210, 217, 218, 225 [7.137], *230, 231, 232*

Hirota, E., 141, *158*
Hirsch, J., 152 [5.214], *160*
Hirthe, W. M., 136 [5.88, 5.89], 150, 151 [5.201], *157*, *159*, 251 [8.108a], *271*
Hörz, G., 92 [4.51], *116*
Hogarth, C. A., 171, *191*
Hollander, L. E., 90, *115*, 135 [5.78], 137, 138 [5.78], *156*, *157*
Holmes, L., 246, 247 [8.97], *270*, *271*
Holtzberg, F., 236 [8.9], 246, 260 [8.190], 264 [8.190, 8.215], 265 [8.190, 8.206], *268*, *270*, *273*, *274*
Honig, J. M., 142 [5.132], *158*, 246, 251, *270*, *271*
Horita, H., 210, *232*
Hosler, W. R., 131, 134, 135 [5.79], 136, 138, *155*, *156*, *157*
Houston, M. D., 258, *273*
Hoy, J., 92 [4.50], *116*
Hu, J.-H., 173 [6.65], *192*
Hufner, S., 246 [8.77], *270*
Hulliger, F., 84, 85, 103 [4.8], 105 [4.98], 106 [4.8, 4.99], 109 [4.100, 4.107], 111 [4.8], 112 [4.100], *115*, *117*, 141, *158*, 201, 208 [7.63], 212 [7.63], 213 [7.63], 219 [7.63], 229 [7.223], *230*, *231*, *234*, 243 [8.58, 8.59], 244 [8.64], 266 [8.58], *270*
Hund, F., 33, *48*
Hung, C. S., 44, 45, *48*
Hurlen, T., 133, *156*, 228 [7.215], *234*
Hyde, B. G., 246 [8.87c], 247 [8.87d], *270*
Hyland, G. J., 130 [5.51], 140 [5.119], *156*, *157*

I

Iandelli, A., 236, 243, 244, *268*, *270*
Ichinokawa, T., 152 [5.209], *159*
Igosheva, T. N., 111 [4.116], *117*, 221 [7.165], *233*, 311, *320*
Iida, S., 141, 145, *158*, 186 [6.164], 190 [6.164], *194*, *195*, 255, *272*
Illarionov, S. V., 241, *269*
Imbert, P., 67 [3.48], *75*, 188, *194*, 222, 225, 226, *233*
Ingraham, J. N., *158*
Inouye, H., 168 [6.17], *191*
Inowaki, S., 154 [5.241], *160*
Ioffe, A. F., 44, *48*, 95, *116*
Ioffe, V. A., 150, *159*
Irie, F., 152 [5.212], *160*

Irkhin, Iu. P., 246, *270*
Irkhin, Yu. P., 280 [9.11], *296*
Ishikawa, Y., 154 [5.241], *160*, 182 [6.133, 6.134], 183 [6.134], 186 [6.133], 187 [6.133], 188 [6.133], *194*, 210 [7.92], *232*
Ishikura, O., 139, *157*
Ito, A., 182 [6.133], 183, 186 [6.133], 187 [6.133], 188 [6.133], *194*, 210 [7.92], *232*
Itoh, F., 92, 93, *116*
Ivakin, A. A., 246 [8.78, 8.83], *270*
Ivukina, A. K., 139, *157*
Izuyama, T., 291, *297*

J

Jaffe, H., 336 [11.27], *342*
Jaffray, J., 126 [5.28], 127 [5.28], 128, 139, *155*, *156*, *157*, 179 [6.108], *193*
James, W. J., 92 [4.49], *116*
Jan, J. P., 281, 282 [9.19], *296*, 305, *320*
Jancovici, B., 207 [7.49], *231*
Janninck, R. F., 147, 150, 151, *159*
Jansen, R., 205 [7.41, 7.43], 207 [7.43], 214 [7.41, 7.43], 221 [7.41, 7.43], *230*
Jarrett, H. S., 293, *298*
Jellinek, F., 196 [7.4], 221 [7.169], *229*, *233*
Johansen, H. A., 225 [7.187], *234*
Johnson, J. R., 147 [5.169], *159*
Johnson, O. W., 139 [5.110], *157*
Johnston, H. L., 173 [6.65], *192*
Johnston, R. W., 171, *191*
Johnston, W. D., 61, 62 [3.32], 75, 174, *192*, 254, *272*
Jona, F., 151, *159*
Jones, O. C., 336 [11.32], *342*
Jones, R. V., 291, *297*
Jonker, G. H., 62, 68 [3.34], *75*
Jonsson, T., 255 [8.158], *272*
Julg, A., 18, *23*
Junod, P., 242 [8.55, 8.56], 243 [8.58], 244 [8.56], 245 [8.56, 8.66], 246 [8.73], 247 [8.73[, 254 [8.55], 257 [8.176], 258 [8.73, 8.176], 262 [8.176], 263 [8.73], 265 [8.176], 266 [8.58, 8.176], 267 [8.73], *269*, *270*

K

Kabashima, S., 224, *234*
Kachi, S., 120 [5.10], 122 [5.10], 126, 127 [5.35], 139 [5.114, 5.116], 140, 150, *155*, *157*

Author Index

Kafalas, J. A., 142 [5.132], *158*
Kalitin, V. I., 267 [8.223], *274*
Kalnajs, J., 132 [5.58], 133 [5.58], 134 [5.58], *156*
Kalvius, G. M., 210 [7.94], *232*
Kamigaichi, T., 205 [7.37], 207, 210, 211, 214 [7.120], *230, 231, 232*
Kaminow, I. P., 336 [11.31], *342*
Kanamori, J., 289 [9.48], *297*
Kang, Y. H., 190 [6.190], *195*, 205 [7.39], 210 [7.39], *230*
Kanter, M. A., 260 [8.194], *273*
Karantassis, T., 173, *192*
Karo, A. M., 18, *23*
Karpenko, B. V., 217 [7.135], *232*
Karplus, R., 305, *320*
Kasha, M., 17, *23*
Kastler, A., 326, *341*
Kasuya, T., 278 [9.4], *296*
Kataoka, S., 134, *156*
Katz, J. J., 251 [8.105], *271*
Kauer, E., 241, *269*
Kawakubo, T., 125, 140 [5.117], *155, 157*
Kawamiya, N., 135 [5.82], *157*
Kawano, S., 120, 122, *155*
Kaye, G., 188, *195*
Kazmierowicz, C. W., 260 [8.194], *273*
Keezer, R., 135, *157*
Kennedy, T. N., 150 [5.195], *159*
Ketelaar, J. A. A., 4 [1.3], 11 [1.3], 15 [1.3], 16, 18 [1.3], *23*
Keys, L. K., 126, *155*
Kharkhanavala, M. D., 255, 256, *272*
Kholoday, G. A., 173, *192*
Kieffer, R., 87 [4.11], *115*
Kienle, P. 246 [8.77], *270*
Kiessling, R., 237 [8.14], *268*
Kikoin, I. K., 111, *117*, 220 [7.156, 7.157], 221, 233, 285 [9.31], 286 [9.31], *297*, 311, *320*
Kimball, G. E., 7 [1.4], *23*
Kirschbaum, A. J., 210 [7.87], *231*
Kistner, O. C., 164, *191*
Kitahiro, I., 140 [5.122, 5.123], *157, 158*
Kjekshus, A. J., 81, 82, 83, *115*, 197, 229 [7.224], *230, 234*
Kleber, W., 135, *156*
Klein, G. P., 154 [5.245], *160*
Kleinman, D. A., 324 [11.6], *341*
Kleinpenning, T. G. M., 139 [5.103], *157*

Klichkova, V. P., 256 [8.167], *272*
Klimashin, G. M., 90 [4.33, 4.34, 4.38], *115*
Klinger, M. I., 63, *75*, 171, *191*
Klixbüll-Jørgensen, C., 51, *74*
Klyuskin, V. V., 225 [7.188], *234*
Knight, H. T., 145, *158*
Knuvers, G. F., 172 [6.51], *192*
Kochnev, M. I., 106 [4.102], *117*
Koehler, W. C., 236 [8.4], 241 [8.47, 8.48], 243 [8.48], 224 [8.48], 245 [8.48], 251 [8.105], *268, 269, 271*
Kofstad, P., 150, *159*
Koide, S., 171, 174, *192*
Kolchin, P. O., 147 [5.171], *159*
Kolina, E. M., 228 [7.217], *234*
Komarova, T. I., 258 [8.185], *273*
Komatsubara, T., 225, *234*
Komoto, T., 205 [7.42], 210 [7.88], 213, *230, 231, 232*
Kondo, H., 216, 225 [7.185], 226 [7.202], 228 [7.213, 7.221], *232, 234*
Kondo, J. 278 [9.6], *296*
Konkov, B. L., 305 [10.17], *320*
Kooy, C., 331, *341*
Kopytina, V. V., 207, *231*
Korshunov, V. A., 99, 101, 102, *117*
Kosanke, K., 336 [11.36], *342*
Kosolapov, T. Ya., 91 [4.44], *116*
Kosuge, K., 120 [5.10], 122 [5.10], 126 [5.31], 127 [5.35, 5.36], 128, 139 [5.31], 140 [5.31], 150 [5.31], *155, 156, 157, 158*, 166
Kouvel, J. S., *75*, 106 [4.101], 107, 109, *117*
Koval'chenko, M. S., 90 [4.35], *115*
Kovelev, L. Ya., 303 [10.12], *320*
Kozlovskii, L. V., 90 [4.33, 4.38], *115*
Kramers, H. A., 58, *75*, 167, *191*
Krasovskii, V. P., 104, *117*, 311, *320*
Krebs, H., 24, 25, 26, 27 [2.5], 28 [2.4], 29 [2.4], 33, *47*, 123, *155*, 197, 230
Krinchik, G. S., 326 [11.12], 329 [11.15], *341*
Kristya, V., 139 [5.109], *157*
Kröger, F. A., 147, *158*
Kruger, O. L., 241, 244, 260, *269*
Kruglov, V. I., 173 [6.57], *192*
Krusse, G., 133 [5.64], *156*
Ksendzov, Ya. M., 62, *75*, 138 [5.100], 139, *157*, 173 [6.57], 174, *192*
Kubota, B., 141, *158*
Kudinov, E. K., 225 [7.195], *234*
Kudintseva, G. A., 240 [8.22], *268*

Kuhrt, F., 315 [10.50], 316 [10.50, 10.52], 318 [10.53], *321*
Kulcke, W., 336, *342*
Kumitomi, N., 225, *234*
Kundt, A., 303, *320*
Kurnick, S. W., 260, *273*
Kushiro, I., 190, *195*
Kusma, J. B., 99 [4.89], *117*
Kusumoto, H., 140 [5.120], *157*
Kutsev, V. S., 98 [4.84], *117*
Kuylenstierna, U., 133 [5.63, 5.64], *156*
Kuznetsov, V. G., 212, 213, 219, *232*, 266 [8.221], 267 [8.221], *274*

L

Lafferty, J. M., 239, *268*
Laffitte, M., 212 [7.99], 213 [7.108], *232*
La Grange, L. D., 147 [5.159], *158*
Lagrenaudie, J., 207 [7.46, 7.47], 208 [7.51], *231*
Lal, H. B., 291, *297*
Lallement, R., 241 [8.32, 8.34, 8.49], *269*
Lambe, J., 240 [8.29], *269*
Landau, L., 61, *75*, 288, *297*
Laplaca, S., 240 [8.26], *269*
Lashkar'ev, G. V., 236 [8.10], 238, 260, 262 [8.208], 264 [8.208], 265 [8.208], *268*, *273*, *274*
Lasker, M. F., 252, *271*
Latysheva, V. M., 62 [3.37], *75*, 174 [6.82], *192*
Laurance, N., 240 [8.29], *269*
Lavine, J. M., 180, 181, *193*, 306, 307, 308, 309, 310, *320*
Lawrence, P. E., 267 [8.218], *274*
Lawson, A. W., 139 [5.115], *157*
Leask, M. J. M., 255, 256, *272*
Leavy, J. F., 260 [8.200, 8.203, 8.204], *273*
Lebedev, A. G., 252, 256 [8.138, 8.169], *272*
Leciejewicz, J., 218, *233*
Lecocq, P., 102, *117*
le Craw, R. C., 331, 332, 339 [11.41], *341*, *342*
Leeuwerik, F. J., 139 [5.103], *157*
Lepinskikh, B. M., 150, *159*
Lepetit, A., 223, *234*
LeRoy, N. R., 257 [8.174], 260 [8.197], *272*, *273*
Levin, R. L., 170 [6.22], 174 [6.22], *191*
Levy, F., 242 [8.55, 8.56], 244 [8.56], 245 [8.56], 254 [8.55], *269*

Lewis, D. C., 181 [6.126], 186 [6.126], *193*
Lewis, G. N., 9, *23*
Lewis, J., 226 [7.198], *234*
Lifshitz, E. M., 288, *297*
Lihl, F., 218 [7.145], *233*
Lin, S. T., 186, *194*
Lindars, P. R., 148 [5.173], *159*
Lindberg, O., 299, 301, 302, 303 [10.1], *320*
Lippmann, H. J., 315 [10.50], 316 [10.50], *321*
Lister, B. A. J., 252, *272*
Lister, M. W., 252, 256, *272*
Llewellyn, J. P., 226 [7.204], *234*
Lofgren, N. L., 260 [8.192], *273*
Lommel, J. M., 142 [5.135], *158*
London, F., 9, *23*
Lorang, R., 138 [5.96], *157*
Loriers, J., 124, *155*
Loshmanov, A. A., 225 [7.188], *234*
Lotgering, F. K., 110, *117*, 183, *194*, 214, 215, 220, 223 [7.118], *232*
Lototskii, B. Yu., 154 [5.238], *160*
Lowell, M. C., 173, *192*
Lundin, C. E., 241 [8.42], *269*
Lundström, T., 79 [4.3], 106 [4.3], *115*
Luttinger, J. M., 305, *320*
L'vov, S. N., 88, 89, 90 [4.29], 91 [4.44], 92 [4.52], 95 [4.29], 97 [4.29], *115*, *116*, *120*, *155*, 241 [8.30], *269*
Lyand, R., 126 [5.28], 127 [5.28], *155*
Lyashenko, V. I., 207 [7.52], *231*
Lyon, L., 190 [6.181], *195*

M

Maaz, K., 318 [10.53, 10.54], *321*
McChesney, J. B., 251 [8.109], *271*
McClure, J. W., 258, 263 [8.183], 266 [8.183], *273*
McGuire, T. R., 216, *232*, 257, 258 [8.179, 8.180], 260 [8.190, 8.206], 262 [8.179, 8.180], 263 [8.180], 264 [8.190, 8.206, 8.215], 265 [8.190, 8.206], 266 [8.180], *272*, *273*, 288 [9.41], *297*, 331 [11.21], *341*
McKeehan, L. W., 283, *297*
Mackenzie, J. D., 150 [5.195], *159*
Mackintosh, A. R., 154, *160*
McNeill, D. J., 95, 97, *116*, *117*
McNeilly, C. E., 251, *271*
McReynolds, A. W., 179, *193*
McTaggart, F. K., 204, *230*

Maeda, S., 205 [7.35], 210 [7.77, 7.78], 217 [7.136], *230, 231, 232*
Magascuva, S., 152, *159*
Magneli, A., 133 [5.63, 5.64], *156*
Maissel, L. I., 154 [5.242], *160*
Makarov, E. S., 252, *271*
Makarov, V. V., 173, *192*
Makovetskii, G. I., 225, *234*
Manoilova, I. G., 258 [8.186], 259 [8.186], *273*
Manakov, A. I., 150, *159*
Manca, P., 42 [2.19], *48, 55, 56, 64, 66* [3.43], 67, 68, 73, *74, 75,* 226, 227, 228, *234*
Mann, W., 180 [6.117], *193*
Mansfield, R., 207, 208, *231*
Maranzana, F. E., 305, *320*
Marchenko, V. I., 258 [8.182], 260 [8.205], 262, *273*
Mariani, J. L., 154 [5.246], *160*
Marinace, J. C., 210, 212, *232*
Marincek, O., 245 [8.67], *270*
Marinov, A., 182 [6.132], *194*
Marion, F., 148, 152, 153, *159*, 168, *191*, 255 [8.157], 256 [8.166], *272*
Marotta, A. S., 278, 279, *296*
Marshall, W., 210 [7.87], *231*
Martin, R. L., 246, *270*
Martinet, J., 179 [6.107], 184 [6.107], 190, *193*
Masumoto, K., 205 [7.36], 207 [7.36], 214, 216, *230, 232*
Mathieu, J.-P., 322, 326 [11.2], 327, 328 [11.2], *341*
Matson, L. K., 236 [8.10a], 244 [8.63], 245 [8.63], 263 [8.63, 8.211], 264 [8.63, 8.211, 8.215a], 266 [8.63, 8.211], 267 [8.63], *268, 270*
Matsumura, Y., 104 [4.93], 105 [4.95, 4.96], *117*, 285, 286 [9.29], *297*
Matthias, B. T., 152 [5.213], 154 [5.233], *160*, 219 [7.150], 228 [7.150], *233*, 241 [8.44, 8.45], 246, *269, 270*
Matveyenko, I. I., 91 [4.43], *116*
Matyas, M., 55 [3.15], *74*
Maurat, M., 256, *272*
Max, E., 336 [11.36], *342*
Maxim, M. S., 218 [7.143], *233*
Mayer, S. E., 97, *116*
Mazumder, B. R., 154, *160*

Mead, C. A., 154, *160*
Mead, W., 210 [7.87, 7.88], *231*
Mekata, M., 90 [4.31], *115*
Melikadamyan, V. R., 240 [8.23], *269*
Men'shikov, A. Z., 198, 214, 221, *230*
Menth, A., 242 [8.56], 244 [8.56], 245 [8.56, 8.67], *270*
Menyuk, N., 205 [7.38], *230*
Mercier, M., 205 [7.41], 214 [7.41], 221 [7.41], *230*, 291, 292, *298*
Merriam, M. F., 260 [8.199a], *273*
Merten, U., 147 [5.159], *158*
Mesnard, G., 251 [8.112, 8.113, 8.117, 8.119], *271*
Methfessel, S., 236 [8.9], 246 [8.68a], 260 [8.190, 8.206], 264 [8.189, 8.190, 8.206, 8.215], 265 [8.189, 8.190, 8.206], *268, 270, 273, 274*
Meyer, W., 252, *272*
Meyer-Schützmeister, L., 173 [6.59], *192*
Michel, A., 53, *74*, 140 [5.126], *158*, 190 [6.184], *195*
Miles, P. A., 181, *193*
Miller, C. F., 147 [5.159], *158*
Miller, E. C., 219 [7.150], 228 [7.150], *233*
Miller, J. F., 236 [8.10a], 244 [8.63], 245 [8.63], 262, 263 [8.63, 8.211], 264 [8.63, 8.210, 8.211], 266 [8.63, 8.211], 267 [8.63], *268, 270, 273, 274*
Miller, R. C., 112 [4.123], 113 [4.123], *118*, 123, *155*, 174, *192*, 260 [8.199], *273*
Minami, K., 135 [5.81], *157*
Minomura, S., 127 [5.34], 139 [5.116], *155, 157*, 226 [7.200], *234*
Mishchenko, L. B., 198 [7.13], 230
Mitoff, S. P., 171, *192*
Miyahara, S., 208 [7.67], 226 [7.200], *231, 234*
Miyata, N., 183, *194*
Mlavsky, A. I., 97, *116*
Moers, K., 90, *115*
Mössbauer, R. L., 162, *191*
Mollard, P., 226 [7.198], *234*
Moody, J. W., 252 [8.133], 254 [8.133], *271*
Moore, W. J., 3, 12 [1.1], 13 [1.1], 17, *23*
Mooser, E., 84, 85, 103 [4.8], 106 [4.8], 111 [4.8], *115*, 201, *230*
Morgan, F. H., 251 [8.115], *271*
Morin, F. J., 62 [3.36], 65, 73, *75*, 119, 120, 125, 126, 132, 139, *155*, 171, 174, 184, 185, 188, *191, 194*, 225 [6.152]

Morozov, Yu. N., 246 [8.78, 8.83], *270*
Moser, J. B., 241, 244, 260, *269*
Moss, R. L., 190 [6.180], *195*
Mott, N. F., 44, *48*, 61, 64, *75*, 277, *296*
Mozzi, R. L., 247 [8.93], *271*
Mudar, J., 135 [5.87], *157*
Müller, J., 229 [7.223], *234*
Münster, A., 90, 91, *115*
Mularz, W. L., 267 [8.221a], *274*
Mulay, L. N., 126, *155*
Multani, M. S., 145, *158*
Munson, R. A., 213 [7.110], *232*
Murakami, M., 172 [6.49], *192*, 210 [7.81, 7.82, 7.84, 7.85], 225 [7.190], 226 [7.84, 7.206], 228 [7.84], *231, 234*, 309 [10.34], 312 [10.34], 314 [10.34], *320*
Murasik, A., 205 [7.41], 214 [7.41], 221 [7.41], *230*
Murat, M., 214 [7.123], *232*
Muromkin, Yu. A., 220 [7.163], *233*, 311 [10.37], *320*
Myers, H. P., 255, *272*

N

Nachman, M., 173, *192*
Nagamiya, T., 183, *194*
Nagasaki, H., 127 [5.34], 139 [5.116], *155, 157*
Nagasaki, S., 205 [7.35], *230*
Nagasawa, S., 151, *159*, 247 [8.88], *270*
Nagels, P., 170, 174, *191*, 255 [8.156], *272*
Nagy, I., 179 [6.112], *193*
Nakada, I., 207 [7.53], *231*
Nakamura, Y., 90 [4.31], *115*, 127 [5.35], *155*, 205 [7.40], *230*
Nakau, T., 186, *194*
Narita, K., 140 [5.120], *157*
Nathans, R., 266 [8.217], *274*
Naya, R., 226 [7.206], *234*
Naylor, B. F., 124, 125 [5.19], *155*
Néel, L., 179, 184 [6.143, 6.144, 6.155], 190, *193, 194*, 208 [7.62], 210, 212 [7.62], 213 [7.62], *231*
Nelson, C. M., 148 [5.177], *159*
Nelson, C. W., 94, *116*
Nemchenko, V. F., 88, 89, 90 [4.29], 91 [4.44], 92 [4.46, 4.52, 4.54], 95 [4.29], 97 [4.29], *115, 116*, 120 [5.7], *155*, 241 [8.3a], *269*
Nemnonov, S. A., 92 [4.53], *116*, 198 [7.13, 7.14], 214 [7.14], 221 [7.14], *230*
Nereson, N. G., 246 [8.70], *270*
Neshpor, V. S., 90 [4.29], 91, 95, 96, 97 [4.29], *115, 116*
Ness, P., 204 [7.25], 213 [7.25], 220 [7.25], *230*
Neth, F., 252, *272*
Neuhaus, A., 27, *48*
Neuimin, A. D., 256 [8.170], *272*
Neuman, C. H., 139, *157*
Newman, R., 171, *191*
Newman, W. A., 104 [4.94], 106 [4.94], 108 [4.94], *117*
Nicolau, P., 124 [5.16], 147 [5.168], *155, 159*, 183 [6.141], *194*
Nikitin, E. N., 97 [4.72, 4.74, 4.75, 4.76], 98 [4.72, 4.74], 99 [4.74, 4.75], *116*
Nitsche, R., 204, 213, 214 [7.26], 220, *230*
Noddack, W., 247, *271*
Nogami, M., 111, *117*, 223, *233*, 311, *320*
Noguchi, S., 90 [4.28], *115*
Nomura, S., 125 [5.23, 5.215], *155, 160*
Norreys, J. J., 241 [8.33], *269*
Novogrudskii, V. N., 221 [7.164], *233*
Nowlin, C. H., 291, *297*
Nowotny, H., 99 [4.89], *117*
Nuralieva, R. D., 329 [11.15], *341*

O

Obolonchik, V. A., 236 [8.10], 238, 262, *268*
O'dell, T. H., 291, *297*
Ofer, S., 182 [6.132], *194*
Ogorodnikov, V. V., 90 [4.35], *115*
Ohashi, T., 140 [5.121, 5.122], *157*
Ohm, E. A., 324 [11.7], 338, *341*
Okada, T., 152, *160*
Okamura, T., 179 [6.102], *193*, 302, 310, *320*
Okazaki, A., 217 [7.141], 218 [7.141], *232, 233*
O'keefe, M., 173, *192*
Oleinik, M. I., 221, 223 [7.167], *233*
Olsen, C. E., 246 [8.70], *270*
Olson, A. R., 18, *23*
Ōno, K., 182 [6.134], 183 [6.134], 186, 187, 188, *194*, 210, *232*
Oppegard, A. L., *158*
Oprea, F., 173 [6.62], *192*
Ordan'yan, S. S., 90, *115*
Orgel, L. E., 49, 51, 52, 53, *74*
Ormancey, G., 105, *117*
Ormont, B. F., 98 [4.84], *117*

Ornatskaya, Z. I., 154, *160*
Osika, L. M., 267 [8.218], *274*
Ovechkin, B. I., 98 [4.83], *117*
Ozerov, R. P., 154, *160*

P

Paderno, Yu. B., 240 [8.23, 8.24, 8.28], 260 [8.207], 265, *269, 273, 274*
Pajares Somoano, J. A., 129 [5.46], *156*
Pal, L., 179 [6.112], *193*
Pallagi, D., 179 [6.112], *193*
Palmer, W., 216, 225, *232*
Pannetier, G., 14 [1.8], *23*
Panova, Ya. I., 139 [5.109], *157*
Pardo, M. P., 236 [8.8], 260 [8.201], 264 [8.201], *268, 273*
Parker, R. A., 130, *156*
Parravano, G., 174, *192*
Pascal, P., 87, 90 [4.21], 91 [4.21], 92 [4.21], 95 [4.21], 102 [4.21], 103 [4.21], 106 [4.21], 109 [4.21], *115*, 120, 123 [5.8], 126 [5.8], 128 [5.8], 130 [5.8], 132 [5.8], 139 [5.8], 148 [5.8], 150 [5.8], 151 [5.8], 152 [5.8], *155*, 168 [6.13], 170 [6.13], 171 [6.13], 178, 179 [6.13], 184 [6.13], 185, 190 [6.13], *191*, 197 [7.6], 204 [7.6], 205, 207 [7.6], 208 [7.6], 212 [7.6], 213 [7.6], 214 [7.6], 216 [7.6], 217 [7.6], 220 [7.6], 223 [7.6], 225 [7.6], 228 [7.6], *230*, 235 [8.1], 240, 243 [8.1], 249 [8.1], 251 [8.1], 252 [8.1], 255 [8.1], 257 [8.1], 260 [8.1], 265 [8.1], *268*
Pascard, R., 241 [8.34], *269*
Pasynkov, V. V., 247 [8.98], *271*
Patrakhin, N. P., 303, *320*
Patrie, M., 236 [8.8], 257 [8.173], 260 [8.201], 264 [8.201], *268, 272, 273*
Patrina, I. B., 150, *159*
Paul, W., 130 [5.50], *156*
Pauling, L., 8 [1.5], 14 [1.5], 16 [1.5], 20 [1.5], 21, *23*, 26, 31, *48, 115*, 201, *230*
Pauthenet, R., 184, *194*, 205 [7.41, 7.43], 207 [7.43], 208, 210, 214 [7.41, 7.43, 7.122], 221 [7.41, 7.43], *230, 231, 232*, 291 [9.65], *298*
Pearson, A. D., 42, *48*, 119, 120, 124, 125, *155*
Pearson, G. L., 313, *321*
Pearson, W. B., 69, *75*, 81, 82, 83, 92 [4.48], 106, 111, *115, 116, 117*, 197, 229 [7.224], *230, 234*

Peibst, H., 135 [5.76], *156*
Penn, T. E., 338 [11.39], *342*
Perakis, N., 139 [5.112], *157*, 173 [6.64], *192*
Perekalina, T. M., 220 [7.151, 7.159], *233*
Pereue, J. H., Jr., 173 [6.52], 174 [6.52], *192*
Peri, G., 173 [6.60, 6.61], *192*
Perny, G., 138 [5.96], *157*
Perrot, M., 173 [6.60, 6.61], *192*
Perthel, R., 212 [7.103], *232*
Pestmalis, H., 133 [5.64], *156*
Peters, C. J., 336 [11.33], 338 [11.40], *342*
Pettus, C., 148 [5.174], 152 [5.174], *159*
Philipp, W. H., 92 [4.47], *116*
Pickart, S. J., 266 [8.217], *274*
Picon, M., 257 [8.173], 260 [8.193], 262 [8.193], *272, 273*
Pinard, J., 337 [11.37], 338 [11.37], *342*
Piper, J., 90, *115*
Pistoulet, B., 299, 301 [10.2], *320*
Pockels, F., 324, *341*
Pollak, H., 255 [8.163], *272*
Pollak, M., 44 [2.24], *48*
Polyakova, M. D., 240 [8.22], *268*
Popescu, F. G., 173 [6.54], *192*
Porter, C. S., 339, *342*
Porter, J. T., 147 [5.159], *158*
Posnjack, E., 177 [6.91], *193*
Post, B., 240 [8.26], *269*
Potter, H. H., 282 [9.26], *297*
Potter, J. F., 251 [8.109], *271*
Pouchard, M., 154 [5.223, 5.224], *160*
Prakash, B., 207 [7.56], *231*
Pratt, G. W., Jr., 291, *297*
Pugh, E. M., 303, 304, 305 [10.14], 306 [10.27], 319 [10.14], *320*
Putley, E. H., 299, *320*
Putseiko, F. K., 213, 214 [7.114], *232*

Q

Quinn, R. K., 214 [7.115], *232*
Quitmann, D., 246 [8.77], *270*

R

Rabenau, A., 204 [7.25], 213 [7.25], 220 [7.25], *230*
Rado, G. T., 53, *74*, 289 [9.44], 291, *297, 298*
Radovskii, I. Zh., 98 [4.82], 99 [4.82], *117*
Radzikivs'ka, S. V., 260 [8.207], *273*
Ramsey, T. H., 267 [8.219], *274*
Rao, K. V., 170, 172, *191*

Rapp, R. A., 252, *271*
Raub, C. J., 154 [5.233], *160*
Raz, B., 255 [8.153], *272*
Rebouillat, H. P., 291 [9.65], *298*
Rebouillat, J. P., 242 [8.54], *269*
Regel, A. R., 150, *159*, 228, *234*
Régnault, F., 207 [7.49], *231*
Reid, F. J., 244, 245, 262 [8.210], 263 [8.211], 264 [8.210, 8.211], 266 [8.211], 267 [8.63, 8.210], *270, 273, 274*
Remeika, J. P., 291, *297*, 331 [11.20, 11 22], 332 [11.22], 333 [11.23], 334 [11.22, 11.23], 335 [11.22, 11.23], *341*
Renon, J., 147 [5.163], *158*
Resnick, R., 91, *116*
Revolinsky, E., 216, 223 [7.129], 224, *232, 234*
Rey, J., 213 [7.108], *232*
Ricci, F. P., 173 [6.67], *192*
Ridgley, D. H., 104 [4.94], 106 [4.94], 108 [4.94], *117*
Rieder, H., 295 [9.74], *298*
Rienäcker, G., 251 [8.104], *271*
Risi, M., 245 [8.66], 257 [8.176], 258 [8.176], 262 [8.176], 265 [8.176], 266 [8.176], *270, 272*
Riste, T., 179, *193*
Robbrecht, G. G., 186, *194*
Robert, J., 173 [6.60, 6.61], *192*
Roberts, B. W., 82 [4.6], *115*
Roberts, L. E. J., 252 [8.135], 253, 255 [8.152], 256 [8.152], *272*
Roche, J., 128, *156*
Rodbell, D. S., 106 [4.105], *117*, 142, *158*, 267 [8.218], *274*
Rodier, G., 143 [5.143], *158*
Rodionov, K. P., 221, 223 [7.167], *233*, 303 [10.12], *320*
Rodot, H., 226 [7.197, 7.198], *234*, 319 [10.55], *321*
Rodot, M., 226 [7.197], *234*, 319 [10.55], *321*
Rogers, D. B., 154 [5.220], *160*
Roilos, M., 170, 174, *191*
Romeijn, F. C., 174 [6.76], 175 [6.76], 179 [6.96], *192, 193*
Romeyn, F. C., 46, *48*, 61 [3.33], *75*
Rooksby, H. P., 82, *115*, 174 [6.74], 179 [6.109], 184, *192, 193, 194*
Roquet, J., 184 [6.145], *194*
Rosenberg, M., 124 [5.16], *155*, 183, *194*

Rosenberg, R., 340 [11.45], *342*
Rosenqvist, T., 212 [7.98], *232*
Rossignol, D., 241 [8.49], *269*
Røst, E., 199 [7.15], *230*
Rostoker, N., 304, 305 [10.14], 319 [10.14], *320*
Roth, R. S., 188 [6.171], *194*
Roth, W. L., 167, 168 [6.15], 170 [6.10], *191*
Roult, G., 205 [7.41, 7.43], 207 [7.43], 214 [7.41, 7.43, 7.122], 221 [7.41, 7.43], *230, 232*
Rubinstein, C. B., 340 [11.45], *342*
Ruby, S. L., 164 [6.7], 168 [6.7], 169 [6.7], *191*
Rudnitsky, V., 305 [10.16], *320*
Rulli, J. E., 254 [8.148], *272*
Rundqvist, S., 79 [4.3], 106 [4.3], *115*
Rupprecht, J., 154 [5.235], *160*
Rustamov, A. G., 119, 121, *155*
Rutter, J., 173 [6.54], *192*
Ryan, F. M., 260, *273*
Ryvkin, S. M., 204, 213, *230, 232*

S

Sadler, A. G., 181, 186, *193*
Sadler, M. S., *158*
Sadofsky, J., 241 [8.35], *269*
Sagar, A., 112 [4.123], 113 [4.123], *118*
Sagel, 90 [4.26], 91 [4.26], *115*
Sakata, K., 135, *157*
Sakata, T., 97 [4.77], 98, *116*, 139, *157*
Sakurai, H., 90 [4.31], *115*
Salam, S. A., 207 [7.54], 208, *231*
Salikhov, S. G., 179 [6.111], *193*
Saltsburg, H., 172, 173 [6.52, 6.53], 174 [6.52], *192*
Salunina, A. E., 133 [5.70], *156*
Salzano, F. J., 241 [8.35], *269*
Samoilovich, A. G., 305 [10.17], *320*
Samokhvalov, A. A., 119, 121, *155*, 180, 181, *193*, 246, *270*, 285, *297*, 305, 306 [10.26], 310, 311, *320*
Samsonov, G. V., 88, 89, 90 [4.29], 91 [4.44], 92 [4.46, 4.52, 4.54], 95 [4.29], 96, 97 [4.29], *115, 116*, 120 [5.7], *155*, 238, 240 [8.21, 8.22, 8.23, 8.28], 241 [8.30], 258 [8.181], 260, 262, *268, 269, 273*
Santoro, A., 173 [6.67], *192*
Santoro, G., 91 [4.42], *116*
Sapet, A., 204, *230*

Sasaki, Y., 97 [4.78], 98 [4.78], 99 [4.85], 101 [4.85], *116*, *117*, 154 [5.241], *160*
Sato, T., 90 [4.28], *115*
Saut, G., 42 [2.19], *48*, 228, 229, *234*
Sauze, A., 173 [6.60], *192*
Sawada, S., 152 [5.215, 5.216, 5.217, 5.218], *160*
Sawamoto, K., 171 [6.40], *191*
Sawaoka, A., 226, *234*
Scatturin, V., 173 [6.67], *192*
Schab, G. M., 247, *271*
Schafer, M. W., 257 [8.175], 258 [8.179, 8.180], 262 [8.175, 8.179, 8.180], 263 [8.180], 265 [8.175], 266 [8.180], *272*, *273*
Schaner, B. E., 254 [8.148], *272*
Schelleng, J. H., 291, *298*
Schieber, M., 186 [6.165], *194*, 246, 247 [8.97], *270*, *271*
Schlamp, G., 90 [4.26], 91 [4.26], *115*
Schlosser, E. G., 174 [6.73], *192*
Schmalzried, H., 247, *271*
Schmid, H., 174, *192*, 295 [9.74], *298*
Schmid, R., 318 [10.54], *321*
Schmidt, H. E., 255 [8.153], *272*
Schneiderhahn, K., 282 [9.25], 284, *297*
Schottky, W., 29, 33, *48*
Schröder, H., 181, *193*
Schröder, W., 135 [5.76], *156*
Schultz, G., 128, *156*
Schwab, G. M., 174, *192*
Schwartz, C. M., 170 [6.25], 171 [6.25], 173 [6.25], *191*
Schwartz, N., 154 [5.244], *160*
Schwob, P., 243 [8.58, 8.59], 260 [8.191], 264 [8.191], 266 [8.58], 267 [8.191], *270*, *273*
Sclar, N., 242, *269*
Scott, E. J., 288 [9.41], *297*
Sears, R. W., 303, *320*
Seel, F., 3, 8, *23*
Segal, E., 182 [6.132], *194*
Seitz, F., 61 [3.29], *75*
Seki, Y., 154 [5.241], *160*
Sekinobu, M., 111 [4.117], *117*, 311 [10.42], *320*
Selwood, P. W., 124, 125 [5.20], *155*, 190 [6.181], *195*, 252, *272*
Sergeeva, V. M., 258 [8.184, 8.185], 263 [8.184] 266 [8.184], *273*
Serre, J., 218, 223, 226, *233*, *234*, 308, 311, *320*

Serres, A., 173 [6.64], *192*
Seuter, A. M., 225 [7.192], *234*
Sewell, G., 63, *75*, 171, 186, *191*, *194*
Shabel'nikova, A. E., 154 [5.240], *160*
Shafer, M. W., 331, *341*
Shanks, H. R., 154 [5.229], *160*
Shavkunov, P. M., 138, *157*
Shavrov, V. G., 287, *297*
Shcherbakova, G. A., 190 [6.182], *195*
Shchipanov, V. A., 221, 223 [7.167], *233*
Shechter, H., 186 [6.166], *194*
Shelton, J. P., 128 [5.37], *156*, 184 [6.151], 186 [6.151], *194*
Shelykh, A. I., 170, 172, 174, *191*, *192*
Shen, Y. R., 329, 330, *341*
Sherwood, R. C., 251 [8.109], *271*, 331, *341*
Shiga, M., 127 [5.35], *155*
Shimomura, K., 213, *232*
Shimomura, Y., 174, *192*
Shinjo, T., 127 [5.35, 5.36], 128, *155*, *156*
Shirakawa, Y., 284, *297*
Shirane, G., 151, *159*, 164 [6.7], 168, 169, *191*, 223 [7.179, 7.180], *233*
Shirn, G. A., 154 [5.243], *160*
Shtrikman, S., 188 [6.173], *194*, 291, *297*
Shull, C. G., 123 [5.12], *155*, 167, 179, 184, 186, *191*, *193*, *194*, 216 [7.131], *232*
Shur, Ya. I., 281, *296*
Shur, Ya. S., 220 [7.161], *233*, 285, *297*
Shuvayev, A. T., 198 [7.11], *230*
Shvaiko-Shvaikovskii, V. E., *191*
Sidles, P. H., 154 [5.229, 5.230], *160*
Sidorenko, F. A., 97 [4.71, 4.78], 98 [4.78, 4.82], 99 [4.82, 4.85], 101 [4.85], *116*, *117*, 218 [7.144], *233*
Siegel, L., 150 [5.185], *159*
Sienko, M. J., 152, 154 [5.219], *160*
Simard, G., 150, *159*
Simmons, R., 214 [7.115], *232*
Simonova, M. I., 246 [8.78, 8.83], *270*
Simsa, Z., 183, *194*
Singh, J., 207 [7.56], *231*
Singh, N., 256, *272*
Siratori, K., 141, 145, *158*, 186 [6.164], 190 [6.164], *194*
Sirota, N. N., 55, *74*, 225, *234*
Skrabek, E. A., 99, *117*
Slater, J. C., 61 [3.28], *75*
Sleight, A. W., 154 [5.221], *160*
Smakula, A., 170, 172, *191*

Smart, J. S., 257 [8.175], 258 [8.179], 262 [8.175, 8.179], 265 [8.175], *272, 273*
Smirnov, F. S., 205, *230*
Smirnov, I. A., 225 [7.195], *234*
Smit, J., 246, 258 [8.178], 260, 262 [8.178], 264 [8.79], 266 [8.178], 267 [8.79], *270, 273*, 305, *320*
Smith, A. W., 303, 304, *320*
Smith, B. A., 63 [3.40], *75*, 171 [6.31, 6.33], 176 [6.87, 6.90], *191, 193*
Smith, D. O., 181, *193*, 335 [11.26], *342*
Smith, R. A., 36 [2.14], *48*
Smith, S. D., 328, *341*
Smith, T., 226 [7.204], *234*
Smith, W. T., 148 [5.177], *159*
Smrcek, K., 188, *194*
Smyth, D. M., 154, *160*
Snetkova, V. A., 90 [4.38], *115*
Snitko, O. V., 207 [7.52], *231*
Snowden, D. P., 172 [6.47], 173 [6.52, 6.53], 174 [6.52], *192*
Sobolyev, V. V., 262 [8.208], 264 [8.208], 265 [8.208], *273*
Soffer, B. H., 132 [5.57], *156*
Soria Ruiz, J., 129 [5.46], *156*
Sparks, J. T., 205 [7.42], 210 [7.87, 7.88], 213, *230, 231, 232*
Sparks, M., 293 [9.68], *298*
Spector, H. N., 341, *342*
Spedding, F. H., 241 [8.37], *269*
Spencer, E. G., 339 [11.41], 340, *342*
Springthorpe, A. J., 63 [3.40], *75*, 171 [6.31, 6.33], 174, 176 [6.87, 6.90], *191, 193*
Srivastava, K. G., 132 [5.59], *156*, 291 [9.55], *297*
Srivastava, R., 291 [9.55], *297*
Stähelin, P., 152, *159*
Stalder, E. W., 289, *297*
Standley, C. L., 154 [5.242], *160*
Stark, G., 316 [10.52], *321*
Steger, J., 150 [5.185], *159*
Steinfink, H., 263, 265, 267 [8.219, 8.224], *274*
Steinitz, R., 91, *116*
Stevens, K. W., 207 [7.58], *231*
Stössel, H., 295 [9.74], *298*
Stone, F. S., 173, *192*
Stoyantsova, Z. P., 266 [8.221], 267 [8.221], *274*
Straumanis, M. E., 92 [4.49], *116*

Strauser, W. A., 123 [5.12], *155*, 167 [6.9], 184 [6.149], 186 [6.149], *191, 194*, 216 [7.131], *232*
Strekalovskii, V. N., 272
Stryel'nikova, N. S., 90, *115*
Suchet, J. P., 18 [1.13, 1.14], 19 [1.13, 1.14, 1.15], 21 [1.15], 22 [1.13], *23*, 28 [2.8], 30, 31 [2.11, 2.12], 32, 33 [2.12], 42, 43 [2.20], 44 [2.21], 46 [2.28], *48*, 51, 56 [3.16], 58 [3.21, 3.22], 59, 60 [3.26], 63 [3.42], 64, 65 [3.42], 66 [3.43], 67 [3.48], 68, 69 [3.52, 3.53], 70, 71 [3.16], 72 [3.53], 73 [3.51, 3.52, 3.53], *74, 75*, 83 [4.7], 96, 111 [4.114], *115, 116, 117*, 134 [5.74], 142, 143, 145, *156, 158*, 188, 189, *194*, 199, 200, 221, 222, 223, 225, 226, 227, 228 [7.175, 7.212], *230, 233, 234*, 279, 280 [9.12], 293, 294, *296, 298*, 308, 311, 313 [10.49], 318 [10.49], 319, *320, 321*
Suchil'nikov, S. I., 97 [4.73], 99 [4.73], *116*
Suhl, H., 53, *74*
Suits, J. C., 246 [8.68a], 260 [8.190, 8.206], 263 [8.213], 264 [8.80, 8.190, 8.206, 8.215], 265 [8.190, 8.206], *270, 273, 274*, 324, 325, 326, 329, 331, 335 [11.8], *341*
Sumarokova, N. V., 147 [5.171], *159*
Sunyar, A. W., 164, *191*
Surney, L., 150, *159*
Suzuki, T., 104, 105, *117*, 134, *156*, 285, 286 [9.29], *297*
Suzuoka, T., 110, *117*
Svirina, E. P., 305, 306 [10.25], 319 [10.25], *320*
Sweedler, A. R., 154 [5.233], *160*
Sweett, F., 186 [6.169, 6.170], *194*
Switendick, A. C., 88, *115*
Swoboda, T. J., *158*
Syrbu, N. N., 262 [8.208], 264 [8.208], 265 [8.208], *273*
Syrkin, Y. K., 17 [1.10], *23*

T

Taichki, M., 289 [9.48], *297*
Taft, E. A., 151, *159*
Takada, T., 126 [5.31], 139 [5.31, 5.114], 140 [5.31], 150 [5.31], *155, 157*
Takahashi, S., 251, *271*
Takaki, H., 90 [4.31], 92 [4.55], 93 [4.55], *115, 116*, 127 [5.35], *155*
Takei, H., 171, *192*

Takei, W. J., 223 [7.179, 7.180], *233*
Talalaeva, E. V., 285, *297*
Tallan, N. M., 147 [5.164], *158*
Tallman, N. M., 249, *271*
Tanabe, Y., 282 [9.21], *296*
Tanner, D. W., 186 [6.169], 190 [6.180], *194, 195*
Tannhauser, D. S., 63, 64, 75, 133, 135, *156*, 168, 169, 170, 171, 181, 182, 186, *191, 193*
Tarama, K., 126, *155*
Tare, V. B., 247, *271*
Tasaki, A., 186, 190, *194, 195*
Taylor, A. W. B., 130 [5.51], 140 [5.119], *156, 157*
Taylor, R. I., 255, *272*
Tazaki, H., 210 [7.80], 211 [7.80], *231*
Tazawa, S., 210 [7.90], 218 [7.90], *232*
Templeton, D. H., 239 [8.16], *268*
Teranishi, S., 126, *155*
Tetenbaum, M., 260, *273*
Teufer, G., 214, *232*
Theodossiou, A., 210, *232*, 313, 314, 315, *321*
Thorn, R. J., 253 [8.146], 254 [8.146], 255 [8.162], *272*
Tipsord, R. F., 95 [4.61], *116*
Tkachenko, E. V., 256, *272*
Tokushima, T., 97 [4.77], 98, *116*
Tombs, N. C., 179, *193*
Tomlinson, J. W., 168 [6.17], *191*
Torizuka, Y., 179 [6.102], *193*, 302, 310, *320*
Tortosa, J., 173 [6.60], *192*
Toyozawa, Y., 282 [9.22], *297*
Trammell, G. T., 242 [8.52], 243 [8.52], 245 [8.52], *269*
Tripp, T. B., 154 [5.243], *160*
Tripp, W. C., 147 [5.164], *158*
Trombe, F., 170, *191*
Trüpel, F., 148 [5.183], *159*
Trusova, V. P., 98 [4.84], *117*
Trzebiatowski, W., 252, *272*
Tsaï, B., 143 [5.138], *158*, 168 [6.14], 173 [6.63], *191, 192*
Tsarev, B. M., 240 [8.22], 241, *268, 269*
Tsubokawa, I., 174, *192*, 213 [7.107], 214, 215, 223 [7.117], 226, *232, 234*, 333 [11.24], *342*
Tsuchida, T., 90, 92 [4.55], 93 [4.55], *115, 116*, 243, 245, 246, *270*
Tsuya, N., 205, 208 [7.32, 7.68], 209, 210 [7.77, 7.78], 217 [7.136], *230, 231, 232*

Turner, C. E., 171 [6.33], 176 [6.90], *191, 193*
Turov, E. A., 246, *270*, 280 [9.11], 287, *296, 297*, 312, *321*
Twose, W. D., 44, *48*

U

Uchida, E., 216, 225, 226 [7.202], 228 [7.213, 7.214, 7.216, 7.221], *232, 234*
Ueda, R., 152 [5.209], *159*, 190, *195*
Umanski, J. S., 90, *115*
Umeda, J., 140, *157*
Ure, R. W., 95, *116*
Uzan, R., 251 [8.117], *271*

V

Vaidanich, V. I., 111 [4.121], 112 [4.127], *117, 118*, 218, *233*
Vainshtein, E. E., 240 [8.28], *269*
Valetta, R., 150, *159*
van Con, K., 221, *233*
van Daal, H. J., 172 [6.51], 176, *192, 193*
van Der Pauw, L. J., 308, *320*
Van Houten, S., 62, 68 [3.34], *75*, 174, *192*, 256, 257, 258 [8.178], 262 [8.178], 265, 266 [8.177, 8.178], *273*
van Oosterhout, G. W., 46, *48*, 61 [3.33], *75*, 174 [6.76], 175 [6.76], *192*
van Peski-Tinbergen, T., 278 [9.7], *296*
Van Raalte, J. A., 132 [5.62], *156*
van Run, A. M., 294 [9.72], *298*
Van Vleck, J. H., 49, 51, 55, *74*, 239, 246 [8.69], *268, 270*
Vasenin, F. I., 133 [5.70], *156*
Vasil'ev, Ya. V., 190 [6 182], *195*
Vasil'eva, L. L., 62 [3.37], *75*, 174 [6.82], *192*
Verkhoglyadova, T. S., 91 [4.45], 92 [4.46, 4.52], *116*, 120 [5.7], *155*
Verleur, H. W., 140 [5.124], *158*
Vernon, M. W., 173, 174 [6.74], *192*
Verwey, E. J. W., 45, 46, *48*, 61, *75*, 124 [5.14], *155*, 174, 175, 177, 179 [6.94, 6.95, 6.96, 6.97], 183, 184, *192, 193*
Vest, R. W., 129, 147 [5.166], *156, 158*, 249, *271*
Vestersjø, E., 199 [7.15], *230*
Veyssie, J. J., 242 [8.54], *269*
Vick, G. L., 135 [5.77], *156*
Vickery, R. C., 148, *159*
Vigileva, E. S., 267 [8.223], *274*
Vihovde, J., 226 [7.201], *234*

Villers, G., 146, *158*, 226 [7.197, 7.198], *234*, 319 [10.55], *321*
Vlasov, V. G., 252, 256 [8.138, 8.169, 8.170], *272*
Vogt, O., 242 [8.55, 8.56], 243 [8.58, 8.59], 244 [8.56, 8.61, 8.64], 245 [8.56, 8.61, 8.66, 8.67], 246 [8.61, 8.68], 254 [8.55], 257 [8.176], 258 [8.176], 260 [8.191], 262 [8.176], 264 [8.191], 265 [8.176], 266 [8.58, 8.176], 267 [8.68, 8.191], *269, 270, 272, 273*
Volger, J., 126, 135, 138 [5.97], 139, *155, 157*, 190, *195*, 287, *297*
Volkenshtein, N. V., 246 [8.78, 8.83], *270*
Volkov, D. I., 305, *320*
Volokobrinskii, Yu. M., 247 [8.98], *271*
Von Ende, H., 168 [6.16], *191*
Von Hippel, A., 132, 133, 134 [5.58], *156*, 181 [6.120], *193*
Vonsovskii, S. V., 220 [7.155], *233*, 281, 287 [9.34], *296, 297*, 303, *320*
Vu Van Qui, 205 [7.41], 214 [7.41], 221 [7.41], *230*

W

Wachter, P., 246 [8.73], 247 [8.73], 258 [8.73], 263 [8.73], 264 [8.214], 267 [8.73], *270, 274*
Wadsley, A. D., 119 [5.2], 133 [5.2], 139 [5.2], 154 [5.2], *155*, 237 [8.12], *268*
Wagini, H., 111 [4.119], *117*
Wagner, C., 45, *48*
Wagner, J. B., Jr., 170 [6.22], 174 [6.22], 176, *191, 193*
Walch, H., 247, *271*
Wallace, W. E., 243, 245, 246, *270*
Wallbaum, H. J., 87 [4.10], *115*
Walter, A. J., 255 [8.152], 256 [8.152], *272*
Walter, J., 7 [1.4], *23*
Wang, F. F. Y., 246, *270*
Ware, R. M., 95, 97, *116, 117*
Waring, J. L., 188 [6.171], *194*
Wasilik, J. H., 130, *156*
Wasscher, J. D., 225, 226, *234*, 312, *320*
Watanabe, A., 140 [5.121, 5.122, 5.123], *157, 158*
Watanabe, I. H., 205 [7.35], 208 [7.32], *230*
Webster, W. L., 303, *320*
Weinreich, O. A., 251 [8.118], *271*
Weiss, E. J., 267 [8.219, 8.224], *274*
Weiss, H., 313, 317, *321*
Weiss, R. J., 278, 279, *296*
Welker, H., 313, *321*
Wells, A. F., 24, *47, 47*, 79, 80, 81, *115*, 148 [5.181], 149, 152, *159*, 196, 201 [7.1], *229*, 235, 236 [8.2], 237, 238 [8.2], 239, 241 [8.2], *268*, 333, *342*
Wertheim, G. K., 161, *191*
Westbrook, J. H., *57*
Westin, R., 255 [8.158], *272*
Westphal, W. B., 132 [5.58], 133 [5.58], 134 [5.58], *156*, 181 [6.120], *193*
Westwood, W. D., 181, 186, *193*
Wey, R., 126 [5.28], 127 [5.28], *155*
Wheeler, M. J., 241 [8.33], *269*
Whitmore, D. H., 132 [5.60], 134 [5.60], 147, 150, 151, *156, 159*
Whitney, E. D., 147 [5.160], *158*
Wickham, D. G., 174 [6.80], *192*
Wiley, J. S., 145, *158*
Wilkinson, M. K., 241 [8.48], 243 [8.48], 244 [8.48], 245 [8.48], *269*
Will, G., 266 [8.217], *274*
Willardson, R. K., 252 [8.133], 254, *271*
Williams, D. E., 241, *269*
Williams, H. J., 251 [8.109], *271*
Willis, B. T. M., 82, *115*, 179 [6.109], 184, *191, 193*, 255, *272*
Wilson, W. B., 170, 171, 173 [6.25], *191*
Wimmer, J. M., 251 [8.108a], *271*
Winslow, G. H., 253 [8.146], 254 [8.146], 255 [8.162], *272*
Wojtowicz, P. J., 207 [7.59], *231*
Wold, A., 205 [7.38], *230*
Wolf, F., 316 [10.52], *321*
Wolf, W. P., 255 [8.152], 256 [8.152], *272*
Wollan, E. O., 123 [5.12], *155*, 167 [6.9], 184 [6.149], 186 [6.149], *191*, 194, 216 [7.131], *232*, 236 [8.4], 241 [8.47, 8.48], 243 [8.48], 244 [8.48], 245 [8.48], *268, 269*
Wolnik, S. J., 267 [8.221a], *274*
Wood, E. A., 291, *297*
Wood, D. L., 331 [11.22], 332 [11.22], 334 [11.22], 335 [11.22], *341*
Wright, D. A., 251 [8.114], *271*
Wright, R. W., 171, 173, *191*
Wucher, J., 124, 125 [5.21], 126 [5.28], 127 [5.28], 139 [5.112], *155, 157*, 184, *194*
Wyart, J., 152 [5.210], *159*
Wyckoff, R. W. G., 79, *115*, 139 [5.111], *157*, 173, *192*, 196, 197 [7.2], 201, *229*, 235, 236 [8.3], 238 [8.3], *268*

Y

Yadaka, H., 225, 226 [7.189], *234*
Yagi, E., 135 [5.82], *157*
Yaguchi, K., 244, *270*
Yahia, J., 125, 135, 138 [5.98], 139, 151 [5.93], *155*, *157*
Yamada, M., 210 [7.90], 218 [7.90], *232*
Yamaguchi, S., 154 [5.234], *160*, 182, *194*
Yamaka, E., 171 [6.40], *191*
Yanagi, T., 125 [5.23], *155*
Yarembash, E. I., 266 [8.221], 267 [8.221, 8.223], *274*
Yasnopol'skii, N. L., 154 [5.240], *160*
Yoffe, A. D., 207 [7.55], *231*
Yonemitsu, H., 135 [5.81], *157*
Yoshimori, A., 143, *158*
Yosida, K., 205, 208 [7.33, 7.69], *230*, *231*
Young, A. P., 170, 171, 173 [6.25], *191*
Young, D. A., 251, *271*
Young, L., 154 [5.236], *160*
Yousef, Y. L., 145, *158*
Yupko, V. L., 95, *116*
Yurkov, V. A., 150, *159*
Yuzuri, M., 190, *195*, 205 [7.39, 7.40], 210 [7.39], *230*

Z

Zachariasen, W. H., 237 [8.11], 241 [8.44], *268*, *269*
Zalevskii, B. K., 262 [8.208], 264 [8.208], 265 [8.208], *273*
Zalkin, A., 239 [8.16], *268*
Zaveta, K., 183, *194*
Zeilmaker, H., 145, *158*
Zhukovskii, V. M., 252, 256 [8.138], *272*
Zhurakovskii, E. A., 98 [4.81], *117*
Zhuravlev, N. N., 240 [8.23], *269*
Zhuze, V. P., 138, *157*, 170, 172, 174, *191*, *192*, 228, *230*, *234*, 258 [8.186], 259 [8.186], 263, 266, *273*
Ziegler, X., 341 [11.46], *342*
Zijlstra, R. J. J., 139, *157*
Zitter, R. N., 340, *342*
Zobina, B. N., 112 [4.126], *118*
Zolyan, T. S., 150, *159*
Zotov, T. D., 181, *193*, 220 [7.161], *233*, 246 [8.78, 8.83], *270*, *285*, *297*
Zvonarev, A. V., 256 [8.167], *272*

Subject Index

A

a_1 or a_{1g} orbital function, 50, 54
Acceptor atom, 53
Acceptor level, 38, 68
Activation energy, 44, 45, 62, 67
Affinity, electronic, 21, 30, 68
Analyzer, 323
Anisotropy, 70, 138, 140, 210, 216, 291, 322, 324
  ferromagnetic, 284
Antibonding (AB) function, level or band, 11, 13, 30, 33, 36, 41, 42, 47, 57, 68, 69, 71–74, 83, 88, 277
Antibonding state, 47, 57, 58
Antiferromagnetism, 57, 88, 110, 225, 288
Antimonides, 64, 82, 83, 103, 109, 245
Apparatus, 162, 163, 288, 289, 292, 308–310, 315, 329–331
Arsenic structure, see A7 (index of Structures)
Arsenides, 82, 83, 103, 106, 244
Arsenopyrite structure, see EO_7 (index of Structures)
Asphericity, 55
Astrov effect, 288, 291, 295, 344
Atmospheric window, 341
Atomic orbital (AO), see Orbital

B

Bachelor electron, 15, 55, 56, 60, 69, 73, 82, 85, 86, 154, 166, 201
Band magnetism, 61
Band model, 33, 37, 39, 41, 43, 44, 63, 68
Band structure, 33, 38–40, 89, 228
Baretter, 293
Bethe curve, 58
Bilz model, 88, 89, 93, 104
Bipyramidal site, 79, 81, 82
Birefringency, 322, 329
Bloch function, 33, 37
Bohr model, 4, 6
Bond angle, 15, 16, 22, 24, 27, 60, 73, 82, 130
Bond energy, 20, 30, 32, 33, 57
Bond resonance, 34–36, 60, 61, 85, 197, 235, 236
Bonding (B) function, level or band, 11, 13, 36, 41, 42, 47, 57, 68, 69, 72, 88, 277
Bonding state, 47, 57, 58
Borides, 88, 89, 239
Brillouin zone, 39
Bronzes, 152, 154
Brookite, 130

C

Carbides, 88, 89, 241
Carbon, 15, 24
Carnot efficiency, 96
Catalytic effect, 148, 150, 204

Cell parameters, 62, 64, 65, 82, 241
Chalcogenides, 68, 196, 236, 257
Charge, *see* specific types
Cohesion, solids, 18, 61
Commutation, 295
Conduction band, 36–39, 42, 97
Corundum structure, *see* $D5_1$ (Index of Structures)
Cotton–Mouton effect, 324, 328, 329
Coulombic interactions, 4, 11, 21, 30, 57, 61
Covalent bond, 9, 14, 18, 21, 28, 57, 58, 60, 268
Covalent charge, 21, 31
Covalent configuration, 19, 30, 31
Covalent crystal, 63, 82, 200
Covalent function, 17–20
Criterion
    Dudkin, 88, 103, 112, 203
    Goodenough, 65, 112, 148, 203
    magnetic, 69, 82
Critical temperature resistor, 293
Crystal field, 16, 49, 51, 73, 88
Crystallochemical model, 41, 51, 88
Cubic arrangement, 29, 235, 236
Curie and Curie–Weiss laws, 55
Curie temperature, 58, 59, 61, 72, 82, 277, 344
Cyclotron resonance, 43
Czochralski method, 112

D

d-electron, 6, 16, 47, 49, 51, 54, 55, 58–60, 63, 65–68, 72, 85, 114, 119, 154, 198, 200, 201, 226, 240, 268
d level or band, 72, 85, 86, 88, 90, 93, 161, 202, 203, 238–240, 246, 256, 280
d_{xy}, d_{yz}, d_{zx} orbital functions, *see* t_2 orbital function
$d_{x^2-y^2}$, d_{z^2} orbital functions, *see* e orbital function
dγ orbital function, *see* e orbital function
dε orbital function, *see* t_2 orbital function
de Broglie generalization, 3
Delocalization, carrier, *see* Localization
Demagnetizing factor, 282, 318, 339
Density, carriers, 5, 22, 26, 38, 55, 60, 63, 64, 66, 69, 72, 73, 82, 83, 90, 260, 264, 268, 280, 344
    minimum, 38, 44, 73, 111, 221, 228, 280, 344

Diagram, *see* Energy level, Phase diagram
Diamagnetism, 43, 55, 154
Diamond structure, *see* A4 (Index of Structures)
Dichroism, circular, 325
Dielectric constant, 132–134, 143, 189
Diluted powder method, 42, 125, 142, 146
Dipole, electric, 21, 31, 277
Directional character, 14, 26, 82
Distortion, lattice, 29, 33, 61–63, 66, 73, 82, 85, 86, 167, 183, 196, 204, 213, 217, 236
Distribution
    charges, 7, 8, 49, 64
    electronic, 6, 21, 36, 37, 60, 63, 66, 71
Donor atom, 53
Donor level, 38, 68
Doppler–Fizeau effect, 162
Double layer, 27
Dzyaloshinskii theory, 186

E

e or e_g orbital function, 7, 8, 50, 52, 54, 65, 69, 82, 83
e_g level or narrow band, 51, 68, 69, 70, 72–74
Earth field, 184, 315
Effective charge, 20, 22, 31, 33–36, 43, 51, 66, 68, 73
Effective mass, electron, 39, 44, 66, 67, 97, 277
Electrical neutrality, 63
Electron cloud, 19, 31, 55, 161, 289
Electron pair, *see* Lewis pair
Electron paramagnetic resonance (EPR), 135, 150, 241
Electron transfer, 21, 22, 30, 31, 46, 61, 62, 65, 82, 200
Electronegativity, 18, 29, 31
Electrooptical effect, 322, 323
Electrostatic interactions, *see* Coulombic interactions
Energy band, 33, 42, 47, 63, 68, 70
Energy gap, *see* Forbidden band
Energy level, 36, 42, 49, 50, 68, 72
Energy level diagram, 41, 49, 68–71, 94, 104, 108, 142, 145, 333, 335
Epitaxy, 142
Ettingshausen effect, 302
Exchange integral, 57

Excitation mechanism, 44, 47, 67, 68, 71, 154, 174, 190, 280
Exhaustion, 38
Extrinsic conductibility, 37, 38, 43

F

f-electron, 6, 55, 58, 59, 240, 268
f level, 238, 239
Factor of merit, 96, 216, 319, 331, 334, 340
Faraday effect, 325, 328
Fermi level, 36, 38
Ferrimagnetism, 56, 57, 106, 109, 110, 179, 205, 210, 212, 217, 218, 221, 229, 244, 312
Ferrites, 179, 184, 190, 284, 285, 306, 307, 310, 331, 332, 339
Ferroelectricity, 135, 143–145, 151, 152, 293, 295, 324, 344
Ferromagnetism, 56–58, 60, 88, 95, 102, 104, 109, 110, 140, 177, 216, 220, 225, 226, 244, 246, 257, 260, 262, 277, 291, 303, 325
  parasitic, 184, 186
Fluorescence, gamma, 162
Fluorite structure, *see* C1 (Index of Structures)
Forbidden band, 37–39, 42, 44, 67, 119, 228, 247, 341
Free carrier, 38, 65, 66, 69, 83, 86, 166, 240, 341

G

Galvanomagnetic, *see* Magnetoelectric effect
Garnets, 325, 326, 329, 331, 332, 339, 341
Germanides, 95, 102
Gibbs free energy, 62
Goodenough theory, 65

H

Haematite, 184, 185
Hall coefficient
  extraordinary, 303, 305, 310, 311, 318
  ordinary, 43, 62, 88, 89, 299, **300**, 304, 310, 311
Hall effect
  extraordinary, 181, 184, 303
  ordinary, 181, 299, 301, 313
Hall generator, 313, 318
Hall "susceptibility," 305, 306
Hall voltage, 43, 299

Hartree–Fock approximation, 8
Haussmannite, 124, 183
Heisenberg model, 58, 60
Heterojunction, 256
Hole, 37, 39, 63, 277
Homogeneity domain, 63, 91, 154, 200, 219, 228, 249, 260, 264
Homotypical representation, 30, 33
Hopping conductibility, 44, 63, 65–67
Hopping frequency, 62
Hopping time, 66
Hund rules, 66, 239, 240
Hybrid orbital function, 8, 15, 16
  s, p, 8, 15, 24, 60, 235, 236
  s, p, d, 8, 16, 29, 52, 69, 70, 73, 82, 83, 85, 88, 198, 201, 208
  s, p, d, f, 8
Hybridization energy, 30
Hydracid, 20, 30
Hydrides, 88
Hydrogen
  atom, 4, 5, 6, 9, 22
  molecule, 9, 10, 12
Hydrogenoid orbital, 7
Hyperfine field, 164, 166, 182
Hysteresis, 144, 214, 306

I

Ilmenite, 188
Impurity band, 61
Incandescent lighting, 251
Indices, optical, 322
Induction, magnetic, 43, 299, 303, 313, 324, 326
Infrared range, 42, 335, 336
Interactions, *see* specific types
Internal field, *see* Hyperfine field
Interstitial atom, 27, 46, 69, 133, 154, 237
Interstitial compounds, 79, 88, 89, 204
Intrinsic conductibility, 37, 38, 42, 43, 55
Inversion of levels, 51
Inverted lattice, 39
Inverted structure, 177, 179, 344
Ionic bond, 18, 19, 21, 29, 60
Ionic character, 18, 21
Ionic charge, 21, 31, 51, 61
Ionic conductibility, 133, 147, 251
Ionic configuration, 18, 19, 30
Ionic crystal, 49, 58, 61, 63
Ionic function, 17

Subject Index

Ionic radius, 26, 61
Ionicity, 19, 22, 30, 43, 56, 60, 61
  equilibrium, 22, 31
  parameter, 18, 20, 31–36, 41, 58, 66, 67, 88, 93, 103, 112
Ionization energy, 21, 30, 68
Isoelectronic substitution, 29, 56
Isomeric shift, 162–164
Isotope, 161, 162

J

Jahn–Teller effect, 73, 85, 183, 214

K

K capture, 161
$K\beta_1$ emission line, 198
Kerr effects, 323, 328, 330, 334, 336
Kinetic energy, 4, 38

L

Langevin theory, 53
Laser, 335, 338, 340, 341
Lattice, see Distortion, Inverted lattice, Polarization
Lattice vibration, 43, 45, 61, 62, 162, 277
Leclanché cell, 143
Lewis pair, 9, 11, 12, 15, 16, 18, 21, 24, 28, 29, 31, 47, 52, 55, 58, 60, 63, 69, 85, 86, 154, 200, 201
Ligand field, 51
Light propagation, 322, 326
Light vibration, 323, 325, 329
Linear combination of atomic orbitals (LCAO), 9, 12, 16, 18, 51, 53, 54
Liquid semiconductor, 150, 212, 213
Localization, carrier, 12, 16, 39, 60, 61, 65–67, 70–72, 114, 163, 203, 226, 242
Lorentz forces, 281, 299

M

Madelung constant, 30
Maghemite, 190
Magnetic interactions, 11, 53, 58, 60, 179, 246, 258, 263
Magnetic moment, 58, 201, 239
Magnetic resonance, 326, 339, 340
Magnetic semiconductor, 146, 147, 179, 184, 190, 210, 217, 246, 257, 262, 268, 281, 287, 333

Magnetite, 177, 190, 310
Magnetization, 57, 284, 339
Magnetobaryte, 331
Magnetoelectric effect, 125, 128, 277, 287, 288, 299
Magnetometer, 315
Magnetooptical effect, 322, 324, 344
Magnetoplumbite, 331
Magnetoresistance, 104, 180, 181, 220, 281, 285, 293, 302
Magnon drag, 225, 344
Manganites, 287
Marcasite, 210
Marcasite structure, see C18 (Index of Structures)
Mass-action law, 137, 170
Maxwell laws, 3
measurement
  current, 316, 317
  Hall effect, 306, 308
  mobility, 137, 138, 151
Mesomer states, 28
Metallic bond, 61, 198
Metallic transfer, 64, 70, 198
Metamagnetism, 243–246, 264, 266, 284, 325
Millerite, 212
Mobility, carriers, 43, 62, 67, 71, 210, 246, 280, 281
  drift or Hall, 137
Modulation, 323, 335, 336, 338, 340
Mössbauer resonance, 102, 127, 140, 161, 162, 169, 182, 186, 210, 226, 228, 268, 344
Molecular orbital (MO), see also π and σ molecular orbital
  approach or method, 12, 16, 17, 33, 39, 41, 47, 49, 51, 61, 72, 88
Molecular orientation, 324
Molecule, heteronuclear, 12, 13
Molybdenite, 207
Momentum, 3, 5
Motors, dc current, 295

N

Narrow band, 44, 47, 69, 72
Neel temperature, 59, 61, 72, 73
Nernst effect, 303
Neutron diffraction, 82, 125, 143, 167, 168, 179, 184, 190, 210, 213, 214, 218, 223, 241, 243

Nitrides, 88, 89, 241
Nonbonding doublet, 15, 18, 31, 69, 200
Noncollinear spins, 105, 143, 146, 223
Nonlinear element, 293
Normalization condition, 9, 19
Nuclear Magnetic resonance (NMR), 140

O

Octahedra
  block, 133, 152
  chain, 130, 131, 138, 149, 151
Octahedral arrangement, 27, 29, 31, 49, 52, 59, 60, 66, 68, 69, 74
Octahedral site, 27, 63, 79, 81, 82, 177, 179
Octahedron
  distorted, 68, 85, 124, 164, 170, 237
  O_6, 124, 129, 130, 139, 145, 151, 237
Orbit, electronic, 4, 6, 55, 58, 326
Orbital function
  atomic (AO), 3, 4, 9, 16, 49, 51–54, 60
  molecular (MO), 9, 11–13, 16, 33, 51–54
Orbital moment, 56, 239
Order
  crystallographic, 63, 64, 82, 179, 200, 206, 207, 214, 217, 218, 344
  magnetic, 56, 57, 71, 72, 74, 82, 104, 277, 282, 344
Overlap
  atomic orbitals, 14, 16, 18, 24, 61, 65, 70, 246, 268
  bands, 41
  integral, 9, 19, 20
Overstructure, 196, 208, 217, 236
Oxides, 45, 53, 61, 64, 65, 67, 73, 88, 119, 161, 246
Oxygen pressure, 133, 150, 167, 170

P

p-electron, 28, 51, 54, 59, 60, 68, 83, 200
p orbital function, 5, 6, 8, 27, 50, 52, 60, 65, 69, 70, 82, 83, 85, 88
Pair of electrons, see Lewis pair
Paramagnetism, 55, 72, 82, 98, 104
  constant, 119, 120
  Pauli, 64, 65, 72, 73, 148, 226, 228
  van Vleck, 243
Pascal constant, 55
Peltier effect, 44, 302
Permeability, magnetic, 318

Perovskite structure, see $E2_1$ (Index of Structures)
Phase diagram, 89, 108, 167, 168, 247, 253, 266
Phonon, see Lattice vibration
Phosphides, 103, 242
Photoelastic effect, 324
Photon, 3, 42, 161, 162
π band or bond, 52, 70, 71
π and σ molecular orbital, 12, 52–54, 65, 333
Piezoelectric effect, 151, 291, 324
Plane arrangement, 29
Pockels effect, 324
Point charge, 20, 49, 51
Polarization
  ion, 31, 51, 68
  lattice, 44, 61, 63, 66
Polaron, 63, 66, 67, 72, 171
  band, 63, 176
  magnetic, 281
  small, 138, 171, 176, 251
Polianite, 143
Potassium bromide, 42
Probability of presence, 4, 22, 163
Proton, 4, 9, 10, 161
Pseudobonding, 61
Pseudometal, 47, 67, 72, 93, 114, 229
Pyramidal trigonal arrangement, 27, 28
Pyrite, 210
  structure, see C2 (Index of Structures)
Pyroelectricity, 151
Pyrolusite, 143
Pyrrhotites, 208, 209, 313

Q

q-electron, 8
Quadrupolar interaction, 164, 168, 186
Quantum mechanics, 8, 9, 20
Quantum numbers, 4, 6, 7, 58
Quantum state, 24, 27, 36, 85

R

Radioactivity, 161, 239
Radius, see Ionic radius, Tetrahedral radius
Rare earths, 34–36, 55, 235
Recoil energy, 162
Recording, magnetic, 190, 315, 317
Rectifier effect, 151
Reduction, chemical, 132, 138, 154

Refractoriness, 91, 237, 241, 251, 257, 260, 262
Regulation, 293
Repulsion force, 21, 50, 51
Resistivity maximum, see Density minimum
Resonance, see specific types
Rest mass, electron, 39
Restitution energy, 31
Reverse shift, 246, 247, 259, 260, 344
Righi–Leduc effect, 302
Rock salt structure, see B1 (Index of Structures)
Rotatory power, 325, 326, 329, 331
Russell–Sanders coupling, 56
Rutile structure, see C4 (Index of Structures)

S

s-electron, 27, 54, 59, 60, 68, 163, 166, 200
s orbital function, 4, 6, 8, 50, 52
Scattering, magnetic, 71, 72, 277, 281, 282
Schrödinger postulate, 3, 4, 9, 20
Screen effect, 164, 268
Seebeck effect, 44, 62, 142, 303, 344
Selenides, 67, 196, 197, 202, 203, 213
Semiconductor, 36, 47, 55, 56, 58, 61, 64, 65, 67, 69, 72, 88, 114, 154, see also specific types
Semimetal, 97
Silicides, 87, 95, 241
Site, crystallographic, see specific types
Skutterudite structure, see DO_2 (Index of Structures)
Space or spin orbital function, 11, 58
Spin moment, 56, 277
Spin-orbit coupling, 104, 170, 289
Spinel structure, see $H1_1$ (Index of Structures)
Standard heat of formation, 30, 32
Stannides, 95, 102
Stark effect, 164
Stoichiometry problems, 33, 69–71, 79, 90, 91, 93, 132, 138, 139, 196, 197, 204, 208, 213, 214, 228, 236, 237, 249, 252, 260, 268, 280
Sublattice, 27, 179, 208, 209, 237
Sublimation energy, 30
Sulfides, 67, 196, 202–204
Superexchange, 58
Susceptibility, magnetic, 55–57, 73
Symmetry, orbital function, 5, 11, 52, 58, 69

T

t_1 or t_{1u} orbital function, 50, 54
t_2 or t_{2g} orbital function, 7, 8, 50, 52–54, 60, 65, 69, 70, 82, 83, 268
t_{2g} level or narrow band, 51, 68, 69, 71–74, 85, 88, 123
Tellurides, 64, 67, 196, 197, 199, 202, 203, 219
Tetrahedral arrangement, 22, 24, 28–30, 49, 60, 68, 69, 73
Tetrahedral "radius," 26
Tetrahedral site, 15, 26, 81, 177
Tetrahedron, distorted, 85
Thermal agitation, 36, 278, 282
Thermionic emission, 240, 251
Thermochemical data, 20, 30
Thermodynamics, 62, 186, 288
Thermoelectric effects, 44, 95, see also Peltier effect, Seebeck effect
Thermoelectric generator, 95, 96, 98, 99, 112, 216, 260
Thermoremance, 184, 185
Thin layers, 154, 242, 253
Thiospinels, 294, 319
Threshold, optical, 42, 67
Transfer mechanism, 44, 47, 61, 63, 64, 66–68, 70, 71, 154, 170, 179, 190, 199, 222, 229, 246, 256, 268, 280
Transition
  electronic, 27, 39, 97, 244, 268, 277, 326, 328, 333
  magnetic, 82, 109, 126, 184, 244, 293
  nuclear, 165
  semiconductor-metal, 65, 73, 120, 126, 129, 139, 281, 293, 344
Troilite, 208

U

Umbrella-shaped, see Noncollinear spins
Ultraviolet range, 42

V

Vacancy, 26, 34–36, 45, 46, 63, 64, 69, 79, 93, 119, 146, 147, 151, 167, 190, 196, 199, 201, 204, 206–208, 210, 214, 217, 218, 225, 229, 235, 236, 249, 260, 344
Valence band, 36, 37, 39, 42, 97
Valence bond (VB) approach or method, 10, 12, 16, 17, 18, 29, 33, 39, 41, 47, 49, 61, 67, 72

Valency induction, 46, 63, 65, 174, 188, 189, 251, 344
van der Pauw method, 308
van Vleck theory, 239, 240
Verdet constant, 326, 331
Visible range, 42
Vitreous semiconductor, 150
Voight effect, 324

W

Wavelength, 3, 38
Weiss model, 58, 72

Wide band, 47, 66, 69, 71
Wüstite, 63, 64, 167, 199
Wurtzite structure, *see* B4 (Index of Structures)

X

X rays, 38, 198

Z

Zeeman effect, 128, 164, 182
Zinc blende structure, *see* B3 (Index of Structures)